Household Ceramics at Port Royal, Jamaica, 1655-1692

Madeleine J. Donachie

BAR International Series 1195
2003

Published in 2016 by
BAR Publishing, Oxford

BAR International Series 1195

Household Ceramics at Port Royal, Jamaica, 1655-1692

ISBN 978 1 84171 560 5

BAR Publishing is the trading name of British Archaeological Reports (Oxford) Ltd. British Archaeological Reports was first incorporated in 1974 to publish the BAR Series, International and British. In 1992 Hadrian Books Ltd became part of the BAR group. This volume was originally published by Archaeopress in conjunction with British Archaeological Reports (Oxford) Ltd / Hadrian Books Ltd, the Series principal publisher, in 2003. This present volume is published by BAR Publishing, 2016.

Printed in England

BAR
PUBLISHING

BAR titles are available from:

BAR Publishing
122 Banbury Rd, Oxford, OX2 7BP, UK
EMAIL info@barpublishing.com
PHONE +44 (0)1865 310431
FAX +44 (0)1865 316916
www.barpublishing.com

ABSTRACT

From 1981 to 1990, the Institute of Nautical Archaeology (INA), in conjunction with Texas A&M University (TAMU) and the Jamaica National Heritage Trust, excavated a portion of the underwater English colonial city of Port Royal in Jamaica. Port Royal, an important international commercial center in the late 17th century, was destroyed in 1692 by an earthquake, which sank over half of the city beneath the waters of Kingston Bay. The INA/TAMU investigation has resulted in an extensive collection of artifacts and other material remains contemporary with the disaster.

This study examines the ceramic inventory of one of the most fully excavated buildings in the heart of old Port Royal. As household wares, the pottery vessels recovered from the site provide important data on the customs and standard of living of the building's occupants. By extension, they reveal certain social aspects of the town as a whole and provide information about the kinds of material goods that were available to New World colonial settlers at the end of the 17th century.

Minimum vessel counts, by ceramic ware, form, and functional classification, are the basis for the analysis. The assemblage is looked at in the general context of all of the ceramics recovered from the Port Royal site, as investigated by INA/TAMU. It is also compared with similarly well-dated groups from two non-Jamaican sites. English pottery inventories from the 17th century and household probate inventories from Port Royal are examined to cast light on ceramic usage and markets. Social commentaries of the period and northern European paintings of interior scenes provide a snapshot of the everyday roles of ceramic vessels.

TABLE OF CONTENTS

LIST OF FIGURES

FIGURE Page

LIST OF TABLES

CHAPTER I

INTRODUCTION

In 1692, Port Royal, an English colonial city on the island of Jamaica, was hit by a devastating earthquake at the peak of its commercial prosperity. The tremors from the earthquake and the ensuing tidal wave submerged over half of the city beneath the waters of Kingston Bay. Some of the buildings sank into the bay in an almost vertical fashion, their walls falling downward and inward, on to the rooms and floors. As a consequence, the Port Royal disaster produced in these buildings a rare event for archaeology: a near-unmixed collection of material and cultural remains of late 17th-century life.

In 1981, the Institute of Nautical Archaeology, in conjunction with Texas A&M University and the Jamaica National Heritage Trust, took advantage of the unique opportunity offered by the Port Royal site and began systematic excavations. The investigation, which lasted for 10 years, focused on the underwater remains of the northern end of Queen and Lime Streets, in the commercial center of the town, where eight buildings yielded thousands of artifacts. The largest structure, a two-building complex, was particularly impressive, revealing the most information to date about life in the port. Objects of iron, pewter, copper, and glass were found, as was a vast array of fragile organic material, such as wood and bone. In addition, significant numbers of ceramic remains, some local but the majority from wares imported from England, were included in the excavated artifact inventory of the complex. It is these ceramic remains that are the concern of this study.

CERAMIC STUDIES

The word 'ceramic,' derived from the Greek *keramos* ('burned stuff' or 'earthenware') is a blanket term that covers all artifacts molded from clay and then rendered durable by firing (Rice 1987:3). Making pots is one of the most ancient of human activities, and the abundant and widespread presence of pottery in archaeological sites is a testament to its durability. That and its versatility in use and form make it one of the most valuable objects available to archaeology.

Excavated ceramic material can be used as a dating tool; it can also illuminate the lives of those who made and used it in the past. Its usefulness for chronology was recognized in the 19th century, when it was concluded that pottery fragments found together in archaeological deposits were likely to have been placed there about the same time and to be parts of vessels used contemporaneously. Collections of these sherds (and intact vessels) were segregated into categories, largely on the basis of shape, style, and/or decorative motifs. "These descriptive categories, and the stylistic changes they incorporated, then formed the basis for assessing temporal relationships of materials found within a site" (Rice 1984:246; see also Willey and Sabloff 1993:96-121).

As the 20th century progressed, so too did familiarity with the archaeological record, and more elaborate schemes of pottery classification were constructed. No longer concerned with mere chronological ordering of a site, archaeologists sought chronological comparisons *between* sites and between cultures in order to elucidate relationships within human societies (Rice 1984:246; Willey and Sabloff 1993:121-141). With the discovery of radiocarbon dating in the late 1940s, which freed archaeology from uncertainties over chronology, this interest developed into examining the 'behavioral significance' of ceramics. Archaeologists began to look at pottery vessels in a wider context, to see them as they might have been *used* in a developing behavioral system (see Binford 1965; Matson 1965a; Van der Leuw and Pritchard 1984).

Modern archaeological analyses of ceramic function are voluminous and range from studies of vessel form, shape, and size (e.g., Henrickson and McDonald 1983; Smith 1983) to the identification on a vessel's surface of physical or chemical 'traces' pointing to use (e.g, Griffiths 1978; Hally 1983; Cackette et al. 1987; Skibo 1992). The physical properties of pottery fabrics are examined to assess their suitability for various roles (e.g., Henrickson and McDonald 1983:631-634; Steponaitis 1984; Bronitsky and Hamer 1986); ethnographic analogy helps to identify the uses of excavated vessels (e.g., Longacre 1991). The context of recovery, it is recognized, may not always provide proof of function. If, for example, a pot is found on a living surface in association with a hearth, the function of that vessel may appear to be clear. But vessels typically have multiple uses during their life spans; the archaeological context is merely their final resting place rather than an incontrovertible indicator of how their use life was spent (see Binford 1962, 1965).

Historical archaeologists working with ceramic assemblages call in aid other data to indicate how vessels were used in the past. Weatherill (1971, 1986) and Turnbaugh (1977, 1983), for example, have examined pottery probate inventories to interpret the meaning of vessels found in archaeological sites. Household probate inventories, which listed and valued moveable goods held at death, may indicate how people used and perceived their wares (see Stone 1970; Brown 1973; Beaudry 1980; Courtney 1986, 1989). Valuations in import and sales documents provide a valuable tool for analyzing expenditure patterns (see Miller 1980, 1991; Spencer-Wood 1987; Di Zerega Wall 1991). Haselgrove and Van Loo (1998) are the latest in a long line of scholars to use customs lists to provide evidence of production and trading of pottery. Catalogues of museum collections are especially helpful to the historical archaeologist for their descriptions and illustrations of quality pieces (see, e.g., Ray 1968; Ayers 1980, 1985; Britton 1982; du Boulay 1984; Rackham 1987; Grigsby 1990; Archer 1997). Another notable source of information is the representation of pottery in art, which has the double advantage of showing actual use and pinpointing date (see examples in Van der Pijl-Ketel 1982; Archer 1997; Gaimster 1997; Schaefer 1998).

RESEARCH OBJECTIVES

This study seeks to show that the analysis of the Port Royal ceramics—their types, their quantities, their forms—can lead to a better understanding of the community that suffered the disaster. The great merit of the Port Royal site as investigated by the Institute of Nautical Archaeology and Texas A&M University is that it is a relatively undisturbed context, with many of the ceramic wares found at or near their original location. The author has concentrated her attention on Port Royal's Building 4/5, a large, two-building brick structure that may have functioned as a private dwelling (occupied by a single family or partially/wholly rented out); as a commercial (eating/drinking) establishment catering to the many merchants and mariners that passed through the town; or as a combination of both. As household wares, the vessels recovered from the site should provide important data on the customs and standard of living of the building's occupants. By extension, they may also, it is suggested, reveal certain social and economic aspects of the town as a whole.

A second study objective is to examine the finds in the context of their manufacturing origins. To this end, the author has attempted to bring together from a variety of sources an account of the historical development of the main ceramic types, so far as these are relevant to the Port Royal site. This panoramic approach opens up the study to a wider consideration of the cultural influences that came to bear on Jamaica and on colonial society as a whole.

Some of the flavor of the viewpoint of this study is summarized by the historical archaeologist James Deetz (1977:46-47): "Pottery is largely utilitarian. Its interpretation relates to the everyday aspects of life rather than the more esoteric. . . . From their abundance or rarity, their place of manufacture, and the purposes they served, we can discern much about the life and times of their owners." Noted ceramic archaeologist Frederick Matson's oft-cited quote is endorsed by the author: "Unless ceramic studies lead to a better understanding of the cultural context in which the objects were made and used, they form a sterile record of limited worth" (Matson 1965b:202).

RESEARCH PROCEDURES

The stratigraphy of the Port Royal site as excavated by the Institute of Nautical Archaeology and Texas A&M University is composed of three natural layers: Layer 3, the deepest layer, consists primarily of sand and contains partially intact floors and rubble from the 17th-century city's collapsed buildings, together with a mixture of in situ artifacts. Layer 2, above Layer 3, encompasses a thick band of coral, believed to have been deposited in 1722, when two earthquakes and a hurricane hit Jamaica. This layer contains 18th- and early 19th-century artifacts in its upper levels and admixtures of 17th-century material in its lower levels. Layer 1, the uppermost layer, consists of a combination of roots and silt and contains post-1722 refuse from Port Royal and Kingston Harbor. While post-depositional disturbance (in the form of earthquakes and hurricanes, the dragging action of ships' anchors, and the looting of bricks from the 17th-century structures) has affected to varying degrees all three layers of the site, Layer 3 is the least contaminated, having being partially protected by the coral barrier in Layer 2.

A total of 7440 ceramic sherds was excavated from all three layers by the Institute of Nautical Archaeology and Texas A&M University during the 1981-1990 excavations at Port Royal. Of these, 2510 were recovered from Building 4/5. Not all of the sherds from Building 4/5, however, belong to vessels associated with the living building. Approximately one-third (n=893) of them, in fact, are known 18th- and 19th-century wares, and were, for the most part, probably dumped over the site by staff at the British Naval Hospital, which lies close to the excavated area and which was occupied from 1743-1905. This one-third of the assemblage, then, is thus disregarded in this study.

The remaining two-thirds (n=1617) of Building 4/5's ceramics have been identified as wares and types already in production in 1692, and the author has listed these individually on a specially constructed data base. The broad details concerning them are thus readily available for further study if required (see Appendix A). They are referred to throughout this study as the 'data base sherds.'

Ceramic classification and attribution of the data base sherds was based on macroscopic (visual) examination of each sherd's fabric (porous vs. non-porous), surface treatment (slip, glaze, etc.), and type of decoration (applied vs. stamped ornament vs. incised motif, etc.). Undecorated and decorated earthenware, stoneware, and porcelain of English, Continental European, Mexican, Chinese, and Jamaican provenance were recognized using this method.

Vessel form was recognized by reference to intact and reconstructed vessels and to diagnostic rim and basal sherds. The identification of vessel form from individual body sherds required to take account of other sherd attributes. An undecorated earthenware bowl body sherd was distinguished from a cooking vessel body sherd, for example, by considering the presence of a lead glaze on the vessel's interior (as in the former) or its absence from the interior (as in the latter). Evidence for fire use and sooting also helped, in some cases, to distinguish cooking pot body sherds from those of other vessels. Several vessel forms were also identified based upon the uniqueness of a body sherd's overall shape, size, and decoration. At least 189 vessels have been notionally constructed by the author for the data base assemblage.

The data base sherds were recovered from all three layers of the Building 4/5 site. Scrutiny of the archaeological record showed that approximately 70 percent of them (n=1155) were found in Layer 3, identified as the 17th-century context. These being the sherds most likely to belong to vessels associated with the living building, they were subjected to close examination by the author. Referred to as the 'analyzed sherds,' they provide the focus of this inquiry.

The analyzed sherds are the foundation for the several analyses, with supporting tables, which are undertaken in this study. They are examined and quantified by ware and type/variety and by the estimated number of vessels they represent. They are also compared with similarly well-dated ceramic assemblages from two contemporary non-Jamaican sites and are briefly looked at in the general context of the ceramics recovered from all eight buildings at Port Royal excavated by the Institute of Nautical Archaeology and Texas A&M University. At least 144 vessels have been notionally constructed by the author for the analyzed assemblage.

As a number of American archaeologists have emphasized, the usefulness of sherd counts on their own is limited (see Stone 1970; Deetz 1977:49–50; Miller 1980, 1991; Beaudry et al. 1983:20). While raw numbers may provide a general impression of a site, they do little to yield information on how ceramics were used. Lumping sherds into a 'kitchen group,' following South (1977:167-171), and looking at their percentage frequencies as opposed to those of architectural- or furniture- or activities-related groups, is not undertaken in this study. Similarly, Miller's (1980, 1991) scaling technique, which uses import and sales documents to create indices of prices of standard forms and wares, will not reveal actual vessel use. Rather, the concern of the author has been directed at the function of a vessel as inferred from its form and the proposition that identifying function may best allow us to understand lifestyle. Following this, the main reference in this study is to Beaudry et al. (1983), whose vessel typological scheme (known as the Potomac Typological System, or POTS) classifies vessel shape by attending to use. These classifications themselves are based upon vessel descriptions found in 17th- and 18th-century probate inventories and other documentary sources. Multiple or secondary uses are not considered in any detail in this study.

The author makes use of all of the methods of inquiry used by historical archaeology as described earlier in this chapter. Both archaeological and documentary data are used to interpret the distributional patterning and meaning of the analyzed sherds, and, as already noted, ceramic assemblages from two external sites provide comparative material. To cast light on ceramic usage and markets, English pottery inventories from the 17th century and household probate inventories from Port Royal were examined by the author. Social commentaries and literature of the period help to show how ceramic vessels were used. Northern European paintings of interior scenes provide a snapshot of their everyday roles.

SIGNIFICANCE OF THE STUDY

The Institute of Nautical Archaeology and Texas A&M University investigation at Port Royal was carried out for 10 years under the direction of Dr. Donny Hamilton. This work represents the only part of the 17th-century city to be systematically excavated using modern archaeological methods. The Building 4/5 complex represents the most intensively investigated structure of the entire underwater site. The significance of its Layer 3 ceramic assemblage lies in its coherence as a single deposit, the contamination of which has been minimal.

It is most important to note that ceramics are only part of the story of personal possessions. To understand the daily life of the inhabitants of Port Royal, we must, if we can, take account of all the objects that have survived and are available for study today. But of the wares of a household, pottery is the medium that is the least likely to be reprocessed. Objects made of metal, for example, could be, and were, melted down for other uses and so are less useful in preserving evidence of their use history.

The author believes this study will add to our understanding of English colonial life in the late 17th century. The task undertaken by the author of providing in Chapter IV a consolidated account of the development of the ceramic techniques that were known in the 17th century should be of use to students in the field.

CHAPTER SUMMARIES

Eight chapters follow this introduction. Chapter II presents a brief history of the European settlement of Jamaica and of the rise and tragic fall of the notorious English colonial city of Port Royal.

Chapter III summarizes the archaeological excavations at the Port Royal site and looks closely at the Building 4/5 complex as the primary unit of analysis. An overview of the building's ceramic inventory, with descriptions of methods used in its analysis, is also incorporated in Chapter III.

Chapter IV presents the ceramic collection in more detail. It discusses the history and development of ceramic wares, so far as they are relevant to the site, in order to place the recovered items in context. Also presented are sherd and estimated vessel counts within each ceramic ware and type and their relative proportions between wares and types. In this chapter, too, the author looks briefly at the ceramics recovered from all eight buildings (Layer 3 only) excavated by Hamilton's team.

In Chapter V, the author steps back from the archaeological record and presents a brief look at the domestic world of the 17th century in order to clarify the place of ceramic vessels in the home. In this chapter, travelers' notes and household probate inventories provide a picture of life as it was in Port Royal at that time. The author also reviews English literature from the period regarding food and drink and its relationship to medicine and health.

The author returns to the Building 4/5 assemblage in Chapter VI, to look at vessel form and function and to discuss distribution patterns within the building.

Chapter VII, which forms an addendum to Chapter VI, compares the ceramic wares and forms recovered from Building 4 with those from Building 5. The author also looks at the assemblage vis-à-vis the Layer 3 ceramics from all eight buildings excavated by Dr. Hamilton.

Chapter VIII compares the analytic findings of Building 4/5's ceramic inventory with similar assemblages from two contemporaneous sites: a governor's house in rural Virginia and an urban inn in Guildford, Surrey, England. The comparison attempts to illuminate the differing requirements of the occupants of the three properties, based upon the proportions of functionally defined vessel forms, and the relative access each property had to pottery markets, as reflected in the range of ceramic wares and types present.

The study concludes in Chapter IX with a summary of the data and a critique of the research objectives as outlined above.

APPENDIX SUMMARIES

A list of the ceramic vessel types and forms recovered from each of the rooms and yards in Building 4 and Building 5 (Layer 3 only) is set out in Appendix A.

Appendix B includes the ceramic data base compiled for the 'data base' sherds recovered from Building 4/5. This record includes includes data on site provenance, ceramic ware, type, form, sherd size, decoration, etc. Keys to the field codes used in the data base are also included.

CHAPTER II

THE ISLAND OF JAMAICA AND THE HISTORY OF PORT ROYAL

JAMAICA UNDER THE SPANISH

When Columbus visited Jamaica in 1494, on his second voyage to the New World, he thought it "the fairest island that eyes have beheld . . . all full of valleys and fields and plains" and took possession of it in the name of Ferdinand and Isabelle of Spain (cited in Black 1965:25). Columbus called the island St. Jago (Santiago), but finding no mineral riches, left soon afterwards.

The first Spanish settlement in Jamaica, Sevilla la Nueva, located on the island's north coast, was established in about 1510 by Diego Columbus, the son of the Discoverer (Figure 2.1) (Phillippo 1970[1843]:14-15; Black 1965:34). It consisted of only a fort, a castle, and an unfinished church, and by 1534 had been largely abandoned (Mayes 1972:5). Two other settlements of note were founded by the Spanish on Jamaica's southern coast, in the vicinity of Kingston Harbor: Caguaya and St. Jago de la Vega (present-day Spanish Town). The latter became the center of Spanish activity, due to its "convenient and healthful situation" and its "ample water supply and fertile country round about" (Black 1965:36).

In 145 years of occupation, the Spanish used Jamaica mainly as a supply base from which provisions and weapons were sent to help in the conquest of the New World. The few colonists who settled the island grew cotton, tobacco, cocoa, and sugar, primarily for local use, and introduced European fruits, such as bananas and oranges, and animals, such as pigs, goats, and cattle (Black 1965:32-33; Gardner 1971:13-14). Despite such constructive activities, however, Jamaica under the Spanish "never prospered, was always poor, and more of a burden than a benefit to Spain" (Black 1965:32). Gold was never found, despite many local Indians and African slaves being worked to death in the search. And so the years passed by, and the settlers, neglected by the Spanish Crown, slowly settled into apathy and indolence (Phillippo 1970[1843]:83, 154; Black 1965:32-33).

ARRIVAL OF THE ENGLISH

A severe shock to Spain's lethargic habits was given towards the end of the 16th century, when an English buccaneer dropped anchor in Kingston Harbor in 1597 and plundered St. Jago de la Vega. In 1643, Jamaica was again raised to the defensive when the rogue ship of Captain William Jackson hovered on the horizon (Phillippo 1970[1843]:19). For 12 more years, the Spanish held their own against numerous attacks by pirates, but in May 1655, the island was taken with ease by the English as a 'better than nothing' tactic after the miserable failure of Oliver Cromwell's 'Western Design' against Spain.

Cromwell, the Lord Protector, had determined to gain an economic foothold for England in the Caribbean (Black 1965:42; Howat 1974:88). So, in December 1654, on his orders, a secret expedition left England to attack Santo Domingo in Hispaniola, capital city of the Spanish Indies. It was, perhaps, the worst equipped and organized campaign ever to leave English shores. Most of the 4000 troops were 'volunteers'—nothing more than "common Cheats, Theeves, Cutpurses, and such like lewd persons" (cited in Black 1965:43)—and in January 1655, they were reinforced by 3000 incompetents recruited from England's Caribbean colonies of Barbados and Nevis. With a further 8 added to the original fleet of 30 ships, the English force arrived in Hispaniola a few months later (Black 1965:44; Gardner 1971:30).

The great plan of attack failed miserably, however, and rather than return empty-handed and face the wrath of Cromwell, the expedition's disheartened leaders, Admiral Penn and General Venables, decided to seize Jamaica to the west, which was known to be weakly held by a few hundred Spanish planters (Black 1965:43-44; Gardner 1971:28; Howat 1974:87). Thus on 3 May 1655, the English fleet rounded the point of Caguaya and anchored off the southeastern side of the island. "No opposition of a serious character was offered, the fire of the few guns were silenced," and it was not long before the capture of Jamaica was complete (Gardner 1971:30).

Disappointed at the failure at Santo Domingo but not the man to lose any advantage, Cromwell determined to make the most of the new colony of Jamaica. He saw its strategic importance for harrying Spain and for coordinating trade (Howat 1974:89). Liberal supplies were thus sent to the infant colony, and active measures were taken to secure planters from New England, Bermuda, and Barbados. A number of Portuguese, including some Jews who had been living with the Spanish on the island, asked to remain as settlers.

PORT ROYAL EMERGES

It quickly became clear to the English that they would have to fortify Caguaya (which they renamed the Point), and after a shaky start, the building of Fort Cromwell began (Gardner 1971:38). By 1657, there was, in the words of Commander Brayne,"the faire beginning of a town upon the poynt of this harbor . . . there I intend all our storehouses and trade shall be, which shall soon make it a flourishing place" (cited in Pawson and Buisseret 1975:10). A great spurt in building activity commenced in 1658, by which time there were at least three rows of private houses, a state storehouse, the beginnings of a church, and a courthouse (Pawson and Buisseret 1975:10-11).

The early years of Jamaica's English occupation were, however, marked by violence. Hostilities continued with Spain, who attempted to regain possession of the colony, and an internal rebellion occurred against the military rule of the island. It was only with the Restoration in1660 of Charles II to the English throne, which formally ended the war with Spain, and the establishment

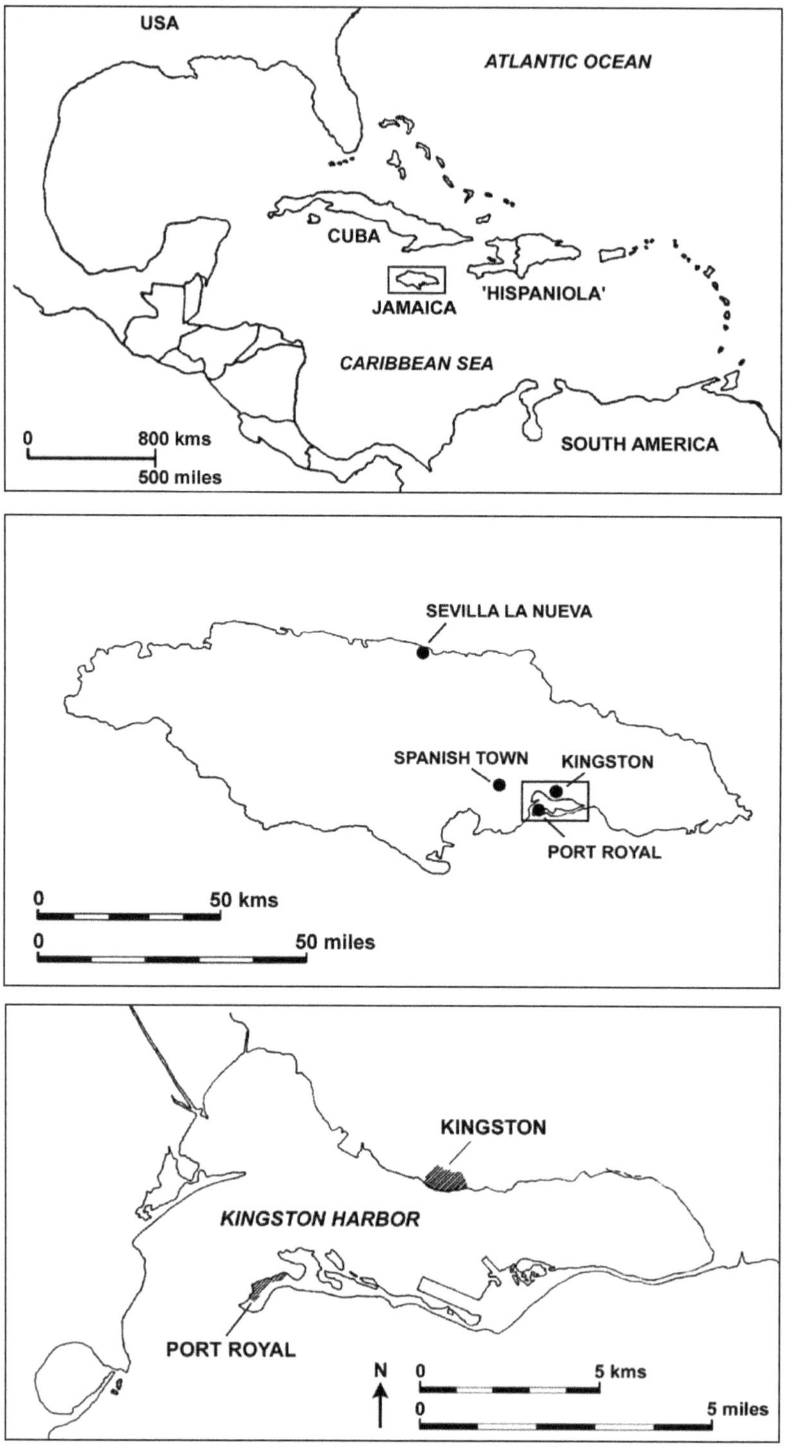

FIGURE 2.1. Location of Port Royal, Jamaica

of a civil government in Jamaica, that the colony's economic and strategic importance began to emerge. Fort Cromwell was refurbished and renamed Fort Charles in honor of the King, and the Point was renamed Port Royal. Ranching and the planting of tobacco, cocoa, and sugar cane for export began, and new settlers arrived in their thousands, attracted by the King's generous promises of land and privileges (Pawson and Buisseret 1975:18-19).

Port Royal in 1660 was described by Sir Thomas Lynch, who later became governor, as "seated on the extreme end of the pointe, containing in itt about 200 houses, all buylt by the English, with some publique houses . . . and the houses where all the stores for fleete and army were kept . . . . This is the place where all merchants, strangers and saylers reside as being the seate of trade and the most healthy place in the island" (cited in Pawson and Buisseret 1975:13). This extraordinary growth of the city has been attributed to its "location within a well-protected harbor, its flat topography, and deep water close to shore [where] large ships could be easily serviced, loaded, and unloaded" (Hamilton 1992:40).

The growing security and opportunities afforded by Port Royal also attracted buccaneers, and the part played, after 1660, by these notorious adventurers is an important feature in Jamaica's colonial history. They were mostly French, English, and Dutch castaways and escaped criminals, who had joined together in their common hatred of Spain. For a while, they lived on Hispaniola's deserted north coast, where they hunted the island's wild cattle and pigs and sold or traded them to the numerous passing ships. Eventually, however, due to frequent attacks on their camp by the Spanish, the buccaneers were forced to move to the nearby island of Tortuga. When assaults by the Spanish continued on Tortuga, the buccaneers took to the seas, where they made a career out of plundering Spanish ships. These raids, by 1650, were proving to be extremely profitable, and the 'brethren of the coast,' as the buccaneers came to be known, were a strong and striking force in the Caribbean (Black 1965:52; Gardner 1971:56-57; Pawson and Buisseret 1975:20).

Tortuga remained a favorite rendezvous of the swashbuckling band of pirates until the capture of Jamaica by Cromwell. Well aware of the island's vulnerability in the heart of Spanish America, Colonel Edward D'Oyley, Jamaica's first civil governor, soon recognized that the buccaneers could protect their new colony and also generate revenue. They were thus encouraged to use Port Royal as their base of operations. In return for their services, England granted them letters of marque, to legalize their activities, and changed their status to that of privateers (Black 1965:51-53; Zahedieh 1986a:574).

Employing the buccaneers as privateers was a resounding success. Their many sorties against the Spanish West Indies, the Spanish Main, and against Spanish, and, increasingly, Dutch, ships were not only lucrative to Port Royal—they made it the wealthiest city in the Carribean—but they also held down the enemy (Pawson and Buisseret 1975:22-24). As a contemporary English historian, Sir Edward Long (1970[1774]:1:300), wrote: "It is to the Bucaniers that we owe the possession of Jamaica at this hour."

The brief but colorful period of sanctioned piracy ended in 1671, with the signing of the Treaty of Madrid (1670), which officially ended the war between England and Spain (Phillippo 1970[1843]:21; Pawson and Buisseret 1975:36). Privateering and/or piracy, however, continued in one form or another into the 18th century (Zahedieh 1986a:574; Hamilton 1992:40). Indeed, by 1671, many of Port Royal's merchants were well supported by "the torrents of money which [the buccaneers] poured" into the city (Long 1970[1774]:1:299). The merchants sent the proceeds to their representatives in London for sale at profit. More goods were then bought in England and transhipped across the Atlantic for sale in Port Royal (Pawson and Buisseret 1975:31).

It was this economic activity that eventually led to England's recognition of Port Royal as a potential *international* trading center in the Caribbean, located as it was at the very heart of the principal islands of the West Indies. Thanks in part to the privateers, who had by this time, in effect, worked themselves out of a job, the stage was set for the new policy of attracting into English hands the trade of the Spanish New World colonies (Black 1965:61). Trade with Port Royal was too attractive to be ignored by Spain's New World colonists, since by this time, poor organization and the corruption of the Spanish fleet system was creating highly irregular trade scheduling and raising prices for European goods (Zahedieh 1986a:573). Direct contraband trade with Jamaica meant imports at lower prices for the Spanish colonists and regular exports of their agricultural staples. As observed by a contemporary, Francis Hanson, on the city's economic success:

> [Port Royal] being most commodiously seated in the midst of the *Spaniards*, so that we drain the benefits of their Gold and Silver Mines without their Labour and expences . . . . is always like a continual Mart or Fair where all sorts of choice Merchandizes are daily imported not only to furnish the Island, but vast quantities are thence again Transported to supply the *Spaniards*, *Indians*, and other Nations, who in exchange return us bars and cakes of Gold, wedges and piggs of Silver, Pistols, Pieces of Eight, and several other Coyns of both Mettles (Hanson 1683:55-56, emphasis original).

At the center of this trade system were the Port Royal merchants, who were now a thriving community, comprising "almost half" of Port Royal's residents, according to probate inventories for the period 1686-1694 (Zahedieh 1986a:570). As noted by Port Royal visitor John Taylor (1688:266), they traded with dealers from England, who brought "Wine, Cloath, Linnen, Stuff, silks, Fruit, Ironwork, Pitch, Tarr, Ropes, and other Things," as well as with dealers from Ireland, who brought "Beef, Pork, Salmon, Cheese, Butter, flower, Beer, &c," and with those from New England, who brought "Sturgeon, Macerell [and] . . . steaves and hoops for cask, with other the like Commodities." In exchange, exotic products, such as "sugar, Indigo, Coco . . . Rum, and Die wood" were loaded on to the ships for their long journey back to Europe.

Many of the exports from Port Royal, in fact, formed part of a re-export trade. Timber, "hides, brazil nuts, cochineal, and bullion" came from Central America and other Spanish New World colonies (Pawson and Buisseret 1975:68). The bullion was particularly important, as it was the measure and standard of riches in 17th-century England.

Jamaica's shipping registers attest to the significance of the Port Royal trade: in only two years, between 1668 and 1670, some 208 ships of 6727 tons "arrived in Port Royal harbor" (Pawson and Buisseret 1975:64). In 1688 alone, 213 ships docked at the port. This can be compared to the 226 ships that in the same year sailed "into all the New England ports combined" (Zahedieh 1986a:570).

The population of Jamaica was all the while steadily increasing. Early in 1673, an official letter to Lord Arlington, the Secretary of State, described it as "containing upwards of 7700 whites; of these 4050 were men . . . more than 2000 women and 1712 children. In addition to these it was reckoned that 800 seamen belonged to the ports. The slaves numbered 9504" (cited in Gardner 1971:60). Port Royal, it was also noted, had about 2000 inhabitants. This last estimate, however, seems a little low, since the letter also states that the town held as many as 800 houses, although the number of 'houses,' if correct, may have included such things as workshops and storehouses. (That the buildings were not in excess of demand is suggested by Taylor [1688:252] when he notes that "Lodgings are here verey deare," comparable, in fact, to London.) A more reliable estimate is for 1688, when Port Royal contained between 1200 and 1500 houses (Taylor 1688:252; Pawson and Buisseret 1975:98). Four years later, this number had escalated to approximately 2000 houses of various kinds, with a total population for the town of between 6500 and 7000 (Pawson and Buisseret 1975:99; Hamilton 1992:40). This was a phenomenal increase. Only Boston, Massachusetts, with a population in 1690 of approximately 6000 people (Henretta 1965:73), rivaled it in size and importance in the 17th-century New World.

The area that comprised Port Royal in the 17th century was never greater than about 50-60 acres, and this lack of space conditioned every aspect of life (Pawson and Buisseret 1975:81; Hamilton 1992:40). While the street plan of the town was, in many respects, ordered and balanced, consisting "essentially of a rough triangle, with right-angled 'blocks' running from its base up to each of the roughly equal sides" (Figure 2.2), the buildings were cramped together in the rather incoherent fashion characteristic of English towns of the period (Pawson and Buisseret 1975:81). Continuous division and subdivision of property meant that large merchant houses intermixed with humble artisan dwellings, and the courthouse and prisons were cheek by jowl with the town's busy streets.

The style of individual buildings reflected the northern European origins of the settlers with some necessary adaptations to the tropical climate (Pawson and Buisseret 1975:97). "The Houses . . . are for the most part Brick, and after the English manner," comments Sir Hans Sloane (1707:47), who lived in Jamaica in the 17th century and who later established the British Museum. The 'cook room' was usually a separate building in the yard: "The Kitchens, or Cook-Rooms here, are always at a small distance from their Houses, because of the heat and smell, which are both noisom and troublesome" (Sloane 1707:47). None of the houses, in fact, had chimneys, since heating was unnecessary (Pawson and Buisseret 1975:94).

John Taylor's (1688:252) account is illuminating: "at least 600 well built brick houses, and as many more built with Timber: ye houses are built four story high, Cellard, covered with Tiles, & glazed with Sash Windows; haveing large shops, and commodious store houses belonging to 'em." Taylor (1688:253) also describes some of Port Royal's civil buildings: St.Paul's Church was "built High with batlements of stone [and] paved with Marbell."[1] The "Kings house . . . where the Governore holds his Courte of Chancery [was] an old Timber work House." The governor's house was "the best built house on Port Royall," and Bridwell prison was "a brick structure . . . for Lazie strumpet of which here are plenty." The "Rigament of Infantrey, consisting of 2500 Men, well armed, and good disiplined soldiers" (Taylor 1688:256), and the five forts and a gun line built around the city, namely Charles, Rupert, Carlisle, James, Walker, and Morgan, afforded the emergent economic center good protection (Pawson and Buisseret 1975:40-41).

Surely it cannot be true that Port Royal at this time was nothing more than "the Receptacle of Vagabonds, the Sanctuary of Bankrupts, and a Close-stool for the Purges of our Prisons," as Edward Ward (1933[1700]:13) suggested in 1698. Nor should the character of the population be generalized as "deplorable . . . the dregs of the three kingdoms, who exhibited every kind of excess, and perpetrated almost every degree of wickedness," as scorned by James Phillippo (1970[1843]:22), a Baptist minister, who lived in Jamaica in the early 19th century. There were individuals of presumed respectability, such as judges and military officers, as well as individuals of various trades and professions. Architects, carpenters, bricklayers, cabinet-makers, shoemakers, hatmakers, and tailors, as well as gold and silversmiths all found their services in demand, indeed at "thrice the Wages given in England" (Taylor 1688:267). The numerous houses of worship in Port Royal included a Protestant church (St. Paul's Church) and meeting places for the Roman

[1] Mayes (1972:14, 64) excavated this church and found the floor to be made of wood. Taylor is probably referring to paved walls, the 'Marbell' of which was actually decorative plaster.

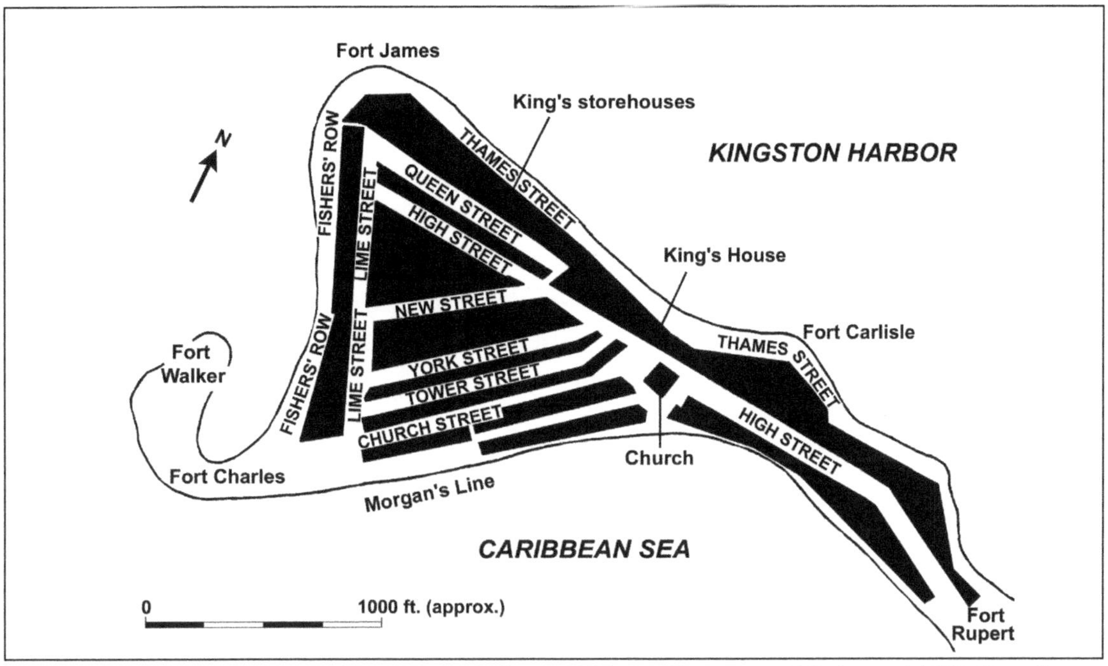

FIGURE 2.2. Street plan of Port Royal before 1692 (after Pawson and Buisseret 1975: Map 8)

Catholic, Quaker, and Jewish communities. The various doctors and at least one pharmacist were all considered by Taylor (1688:268) to be "Learned, Skilfull and Men of known integrity." The city's bad reputation, it seems, came from its considerable energy and enterprise, its "laissez faire attitude toward life," according to Hamilton (1992:40), which those living under the rules of puritanical England or its North American colonies would have found difficult to understand. Everyday life at Port Royal, in fact, seems to have been very similar to what might have been observed in a contemporary seaport of similar size involved in the dynamic, competitive world of Atlantic commerce. Henretta's (1965:75) description of Boston, for example, can be easily applied to Port Royal, in which "the disruptive pressure of rapid economic growth . . . made the social appearance of the town more diverse, more complex, more modern."

PORT ROYAL FALLS

Unlike Boston, however, Port Royal's growth was dramatically curtailed. On 7 June 1692, the business of the day had begun just like any other, early, to avoid the midday heat. The legislative council had already met in session, and the wharves were probably still laden with merchandise unloaded from the many ships docked in the busy harbor. The markets and stores would be proudly displaying all manner of exotic goods, and the streets were likely crowded with people. Then, almost without warning:

> At twenty minutes to twelve o'clock a noise not unlike thunder was heard in the hills of St. Andrews, to the north of Port Royal. Then three shocks were felt: the first comparatively gentle, the next more severe, the last so violent as to fill all with alarm, and only too plainly showing the true character of the calamity. Not only did the earth tremble, and in some parts open beneath the feet of the [city's] terror-stricken inhabitants, but the horrors of the event were intensified by the mysterious, awful sounds, that one moment appeared to be in the air, and then in the ground. Houses . . . were the first to fall; then the church and tower. [Fort James], to which many fled as to a place of security, was next observed to disappear, the sea rolling completely over the place where it stood. And now on every hand nothing was to been seen or heard but the crash of falling houses and walls . . . . [which] sunk into the harbor and were completely overwhelmed (Gardner 1971:75-76).

The human and economic consequences of the earthquake were devastating. In a few short minutes, Port Royal was reduced to approximately 25 acres, less than half its original size. All waterfront buildings, all of the storehouses, the King's House, the Anglican church, and four of the five forts were engulfed in the huge tidal wave that followed the disaster. Over 2000 of Port Royal's citizens died immediately; another 3000 would perish in the following days and weeks, of exposure, injuries, and disease (Mayes 1972:7-8; Pawson and Buisseret 1975:121-123).

Many of Port Royal's survivors were relocated to the new settlement of Kingston, across the harbor. Many, however, stayed in the city, intent upon restoring the "store house or treasury of the West Indies" (Hanson 1683, cited in Mayes 1972:7). The government, noting that "the fort [Fort James] and many of the houses were still left standing; that it was a place excellently adapted to carry on trade, and of great strength to resist an enemy," helped with the resurrection (Long 1970[1774]:2:143).

The city did, indeed, rise again from its ruins; agriculture and commerce were re-established, and the appearance of wealth and splendor was revived. But the new good times were brief. In 1703, Port Royal was almost destroyed by fire, and in 1712 and again in 1722, it was devastated by hurricanes. The decision was then taken to transfer the seat of commerce to Kingston. Even the establishment of a dockyard by the British Royal Navy in 1735, with a coaling station, administrative center, and hospital complex, could not change Port Royal's fortunes. It continued to be battered by countless hurricanes and earthquakes, which left the city a shadow of its former self (Pawson and Buisseret 1975:123-124; Hamilton 1992:41). The glory days of Port Royal were over forever.

CHAPTER III

ARCHAEOLOGICAL RESEARCH AT THE PORT ROYAL SITE

The geology of the part of Jamaica on which Port Royal was built contributed to the magnitude of devastation. Situated on the western end of a long sand spit, at the mouth of Kingston Harbor, the area was noted by Sir Edward Long (1970[1774]:2:139) in the 18th century to be "extremely incommodious in many respects. It had neither earth, wood, nor water; the soil was nothing more than loose sand, lodged at first upon a ridge of rock, and gradually extended in breadth, as well as depth, in a long series of time, by the renditions of the sea." In 1688, this spit, known as the Palisadoes, while "not above a quarter of a mile broad," extended at least 12 miles (Taylor 1688:251).

Archaeologist Philip Mayes (1972:3) is more precise in his description of the site: "[it] is, and was, surrounded by very steep submarine contours, and the whole length of Palisadoes exhibits geological instability . . . coralline limestone occurs at depths varying from 140 ft. to 180 ft.; superimposed on this are sands, gravels, silts and coral." Mayes (1972:3) suggests further that the earthquake shock was transmitted from this underlying limestone into the precariously piled superimposed deposits, which caused what he refers to as a "submarine slump." It was this slump, or slight shift in the position of the earth's plates, he argues, that resulted in 66 percent of Port Royal sinking to as much as 35 ft. below water. This hypothesis has the submarine landslide causing the town to slip outwards rather than topple over (Mayes 1972:3; Pawson and Buisseret 1975:2).

A more recent interpretation of the event has been given by geologist George R. Clark (1995), who attributes the destruction of much of the city to the process known as liquefaction, whereby ordinary sand is transformed into something like quicksand. Clark (1995:37) takes the view that when the vibrations from the earthquake passed through the sedimentary deposits underlying Port Royal, the sand grains failed to stick together, creating a treacherous fluid. Buildings were swallowed up, with brick structures suffering the most. In some streets, Clark propounds, especially those nearest the harbor, *every* building was sucked down into the sand.

Certainly, the liquefaction explanation does seem to fit the archaeological remains of intact buildings and floors in some parts of the city. The walls of some of these structures toppled downward and inward on to the rooms and floors, sealing the contents of each as if in a tomb. For nearly 300 years, the 17th-century site of Port Royal lay relatively undisturbed in its watery grave. It is one of only a few such catastrophic archaeological sites known in the Americas.

HISTORY OF ARCHAEOLOGICAL RESEARCH TO 1981

For many years after the disaster, the city's sunken buildings remained clearly visible in Kingston Bay. Indeed, many lay in water so shallow that salvaging and looting of their contents began almost immediately. "The Earth hath still fits of Shaking, with very much Thunder and Lightning," eye-witness Captain Crocket (1692) reported, "yet, this had so little effect upon some People here, that the very same Night they were . . . Pillaging and Stealing from their Neighbours, even while the Earthquake lasted." In time, however, the heavier, smaller objects sank deep into the seabed, and buildings came to be obscured under layers of sand, silt, and coral. By 1859, when British Royal Navy helmet diver Jeremiah D. Murphy investigated the submerged town, only the tops of the highest buildings were visible (Pawson and Buisseret 1975:144-145).

Although various explorer divers surveyed the site in the following years, it was not until the 1950s and the development of SCUBA (self-contained underwater breathing apparatus) that extended diving became feasible. In May 1954, Mr. and Mrs. Alexis Dupont of the United States dived in the area around Fort James and discovered an arched doorway with a flight of steps, as well as various bottles, bricks, and roof tiles (Mayes 1972:9; Pawson and Buisseret 1975:145). (The reader is referred to Figure 3.1 for the location of both this and the other excavated areas here discussed.) Two years later, American engineer and underwater explorer Edwin Link discovered and raised a cannon from the same location. In 1959, Link and his wife Marion returned to Port Royal, where they led the first serious attempt to explore the sunken city in an expedition jointly sponsored by the Smithsonian Institution, the National Geographic Society, and the Institute of Jamaica. The team excavated in the area of Fort James and the King's Warehouse and mapped what was believed to be the outline of the area of the city that had sunk in the earthquake (Link 1960; Mayes 1972:9). This fairly accurate pre-1692 map, which was reconstructed from analysis of the real estate records from 1661-1670, has been used by all subsequent researchers. Unfortunately, Link never prepared a final report of the excavation, and the artifacts he recovered were never illustrated or described.

After a brief visit to Port Royal in 1960 by Norman Scott, which resulted in the recovery in the area of Fort Carlisle of roof tiles, glass bottles, clay tobacco pipes, and a wooden wheel believed to have been part of a gun carriage (Mayes 1972:9; Pawson and Buisseret 1975:145), the second major underwater project at the site was undertaken by Robert Marx, an American marine archaeologist, who was hired by the Government of Jamaica to explore an area threatened by plans to construct a deep-water harbor and pier. Between 1966 and 1968, Marx excavated two acres of the southwestern part of the devastated city, near the old naval hospital, where he found the remains of fish and meat markets, turtle crawls (pens), at least two taverns, three ships, and

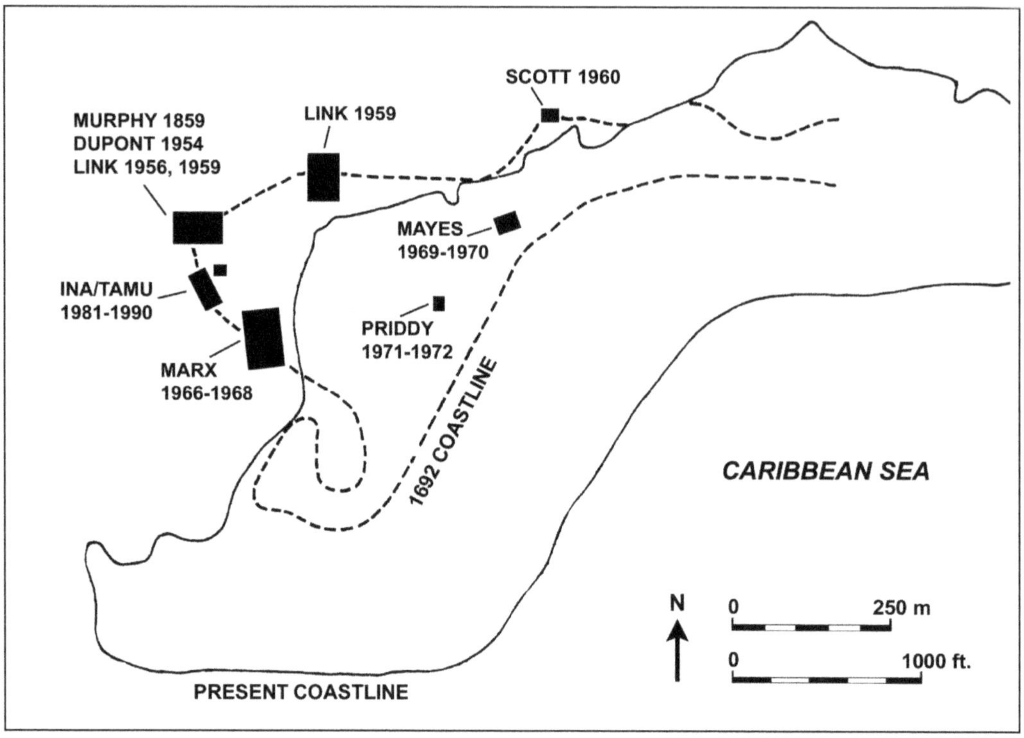

FIGURE 3.1. The coastlines of modern and pre-earthquake Port Royal, showing areas of archaeological investigation

various dwellings, as well as pewter, brass, and silver objects, clay tobacco pipes, glass and ceramic bottles, and mounds of bricks (Marx 1967, 1968a, 1968b, 1968c, 1969, 1971). Much of the area investigated by Marx consisted of jumbled buildings, and it is difficult to correlate his excavations, since none of the mapped structures is aligned with any known streets of the submerged ruins and much of the area lay outside the city boundaries. The site provenance of many of the recovered artifacts is also highly suspect due to excavation techniques. Marx excavated the most extensive area of the city to date, but the artifacts were never throughly studied, and no final report was published (Hamilton 1984:15; Hamilton and Woodward 1984:42). Moreover, sad to say, over the years, many of the artifacts have been lost.

Given the problems of the site's location, in silt-laden waters of poor visibility, the excavation tools then available, and the huge backlog of artifacts awaiting conservation, the Jamaican government in 1968 decided to suspend the underwater investigation of Port Royal until more satisfactory conditions could be created (Mayes 1972:10). They turned their attention, instead, to the commercial development on the site of the post-earthquake naval dockyard, and secured the services of Philip Mayes, an English archaeologist, whose interests lay in land rather than in marine archaeology. From 1968 to 1970, Mayes (1972) investigated a number of 'dry' sites, including a large part of the old dockyard, where he uncovered mostly 18th- and 19th-century components, and the amazingly well-preserved 17th-century St. Paul's Church in a below water table, land-locked environment (Mayes 1972:45-66). He also plotted the position of many features in the submerged city (which allowed him to create a comparative sequence of scaled maps), developed techniques to improve below water table excavation, and established badly needed on-site conservation facilities in the old naval hospital (Mayes 1972:11, 133).

Mayes was followed, in 1971, by another English archaeologist, Anthony Priddy, who for four years conducted another land excavation, this time on New Street, located near the town's central business district (Priddy 1975). This area was densely packed with 17th-century buildings that had survived the earthquake and had been occupied continuously into the late 18th century. The investigation yielded thousands of artifacts and architectural features, but the final site report was never completed, and it is only in recent years that other researchers have used Priddy's data and collections (see Cox and Cox 1987; McClenaghan 1988; Cox 1992; Heidtke 1992; Brown 1996).

HAMILTON'S EXCAVATIONS, 1981-1990

In 1978, at the request of the Jamaican government, the Institute of Nautical Archaeology (INA), in cooperation with Texas A&M University (TAMU) and the Jamaica National Heritage Trust, resumed the archaeological investigation of the underwater site of Port Royal (see Figure 3.1). The work, which continued from 1981 to 1990, was directed by Dr. Donny Hamilton, Associate Professor of Anthropology at TAMU. The main objectives of the project were to identify and excavate a series of submerged 17th-century buildings in the commercial center of the town. It was hoped that the findings would create a detailed picture of life in the city up to the time of its destruction in 1692 (Hamilton 1992:41).

The area investigated by Hamilton's team is located southwest of Kingston Harbor, at the intersection of the northern ends of Queen and Lime Streets. The buildings in this part of the city were built on a relatively flat area close to the harbor and sank into the water almost vertically due to the effects of liquefaction. Hence there is very little horizontal disturbance of the 17th-century components. Unlike the majority of terrestrial archaeological sites, which usually contain refuse of variable date divorced from its original context, the site offered the chance to study the recovered material in the context of contemporary use. With the extensive historical records pertaining to the city that are available, it has been also possible to illuminate less tangible data, such as social/economic class-role differences, ethnic-religious disparities, and the emergence within Port Royal of several distinct communities. Indeed, the 10-year INA/TAMU investigation has already generated a vast amount of written material, including journal articles (Hamilton 1984; Hamilton and Woodward 1984; Meyers 1998), chapters in edited books and conference proceedings (Hamilton 1986a, 1986b, 1988a, 1991, 1992), masters' theses (Wadley 1985; McClenaghan 1988; Gotelipe-Miller 1990; Franklin 1992; Heidtke 1992; Thornton 1992; Clifford 1993; Darrington 1994; Brown 1996) and doctoral dissertations (Hailey 1994; Smith 1995; Dewolf 1998; Fox 1998), as well as the Institute of Nautical Archaeology's quarterly newsletters (Hamilton 1987a, 1988b, 1990a, 1990b). A final report is in preparation.

When Hamilton's team first explored the site, previous reports of zero visibility, difficulties in reaching the 1692 context, and dangers from collapsing walls, jellyfish, and sharks all proved to be unsubstantiated. Visibility was normally between 4 and 10 ft., clear enough for the divers to maintain good stratigraphic control, plot the location of artifacts within an established metal grid framework, prepare detailed drawings of architectural features, and take underwater photographs. Due to the short distance of the site from the shore and the shallow water (most of the area investigated was in depths of less than 15 ft.), SCUBA was not necessary. The divers, instead, were connected by air lines to an air compressor that was itself situated on a barge anchored over the site. A water dredge was used to expose the archaeological deposits from the overlying sediments; standard archaeological tools, such as trowels, line levels, and measuring tapes, were used to maintain proper controls, as floors and walls and artifacts were slowly and carefully revealed (Hamilton 1984:16; Hamilton and Woodward 1984:42-43).

As noted in Chapter I, the excavation techniques allowed the submerged portion of the site investigated by Hamilton's team to be separated into three natural layers (Figure 3.2) (Hamilton 1984:22-23). Layer 3, the deepest layer, and the layer upon which the city was built, consists primarily of sand. It contains partially intact brick and plaster floors of the 17th-century structures, rubble from the collapsed buildings, and a mixture of 17th-century artifacts. This layer is the least contaminated in the whole site. Layer 2, above Layer 3, contains 18th- and early 19th-century artifacts in its upper levels and admixtures of 17th-century material in its lower levels. It consists of fragments of finger and elkhorn coral believed to have been deposited in 1722, when a hurricane and two earthquakes hit Jamaica. This dense coral layer is particularly significant, since it serves as a barrier between the 17th-century layer below it and later material deposited above. It has thus helped maintain the almost pristine nature of Layer 3. Layer 1, the uppermost and most recent layer, consists of a combination of roots, turtle grass, and loose silt, as well as post-1722 and 20th-century refuse from Port Royal and Kingston Harbor.

Port Royal was surveyed and built in the 17th century using the English measuring system. While metric measurements are generally the standard in modern archaeological excavations, it was decided in this case to use the original system. This has facilitated the delineation of street alignments and building locations when comparing the submerged remains with historical maps of the city, all of which use scaled English units. All of the modern maps available from the Jamaican Survey Office are likewise in English units. The excavation grid was thus divided into 10-x-10-ft. squares. Within the 17th-century layer (i.e., Layer 3), each square was further divided into four 5-x-5-ft. squares, which were themselves subdivided into four 2½-ft. quadrants. Each of the divisions (10ft., 5 ft., and 2½ ft.) was given an individual lot number, and all architectural features and associated material remains were mapped and recorded according to the division from which it was retrieved.

The team relied on computer-assisted drawings programs to plot architectural features and the various artifact distributions on to computer-generated excavation plans (Hamilton 1987a:5-6). Indeed, the project was a first in archaeology for its use, from 1987, of SHARPS (Sonic High Accuracy Ranging and Positioning System), an underwater three-dimensional mapping system that uses high-frequency sound to measure precise distances. This system allowed for exact comparisons between pre-1692 maps of Port Royal and the present underwater topography (Hamilton 1988b:7).

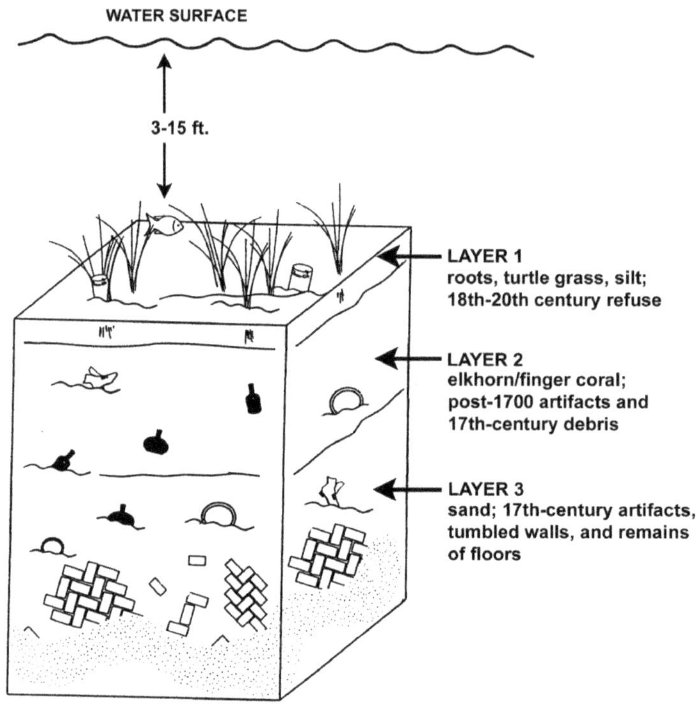

FIGURE 3.2. The stratigraphy of the portion of the Port Royal site excavated by Hamilton from 1981-1990

Hamilton's Excavated Buildings

As noted in Chapter II, many of Port Royal's buildings were fashioned after those in contemporary London: many were made completely of brick, or were a combination of brick and timber, while others were hastily built, simple, earth-bound frame structures (Hamilton 1992:44). The buildings excavated by Hamilton have further revealed that they were usually multi-storied and, in most instances, had wood shingle roofs (some clay roof tiles were also found); floors of brick, plaster, or dirt; and interior walls of wood or brick, the last often plastered. Many of the buildings, too, yielded a vast amount of associated artifacts and have a complement of historical records that pertain either to the structures' owners or occupants or to the makers of the recovered materials (Hamilton 1992:44). (The reader is referred to Figure 3.3 for the location and architectural layout of all of the excavated buildings here discussed.)

Building 1

Building 1, a well-built structure measuring 53 ft. wide and 47 ft. deep, was excavated during the 1982-1985 field seasons. It consists of six ground-floor rooms divided into three separate two-room units, and each unit appears to have housed a distinct business or activity. The volume of fallen bricks on the floors and the remains of components for a staircase indicate that there was also at least one upper story, which probably held living quarters. The three units that comprise the structure are aligned with the south side of Lime Street. One room in each of the units faces directly on to Lime Street; the second room in each is located behind the first. The rooms in each unit are connected by an interior wooden doorway. Plastered, whitewashed walls and herringbone-patterned brick floors comprise the front three rooms. The bricks of the back-room floors, in contrast, were laid out end to end (Hamilton 1984:21, 1986a:105, 1992:44; Hamilton and Woodward 1984:44).

Given the large assortment of leather scraps and shoe soles that were found, as well as a wooden lathe and some planks, it appears that one of the units (Rooms 1 and 2) housed a combination cobbler/wood turner's shop. Large quantities of cut animal bones and sea turtle shells suggest also that butchering and/or food preparation occurred in the unit's back area. The large number of recovered artifacts associated with the selling and consumption of alcohol suggests that Rooms 3 and 4, which comprise the middle of the building, were used as a tavern. Together with tankards and kegs, large ceramic jugs, and at least 60 dark-colored glass liquor bottles were recovered from this area. The artifact assemblage in Rooms 5 and 6 consisted of masses of new white clay tobacco pipes, glass bottles, and pewter plates, suggesting that this unit was used as a combination pipe/wine shop (Hamilton 1984:21, 1986a:108, 1992:44; Hamilton and Woodward 1984:45).

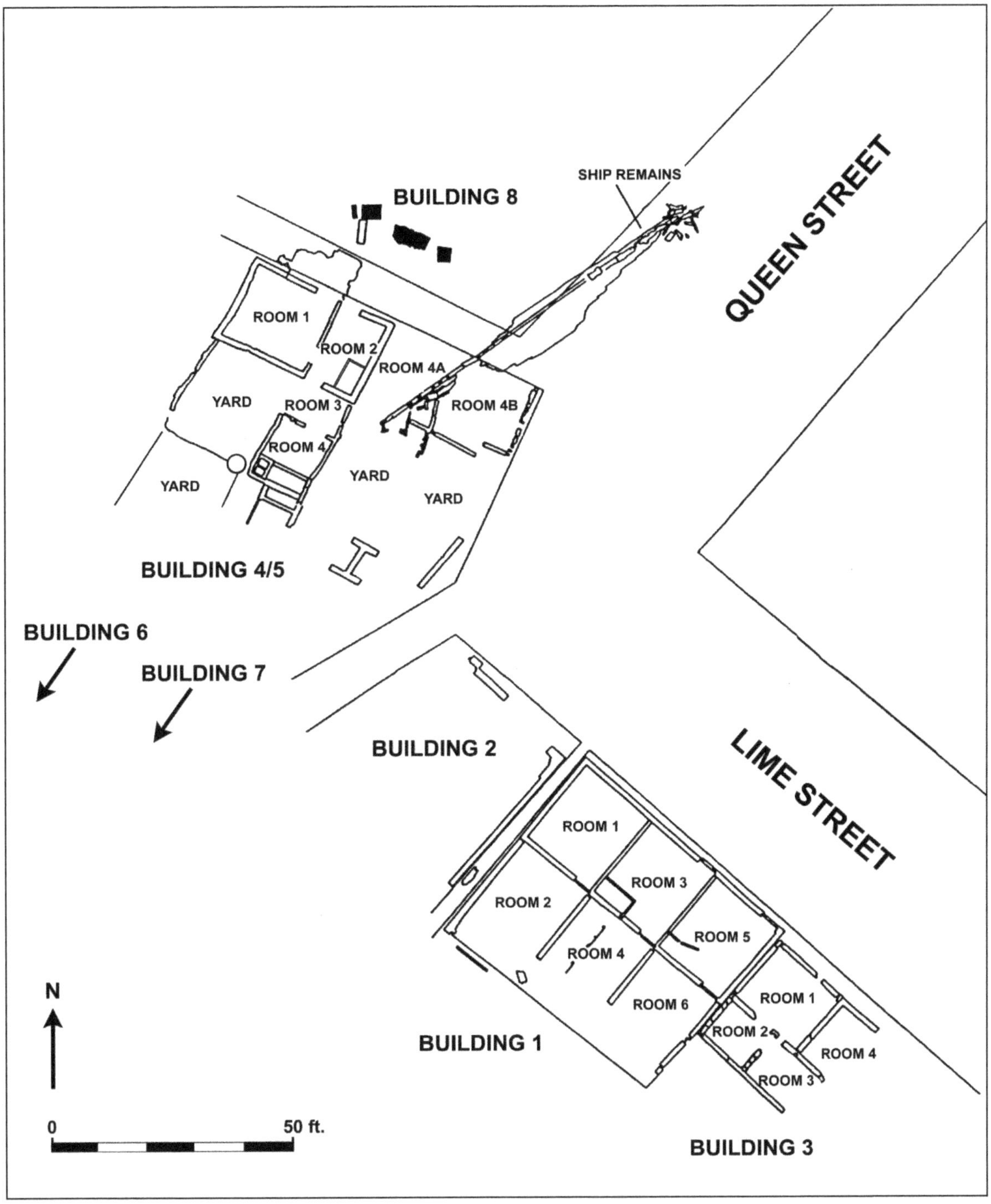

FIGURE 3.3. Plan of the eight buildings at Port Royal excavated by Hamilton (1981-1990), showing locations of walls and possible architectural layout

Buildings 2 and 3

The excellent preservation of Building 1 was not, unfortunately, mirrored in either Buildings 2 or 3, both of which were excavated during the 1985-1986 field seasons. While Building 2, like Building 1, appears to have faced on to Lime Street, its poorly preserved wood frame structure allowed for neither its exact size nor function to be determined. Its only cohesive internal remains were part of a plastered floor in one room and an array of floor planks in another (Hamilton 1988b:6, 1992:44).

Building 3, an earthfast, interrupted-sill, frame building measuring about 38 ft. wide and about 27 ft. deep

(Darrington 1994:44), is slightly more intact than Building 2. Two rooms in this building face on to Lime Street, and two rooms, or possibly yard areas, are located in the back. The front rooms each have crumbled remnants of plastered floors, and the yard areas show remains of brick paving (Hamilton 1992:44). The function of Building 3 could not be determined from the architectural remains. A large number of new white clay tobacco pipes, two corked wine bottles, and two steelyards with associated weights suggest that it was possibly used as a storage area for the various activities in the adjacent buildings and for the nearby outdoor market (Hamilton 1988a:9, 2001, pers. comm.).

Buildings 4 and 5
Buildings 4 and 5 were excavated in the 1987-1990 field seasons. These two distinct, yet attached, brick buildings form a large and rambling complex, measuring approximately 65 ft. wide and over 40 ft. deep, located along a narrow extension of Lime Street (Hamilton 1990a:5, 1992:44). Representing at least two, and possibly three, residences or commercial establishments, or more likely a combination of both, the Building 4/5 complex and its associated yard areas forms a well-defined discrete unit. It is from Building 4/5 that the ceramics examined in this study were recovered. Its architectural layout and artifact assemblage are discussed in more detail below.

Buildings 6 and 7
During 1987-1990, remains of Buildings 6 and 7, located southwest of Building 4/5, were also excavated. These remains included part of a brick-paved yard at the rear of Building 6 and part of a brick-paved yard and a hearth at the rear of Building 7. While the function of Building 6 cannot be determined from the yard deposits, Building 7 may have housed a blacksmith or tinker, since heavily burned piles of scrap metal and numerous tools and locks were found around the hearth (Hamilton 2001, pers. comm.).

Building 8
Building 8, the remains of which were located on the north side of Queen Street, near its intersection with Lime Street, was excavated in the summer of 1987. Although no clear layout of form was evident, with only the remains of a plastered floor and disassociated timber boards and window frame elements recovered, the structure contained a significant amount of Chinese porcelain (see Dewolf 1998). It is suggested that this assemblage could represent either the inventory of a tea/coffee house, an individual's private collection, or, more likely, "in the face of the entrepreneurial spirit of Port Royal in 1692," a commercial establishment specializing in the sale of fine-quality tableware (Dewolf 1998:144).

THE BUILDING 4/5 COMPLEX

Description of Building 4/5

The fourth building excavated by Hamilton's team (Building 4) was later found to be associated with the fifth building excavated (Building 5) (Figure 3.4). Architectural evidence revealed that Building 5 was the first of the two structures to be built, beginning as two ground-floor rooms (Rooms 1 and 2) with 1½-brick-wide exterior walls. Each room had its own front door opening on to a narrow extension of Lime Street. The rooms were themselves connected by an interior door. Excavation of Room 1 showed the remains of plastered walls and a plastered floor, while the floor of the smaller Room 2 was paved with brick laid in a decorative herringbone pattern. The thickness of the exterior walls and the remains of a staircase in Room 2 indicate there was at least one upper story in Building 5, which was probably used for living quarters. At a later date, as indicated by abutting walls, two more brick-paved rooms (Rooms 3 and 4) were added, as was a walled, brick-paved yard (Yard 5) with a cistern in its southeast corner and a privy in its northwest corner (Hamilton 1990a:4, 1990b:14). The purpose of Room 3 appears to have been to provide access to the yard and to join to the building an exterior kitchen (the 'cook room'), represented by Room 4, which contains a large hearth and oven (Hamilton 1990a:5, 1990b:14, 1992:44). It appears, from its location, that the cistern was also used by Buildings 6 and 7.

The presence in Building 4 of one-brick-wide exterior walls and half-brick-wide interior walls clearly show that it was a much less substantial one-story structure. Attached by a common wall to the eastern side of Building 5, it consists of two rooms (Rooms 4A and 4B), which are divided by an interior wall. Each room in Building 4 has its own back yard (Yards 4A and 4B), and the yards contain the remains of two hearths constructed back to back astride a dividing wall. While both rooms in Building 4 show remains of brick paving, the floor of Room 4B appears to have been the only one originally plastered. Both of the yards were unpaved. The architectural layout of Building 4 was disrupted by the earthquake, which badly affected several areas of the structure, including remains of doorways. Horizontal displacements, seen most readily at the east end, in Room 4B, have also skewed the floor and walls several feet. Interpretation of Building 4 is further complicated by the remains of an ca. 70-ft.-long ship that washed over from the harbor in the tidal wave that followed the earthquake. The ship ploughed through the building's front wall, heeled over on to its port side, and came to rest in the middle of the rooms (Hamilton 1990a:5, 1990b:16, 1992:44; Clifford 1993).

For several reasons, the ceramics from the rear of the Building 6 yard (Yard 6) and the rear of the yard and hearth belonging to Building 7 (Yard 7), as well as the ceramics from an excavated alley located east of Building 4, are included in this study as part of the Building 4/5 complex: The direction of the fallen walls indicates that Building 4/5 collapsed to the south and east. Thus objects that may have been in Building 5 also probably

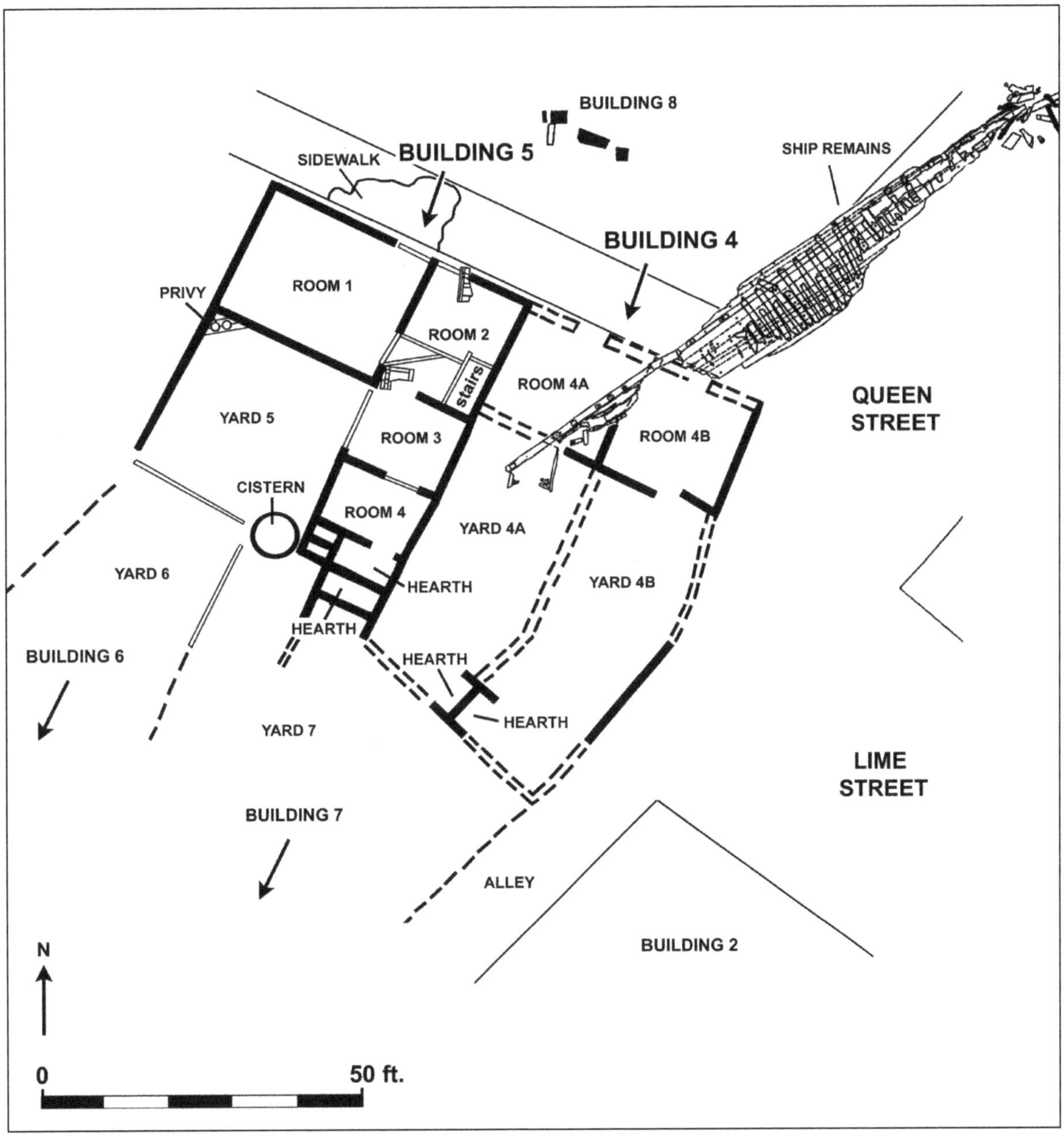

FIGURE 3.4. Plan of Building 4/5, showing locations of excavated walls (represented by solid black lines), suggested locations of walls/subdivisions (represented by dashed lines), and possible architectural layout

moved down in this direction, spilling out not only on to Yard 5 but also across a now destroyed wooden fence into the backs of Yards 6 and 7. Similarly, some of the items recovered from the alley, in all likelihood, came from Building 4.

Recovered Artifacts and Suggested Function of Building 4/5

Just as interesting as the structure's architectural remains were the artifacts recovered. Unlike most terrestrial sites, the underwater environment allowed a considerable amount of perishable, organic materials to be preserved. Wooden handles of axes and hammers, wooden barrels, a bucket and a table, a wooden mortar and a large wooden bowl, as well as leather shoe fragments were recovered from Building 4. Two bone combs, a wooden salt holder, a wooden window frame with lead caming, a calabash gourd dipper (possibly to hold fresh water, which had to be shipped from the main part of the island), and even textile fragments were just some of the organic remains found associated with Building 5 (Hamilton 1988b:6, 1990a:6, 1992:44).

Building 5's architectural layout and overall artifact assemblage suggest it was a reasonably comfortable private residence and/or commercial outfit dealing with food and drink. Room 1, being at the front of the structure and having plastered walls and a plastered floor, may have served as a best room or 'parlor'; alternatively, its separate entrance to the street and range of tableware may point to it being used as a public room in which meals were served to patrons. Room 2, also with its own entrance, provided access to the rear of the building, as well as to an upper floor. Stacks of 22 pewter plates found under the remains of the staircase, and 18 new white clay tobacco pipes and 10 corked glass bottles located near the interior door indicate Room 2 may have also functioned in part as a storage space. Room 3, towards the rear of the building, contained artifacts associated with food preparation, such as cast-iron and copper cooking pots, as well as a large brass strainer and a silver spice grater. Room 4, behind Room 3, was obviously the cook room with its hearth and oven (Hamilton 1990b:14, 1992:44-46; Dewolf 1998:117-118). A large cast-iron cooking pot was recovered from the middle of this room, and two long-handled skillets (still atop charcoal) were found in the hearth. Several measuring weights were also found associated with the hearth in Room 4. These weights are based on the Old English Bread System, indicating that they may have been used in bread production (Smith 1995:113-115). Yard 5 also revealed food-preparation items. A stone *metate* (three-legged grinding stone) and *mano* (hand stone) of Middle American design were the most unusual finds recovered from this location (Hamilton 1990b:14).

It is more difficult to ascribe a specific activity to either of the two rooms that comprise the virtually destroyed Building 4. The yards and hearth areas contained cut animal bones and artifacts indicative of food preparation (two cast-iron cooking pots, a cast-iron trivet, and a brass mortar), while the barrels, the handles of tools, and the bucket and table noted above, as well as scrap pewter and brass, indicate also some sort of domestic/manual activity in this area (Hamilton 1990b:16; Dewolf 1998:122). A few plates, bowls, and mugs suggest that the more poorly constructed Building 4 was also some kind of residence built by Building 5's owner. Perhaps it housed people who worked for the owners/occupants of the larger and structurally more substantial Building 5. Alternatively, it may have been leased out. There is always the possibility, given the location of the structure in the central business district of the city, that there was some commercial activity in Building 4 as well.

Possible Owners/Occupants of Building 4/5

Documentary evidence in the form of two land plats reveals that the plot of land on which Building 4/5 was located was originally owned by John Man, who, in 1665, subdivided it and sold a portion to Anne Thorne and a portion to James Galloway (see Pawson and Buisseret 1975:94, Map 9). No records have been found to indicate that the land was resold before 1692. Archaeological evidence in the form of ownership marks on pewter plates and on silver forks and spoons and on the silver spice grater recovered from the building complex reveal its possible occupants in 1692. Some of the plates (and the spice grater) bear the monogram '$N^{C}I$,' while others show the initials 'IC.' By convention, these plates would have belonged to a married couple, whose surname initial was 'C.' N is the initial for the husband's Christian name, I or J indicates the name of his wife (Hamilton 1990a:7, 1992:46). Nicholas Cransbrough, a vintner, who died some time before the earthquake, and whose probate inventory is on file at the Jamaica Public Archives, is a possible match for these initials, although there is no mention in his inventory of a wife (Jamaica Public Archives [JPA] 1693, 3:383-385). Nathaniel and Jane Cook, identified in the records as living in Port Royal prior to the earthquake, are other possible candidates (Island Record Office 1683, 15:166). Other initials on pewter items found in the dwelling are '$F^{S}D$,' '$I^{P}R$,' '$W^{C}E$,' '$W^{S}F$,' 'AC,' 'AM,' and 'HD.'

CERAMICS AT BUILDING 4/5: AN OVERVIEW

It is important to note that Building 4/5 is the least contaminated and also the most intensively investigated of the structures excavated by Hamilton's team. It was the only building where the fronting street, all of the rooms, the yards and associated hearths, and a cistern were completely uncovered. Further, more in situ artifacts were recovered from the complex than from any other single location in the submerged portion of the city. As was noted above, the archaeological evidence points to its possible function(s), and there is both archaeological and documentary evidence that sheds light on its possible owner(s) and/or occupants.

The author selected the ceramic assemblage from Building 4/5 for this study for several reasons. First, as noted in Chapter I, pottery has long been considered a valuable tool for reconstructing the past. Second, ceramics' technological and stylistic developments are often well established (and were especially so for the period with which this study is concerned), and the Port Royal material is particularly securely dated. Third, ceramics were used in the past for both utilitarian and social purposes. Their analysis thus enables the researcher to identify past habits and to speculate on the socio-economic status of their users.

Condition and Conservation

The condition of the ceramics examined for this analysis differs in several respects from those from terrestrial sites. Most noticeable is the attrition of colors and decoration from some of the finer wares. This is due to chemical alteration of the glazes, the detrimental effects of the absorbed salts, and necessary conservation treatment. Some of the more porous earthenware is friable and worn from the action of waves and currents. The particular nature of the Port Royal event also resulted in a greater number of large sherds than would be expected from a land context.

In common with the rest of the collection of ceramics from Port Royal, the data base sherds underwent mechanical cleaning before shipment to the Conservation Research Laboratory at Texas A&M University. The majority merely required several fresh water rinses to remove soluble salts. If a ceramic artifact is not rinsed immediately after it is removed from sea water, these salts will begin to crystallize within the clay fabric upon contact with the air, causing the object's surfaces eventually to exfoliate. This may lead to sherd, or even total vessel, destruction (Buys and Oakley 1993:23-24; Hamilton 1996:19-20). Some of the sherds in the assemblage required a more intensive cleaning to remove insoluble salts, such as calcium carbonate and calcium sulfate. The use of dental tools and surgical scalpels was sufficient in most cases to remove such marine growth. Some of the finer wares contained insoluble salts too difficult to remove and were submerged for a few minutes in a five percent hydrogen peroxide solution to bleach out the black sulfide staining. All ceramic sherds were air dried before they were sorted, catalogued, and stored in labeled plastic bags. Some have since been cross-mended as complete or partially complete vessels.

Catalogue Procedure

A general data base for all of the Port Royal artifacts has been established by Dr. Hamilton. This data base catalogues lot numbers, recovery layers, type of material culture, and other miscellaneous data, allowing a researcher to apply selected criteria across the entire assemblage and enabling the author to isolate the ceramics used for this study. The author's initial survey of this general data base, however, revealed inconsistencies in the ceramic data at Building 4/5, with some sherds incorrectly catalogued or missing. This was not unexpected, since numerous students have been involved in inputting artifact information. (Indeed, the general Port Royal data base is continually being updated.) More importantly, from the author's point of view, since this general catalogue was not designed to address specific questions of the data set to be studied, there was no information regarding vessel form, cross-mended sherds, or reconstructed vessels. The original artifact records, which are also housed in the Conservation Research Laboratory, are also undifferentiated.

To facilitate her research, the author, therefore, devised a fresh data base for the ceramics she examined at Building 4/5. In this, she has re-recorded the sherds individually and added amended data on the reconstructed vessels. Information on decoration, surface attrition, and cross-mending is also incorporated. This standardization allows detailed comparison of wares, types, and forms, and analysis of distribution patterns within the building complex and its associated yards. This new index is on file at the Conservation Research Laboratory. It is also attached to this study as Appendix B.

As already mentioned in Chapter I, the analysis of the collection by the author included sorting and cataloguing each sherd according to site provenance, ceramic ware and type, individual vessel form (where possible) and part, and sherd size. Several analytical coding techniques were used to assist with data management. These techniques fall in the following categories:

Ceramic Ware and Type

All 7440 sherds recovered from the portion of the Port Royal site excavated by Hamilton's team were catalogued according to a typological classificatory system known as the Port Royal Ceramic Typology (Hamilton 1986c, 1987b). This system organizes historical ceramics into groups according to a sherd's technological and stylistic attributes. In the entire Port Royal assemblage, nine general ceramic wares were identified according to manufacturing technology. These are coarse ware, slipware, tin-glazed earthenware, refined earthenware, stoneware, porcelain, creamware, pearlware, and whiteware. Each ware was then further categorized according to more specific criteria, such as ceramic type (and variety), date of manufacture, presence/absence of glaze, decoration, applied ornament, etc. Of the ceramics recovered from Building 4/5, only the sherds relevant to this study (i.e., the data base sherds, which includes those ceramic wares/types already in production prior to 1692, namely coarse ware, slipware, tin-glazed earthenware, stoneware, and porcelain) were examined by the author and that in a slightly revised and re-organized format (see Appendix B).

Ceramic Part

Having identified ceramic ware and type/variety, the author recorded each of the data base sherds by its vessel part, i.e., rim, base, body, or attachment (handle, spout, lid, lug, etc.). Many sherds in the assemblage include a

TABLE 3.1. Building 4/5's Vessel Forms, Descriptions, and Functional Divisions

Function	Vessel Form	Description
Food Processing	Colander	pan-like utensil with a perforated base
	Cooking Pot	large vessel, usually hemispherical in shape
	Pudding/Pastry Pan	cooking vessel, roughly in the shape of an inverted, truncated cone, less than 25 cm in diameter
	Pipkin	small, bulbous cooking pot with three legs and either one horizontal handle or two ears rising from the shoulder
Food/Beverage Storage	Bottle	bulbous-bodied vessel with a narrow neck, with or without a handle
	Storage Pot	thick-walled, cylindrical or slightly convex-sided vessel, taller than wide
Food Consumption	Bowl	open vessel with convex sides that terminate in a plain/everted rim
	Dish	serving/display vessel, larger than 25 cm in diameter
	Plate	eating vessel from 15-25 cm in diameter
	Porringer	hemispherical vessel with at least one handle, shallower in relation to its diameter; used for eating porridge, stew, and soup
	Salt Stand	pedestaled serving vessel
	Saucer	vessel less than 15 cm in diameter, used for serving condiments (hence, sauce-r) and perhaps as small plates
Beverage Consumption	Costrel	bulbous-bodied drinking and carrying vessel with a narrow neck, similar in form to a bottle but with two ears or strap handles rising from the shoulder
	Cup	(handled) drinking vessel of less than a pint in capacity
	Drinking Pot	one or multi-handled vessel with/without a spout/cover, ranging in capacity from at least one pint; known variously as a posset pot/ syllabub pot/wassail pot
	Ewer	handled, bulbous-bodied serving vessel, similar in shape to a jug but with a narrower, elongated neck with a gutter or spout
	Mug	single-handled, straight-sided drinking vessel, taller than wide
	Punch Bowl	hemispherical vessel with a plain rim, ranging in capacity from a half pint to several gallons
	Tea Bowl	as 'punch bowl' above but smaller
Health and Personal Hygiene	Apothecary Pot	cylindrical vessel with slightly flared rim and base; all sizes; used for drugs, ointments, cosmetics, and condiments
	Chamber Pot	handled vessel with convex sides and a sturdy flared rim
	Washbasin	open vessel with convex sides, of greater width than depth
Other	Flower Vase	stemmed, trumpet-footed vessel with a globular upper section

Source: Beaudry et al. (1983:29-37)

combination of these parts. Sometimes a 'sherd' is, in fact, a complete vessel. Sherd wall thicknesses and the height of intact or near-intact vessels were measured using metric slide calipers; where possible, a diameter chart divided into 1-cm increments was used to determine the diameter of vessels from partial rim and basal sherds.

Ceramic Form

Identification of vessel form, as already noted, was based upon a classificatory system outlined in Beaudry et al. (1983) (the Potomac Typological System, or POTS), in which ceramic vessels are described and classified according to shape and function, using terms from probate inventories and other 17th- and early 18th-century documents. Vessels, once identified, are then grouped under the system into analytical categories designed to distinguish among those used for food processing, food and beverage consumption, food and beverage storage, and health/personal hygiene. A total of 23 classifications of identifiable vessel forms was established for use in this study. They are presented with their descriptions and functional classification in Table 3.1. The author recognizes that these classifications provide only a general framework of the *intended* use of a vessel, and that it is always possible that a household used its vessels in a different way from that for which they were originally made. However, as Beaudry et al. (1983:28) note, "these categories represent the best fit achievable between the multifarious uses suggested by the documents and employable archaeological categories."

Slightly over 80 percent (n=1329) of the data base sherds (n=1617) were identified to form using this classificatory system. The remaining sherds (n=288) were too small to identify specifically and have been recorded as of 'unknown' form.

An important element of the analysis has been to estimate the minimum number of ceramic vessels that can be notionally constructed from the data base sherds. There are, of course, inherent difficulties in making such a judgement with simply a collection of broken pieces of pottery. "Sherd counts may be misleading as estimates of numbers of whole vessels," ceramicist Prudence Rice (1987:291) cautions, "because there is no simple relation between a vessel's size or shape and the number of sherds it breaks into: large vessels typically break into more pieces than small vessels; thin-walled pots may break into more sherds than thick-walled pots; and low-fired pieces may break into more fragments than high-fired objects." Accurate determination of vessel frequencies is also dependent upon correct identification of sherds from the same vessel, and upon a sufficient representation of such sherds to allow successful reconstruction of the vessel (Rice 1987:292). Fortunately, several of the identified vessels from Building 4/5 are near- or completely intact or have been partially reconstructed. In other instances, the author made informed judgements from diagnostic rim, base, and body sherds, and sometimes distinct attachments, such as handles and reconstructed lids.

Preliminary Results

The reader is reminded that the estimated number of vessels identified in this study as being used at Building 4/5 is only of clay containers. Household utensils made in wood, pewter, brass, and iron were also found in significant quantities in and around the building complex. Nevertheless, the ceramic assemblage is one of the structure's more rewarding artifact groups, with sherds and near-intact vessels found throughout the site The fragments represent a wide variety of ware types and vessel forms that range from coarse, heavy pots, simply finished and undecorated, to finer table wares, fragile and carefully painted. More impressive than the variety of types and forms is the numbers themselves, with 1617 sherds recovered, representing at least 189 vessels. The analyzed sherds (n=1155), it should be noted (i.e., those from Layer 3), represent slightly over 70 percent of the data base sherds and 76.2 percent (n=144) of the total number of vessels.

Table 3.2 presents the data base sherd count, by recovery layer and ceramic ware. Table 3.3 presents the estimated total minimum number of vessels, also by layer and ceramic ware. The remainder of this study concentrates on those sherds, and the vessels notionally constructed from them, found in Layer 3 (the identified 17th-century context). Figure 3.5 shows the relative percentages of wares represented by these 'analyzed sherds.' Figure 3.6 shows the relative percentages of wares by estimated minimum number of vessels. (While, as you can see, there are few differences in the ratios between sherd and vessel counts within each ceramic ware, the advantage of minimum vessel analysis is that it can be used to "obverse variations in the cultural organization of activities . . . both at a functional level [the concern of this study] and a symbolic level" [Yentsch 1990:25].)

So, what do the analyzed sherds signify? Are they fragments of the very latest and fashionable ceramics brought over from Europe in the many ships that docked at Port Royal's harbor? Indeed, are they of the quantity and variety that one would expect to find in a large residential dwelling(s) and/or commercial establishment(s) located in the heart of an important colonial maritime trading center? Or, do they, instead, comprise more common types, none too different from those in homes and commercial properties throughout the city, in other North American colonies, as well as in Europe?

And what of the identified vessels represented by these sherds? The latter part of the 17th century was a time of great change in the pottery industry, both in manufacturing technology and types of forms produced. Do the Building 4/5 ceramic vessels reflect these developments?

The following chapters address both these and related

questions. Chapter IV opens the discussion by looking at the history and development of ceramics, so far as they are relevant to the Port Royal site.

TABLE 3.2. Ceramic Wares by Data Base Sherd Counts (n) and Relative Percentages (%) from all Layers at Building 4/5

Ceramic Ware	Layer 3 (17th-century layer)		Layers 1 and 2 (post 17th-century layers)		Total	
	n	%	n	%	n	%
Coarse Ware	396	34.4	343	74.2	739	45.7
Slipware	80	6.9	9	1.9	89	5.5
Tin-Glazed Earthenware	588	50.9	56	12.1	644	39.8
Stoneware	76	6.6	36	7.8	112	6.9
Porcelain	15	1.3	18	3.9	33	2.0
Total	**1155**		**462**		**1617**	

TABLE 3.3. Ceramic Wares by Estimated Minimum Number of Vessels (MNV) and Relative Percentages (%) from all Layers at Building 4/5

Ceramic Ware	Layer 3 (17th-century layer)		Layers 1 and 2 (post 17th-century layers)		Total	
	MNV	%	MNV	%	MNV	%
Coarse Ware	52	36.1	16	35.6	68	36.0
Slipware	11	7.6	3	6.7	14	7.4
Tin-Glazed Earthenware	59	40.9	7	15.6	66	34.9
Stoneware	16	11.1	15	33.3	31	16.4
Porcelain	6	4.2	4	8.9	10	5.3
Total	**144**		**45**		**189**	

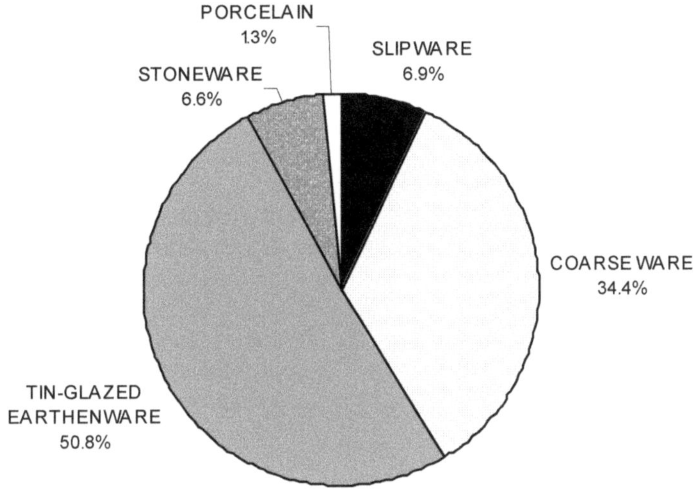

FIGURE 3.5. Relative percentages of ceramic wares, as represented by the analyzed sherds from the 17th-century context at Building 4/5

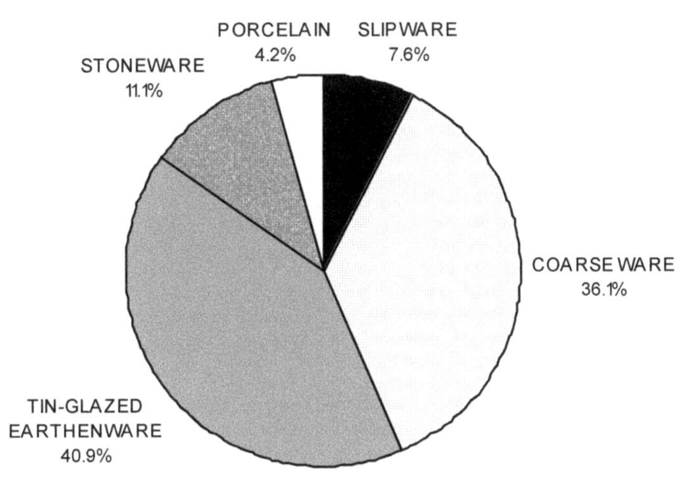

FIGURE 3.6. Relative percentages of ceramic wares by estimated minimum number of vessels, as represented by the analyzed sherds from the 17th-century context at Building 4/5

CHAPTER IV

HISTORICAL CERAMICS AND THE ANALYZED SHERDS

In the early 17th century, pottery making in England and throughout Europe could be very simply organized:

> A man needed a place to work, which, provided he had a house with some kind of yard or a smallholding attached, was not a great problem. He would build sheds to serve as stores and workrooms, and would also construct his own kiln. The few tools he used were easy to make, and his materials could usually be obtained locally. He needed no power supply other than his own labour . . . and would do most tasks himself, from digging clay and cutting fuel to transporting the wares to market and selling them (Grant 1983:47).

But as the market for domestic pottery increased so too did the scale and efficiency of ceramic manufacture. With the advance of technology came an explicit division of labor, and specialist positions emerged, such as the master potter, the thrower, and those who prepared the clay (Grant 1983:65-66; Rice 1987:184). By the late 17th century, the pottery industry in Europe, while still very small, was highly diversified. Its products, which reflect both the everyday and fashionable aspects of life, are one of the most tangible remnants of the changing material culture of the period.

This chapter discusses in detail the ceramics from Layer 3 (the 17th-century context) at Building 4/5 (i.e., the 'analyzed sherds'). The wares are organized according to clay fabric and manufacturing technology, i.e., under the categories of earthenware, stoneware, and porcelain. For the purposes of this study, the earthenware category is subdivided into three groups according to surface treatment, and proceeds from the relatively simple technologies to those which are more sophisticated. These groups are (a) coarse earthenware, or coarse ware, which may be unglazed or partially glazed, and which has minimal, or no, decoration; (b) slipware, which is a thinner, more refined decorated earthenware; and (c) tin-glazed earthenware, a second parallel development in decorated earthenware. Stoneware, the second category to be discussed, is recognized by its hard-bodied, impervious fabric. Porcelain, the third category, is a highly vitrified, translucent ceramic made from white kaolin clay.

The author discusses these three categories in turn by reference to the following: (a) general technology and development, (b) the examples found in Building 4/5 (and to a lesser extent, those found in the other buildings at Port Royal which were excavated by Hamilton), and (c) the manufacturing sources of those examples. A summary of the collection is presented at the end of the chapter. Functional classifications are the concern of Chapter VI.

COARSE WARE

Technology and Development

Coarse ware, as used here, refers to a soft, relatively coarse, porous earthenware that has been fired at temperatures ranging from about 900°-1200°C (Draper 1984:10; Rice 1987:5). Coarse wares are the most basic of all ceramics and have been produced for millennia in all areas of the world. They also comprise the bulk of pottery excavated from most sites that date to before the early to mid 19th century. Coarse ware vessels were used for common, everyday tasks. When broken, they were simply discarded, since they were cheap and could be easily replaced. Coarse wares offer a good starting point for exploring the process whereby clay is transformed into a finished product. Understanding the potting process, and how it was modified for different wares, is a necessary precursor to the examination of those products and their uses.

As noted, pottery manufacture by the 17th century had developed in many ways. But for the majority of producers of cheap, coarse utilitarian vessels, the advantages of low capital outlay and relatively simple technology remained (Weatherill 1971:32; Grant 1983:47; Draper 1984:8). Potters worked out of small, rural potteries, the appearance of the wares' fabric being dependent upon the color of the local clays. Tempering agents, such as sand or gravel, were often added by the potter to modify the clay's workability and to reduce shrinking and/or expansion of the vessel during the firing process. Most European coarse ware pottery of the 17th century, as was the case with pottery in Europe in general, was thrown on kick wheels, or 'treadles.' Wares that could not be thrown were molded or modeled in separate pieces (Weatherill 1971:32; Grant 1983:47; Draper 1984:10).

If coarse wares (and, indeed, all earthenwares) were to be glazed, which made them both waterproof and more attractive, this was done only after they were cut free from the wheel and were completely dry. The glaze, usually made from powdered galena, a lead ore, was sometimes dusted on to vessel surfaces. A better and more common method was to dip the vessel in a liquid mix of galena and finely ground clay suspended in water (Grant 1983:49; Draper 1984:10; Weatherill 1986:57). (In the later 17th century, 'litharge,' or lead oxide, which gives a shinier finish, began to replace galena [Draper 1984:10].) The color of the underlying clay body—or in the case of slip-decorated earthenware, the color of the slip coating—affected the appearance of the lead glaze, which, in its pure form, was virtually translucent. Impurities in the glaze invariably gave it some color, or minerals could be added, changing the color deliberately. For example, a dark brown speckled and streaked effect is the result of either iron or manganese being added directly to the glaze. Varying amounts of copper fillings in the glaze produce numerous shades of green. The thickness with which the glaze was applied also changed its characteristics. A clear lead glaze, for example, appears yellow on a buff fabric but can range from a lemon color through amber to a dull sandy shade, depending upon its thickness. On red-colored earthenware, the glaze can appear reddish brown to orange (Draper 1984:11-12; Pearce 1992:5-6).

Firing hardens a clay vessel and turns the glaze shiny. As Weatherill (1971:36, 1986:60) notes, firing of clay vessels in the 17th century was largely undertaken on a 'hit-or-miss' basis, since methods to ascertain kiln temperatures were not invented until the late 18th century. Indeed, little is known of early kilns and firing practices. That English kilns were quite low and rather small, approximately 6 ft. wide and slightly over 8 ft. tall, and surrounded by a wall of clods to retain the heat, is indicated by contemporary observer Dr. Robert Plot (1686:123). The reconstructed foundation of an early 18th-century kiln at the Albion site in Staffordshire reveals that firemouths were neither evenly spaced nor of the same size or shape (Weatherill 1971:Plate 7, 1986:61; Grant 1983:50).

Coarse wares, unlike finer earthenwares, were fired once at relatively low temperatures, using a variety of fuel, such as wood, peat, or coal (Grant 1983:42-43; Draper 1984:10; Weatherill 1986:57). The firing took a long time, "up to three days," and "was not an easy process, the construction of the kiln and the type of fuel used had to be thoroughly understood by the potter; weather conditions had to be allowed for, as uneven draught could lead to overheating of some wares and underheating of others" (Grant 1983:51). The vessels were packed tightly in the kiln, with the heavy, glazed pots positioned at the bottom, where it was hotter and where drips of glaze would not spoil anything below. (Some drips were unavoidable, however, as shown by some of the Port Royal ceramics.) Occasionally, pottery vessels were protected in the kiln in fireclay containers called saggars, sometimes known as shiggs or slugs (Edwards 1974:21; Britton 1982:17, 1987:190), or, according to Plot (1686:123), "shragers." These protected the vessels from "the vehemence of the fire, which else would melt them downe, or at least warp them" (Plot 1686:124). Firing vessels in saggars also protected the colors on the vessel surfaces.

When the potter decided that it was time to cool the kiln, he had all of the holes in the oven wall closed, cutting off the oxygen supply to the fire. The cooling period for pottery is particularly critical, for too rapid cooling can cause vessels to split and flake. The cool down, like the firing, took several days.

Coarse Ware at Building 4/5

In the Building 4/5 analyzed ceramic assemblage, the coarse ware group can be generally included in a utilitarian, or non-tableware, category, and vessels can be defined by their use in food processing (i.e., cooking and preparation) and food and beverage storage. Some coarse ware chamber pots were also recovered in the excavations. Certain types of coarse ware overlap the tableware and utilitarian categories. This aspect of vessel form and function is discussed more fully in Chapter VI.

As shown in Figure 4.1, most of the analyzed coarse ware sherds were found in the yards at the back at the building complex. It is probable that many of these fragments are parts of vessels that were in active use at the time of the earthquake. It is also possible that at least some of the fragments in these areas were intentionally dumped there as trash, the vessels of which they were a part having broken and become useless during the life of the building (see South 1977:47-80). The coarse ware distribution pattern is also discussed more fully in Chapter VI.

A total of 396 sherds, or 34.3 percent of the analyzed sherds, was identified as coarse ware. (This can be compared to the 1048 coarse ware sherds recovered by Hamilton from all of the other excavated buildings combined [Layer 3 only].) These 396 sherds represent an estimated minimum of 52 vessels, or 36.1 percent of the total minimum number of vessels taken to be associated with the building prior to 1692. The coarse ware represented is a somewhat eclectic group, with English, Continental European, and local African slave wares. The European wares consist of wheel-made, kiln-fired, thick, fine-grained sherds, often with colored glazes. The African-Jamaican wares, in contrast, are hand-coiled, are often heavily tempered with sand, and show evidence of open-hearth firing (sherds often have distinct cores).

English Coarse Ware at Building 4/5

Some 215 sherds, or slightly over half (54.3 percent) of the analyzed coarse ware, are almost certainly of English provenance. These sherds represent an estimated minimum of 30 vessels, or 57.7 percent of the total estimated minimum number of coarse ware vessels (Figure 4.2). Seven additional sherds, representing at least two vessels, are also possibly English, based upon fabric color and surface treatment. The fragments are, however, too small to identify with certainty, so they were recorded in the appropriate 'unidentified' (i.e., glazed and unglazed) categories in the coarse ware typology created and used for this study (see Appendix B).

Redware. The majority of the English coarse ware sherds are wheel-thrown, undecorated Redware (n=153), which taken together represent at least 20 vessels (see Figure 4.2*a-c* for examples). Sherds were recognized by their fabric color, which is generally uniform on both surfaces and which ranges from a light red (Munsell 2.5YR 6/6) through red (2.5YR 4/ -5/6)

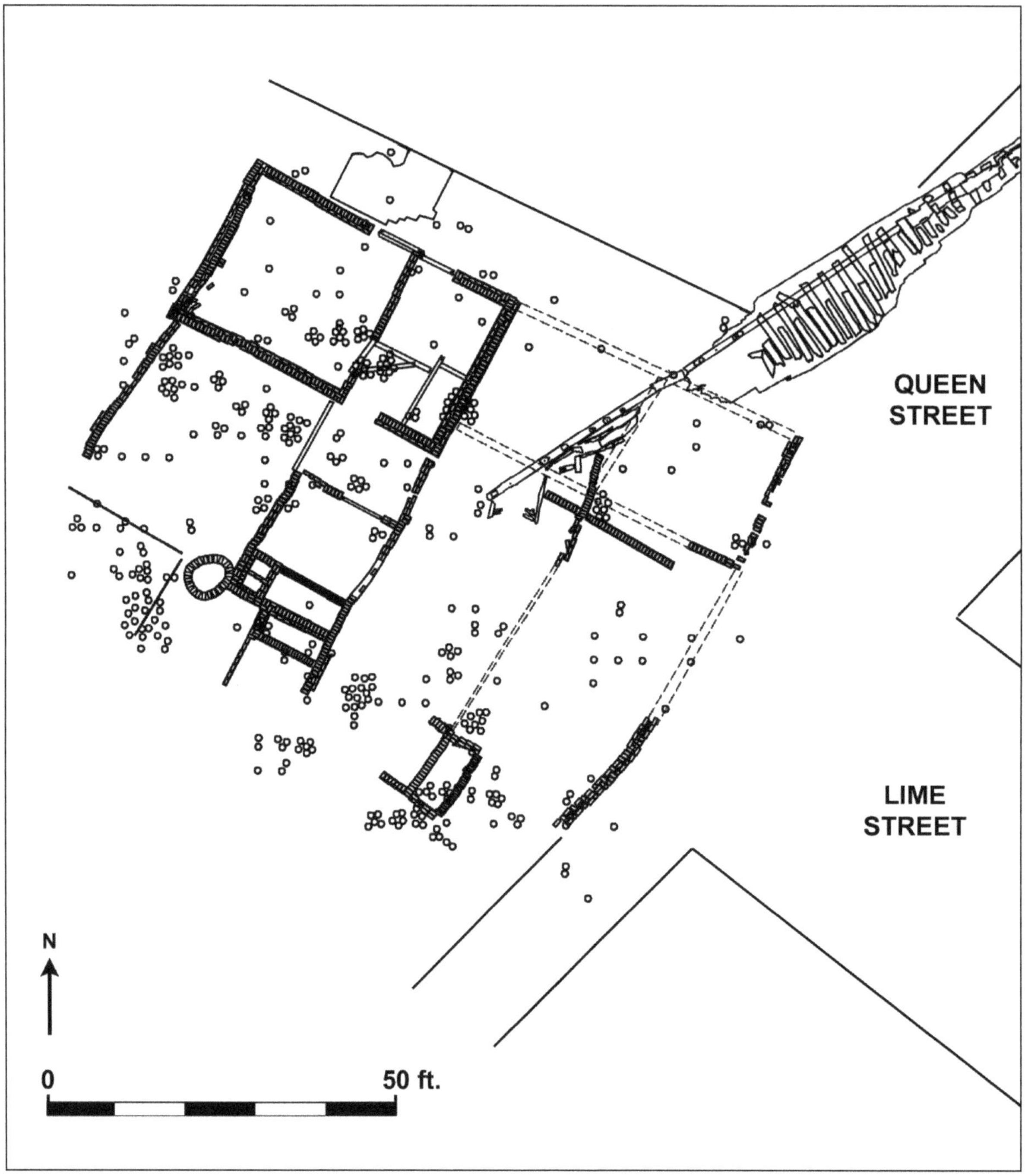

FIGURE 4.1. Distribution of coarse ware ceramic sherds in the 17th-century context at Building 4/5

FIGURE 4.2. Examples of English coarse ware recovered from the 17th-century context at Building 4/5. *a*, intact Redware storage pot (interior glaze); *b*, intact Redware cooking pot (interior glaze); *c*, neck and shoulder of Redware storage pot (interior and exterior glaze); *d*, intact red Border ware chamber pot (interior and exterior glaze); *e*, intact red Border ware storage pot (exterior glaze); *f*, intact red Border ware porringer (interior and exterior glaze)

through a reddish brown (5YR 5/4). Consideration of other criteria, such as construction of bases (usually flat), body composition (minimal mineral tempering), and surface finish (smoothed and usually glazed) facilitated the analysis. Nearly 30 percent of the Redware sherds (n=45) were unglazed on their exterior surfaces. This is not to assume, however, that the vessels to which these sherds originally belonged were themselves glazed on only one side. English potters in the 17th and 18th centuries often glazed only part of a vessel's exterior surface, if they glazed them at all. It was considered unnecessary to incur the high costs of glazing when most of these vessels were produced solely for utilitarian/kitchen use.

Despite the comprehensive information available for identifying most historical ceramics, documentation is notoriously sparse for English plain red coarse wares. Little specific information exists in primary or secondary historical records on their origins or periods of manufacture or even on their cost. That they were an inexpensive commodity compared to other wares is suggested by their near absence in 17th- and 18th-century household probate inventories and colonial bills of lading. Cheap to produce and easy to replace, these domestic utilitarian wares were obviously not worthy of specific itemization. [1] Further, identifying specific production centers is hampered by the lack of distinctive visual features of these wares, which were mass-produced throughout the 17th and 18th centuries. The southeast of England and the West Country (largely Devonshire and Somerset) were known to produce Redwares throughout this period (Noël Hume 1970:102; Edwards 1974:4-7), but they were also manufactured in small quantities from at least as early as the 1620s by English colonists in North America (see Watkins 1950; Noël Hume 1963; Kelso and Campbell 1974; Turnbaugh 1983, 1985).

The classificatory system used in this study, together with a consideration of documented vessel form and function, provided a strong and consistent means of tackling Building 4/5's Redware identification and description, albeit on a general level. Unglazed and partially glazed (interior and/or exterior) cooking, storage, and chamber pots, as well as a few bowls were identified. These vessels are discussed more fully in Chapter VI.

North Devon Ware. Other types of English coarse ware are more thoroughly documented in the literature. These include North Devon ware, which was manufactured in the southwest of England, and Red and White Border wares (see below), which were made near London. Both types are ubiquitous on early to mid 17th-century English colonial sites in America, and both types, in limited numbers, were recovered from Layer 3 in the Building 4/5 complex.

Identified North Devon coarse ware sherds (n=38) account for 17.7 percent of the analyzed English coarse ware. These sherds represent at least two interior-glazed storage pots and one unidentified vessel form. All but one of the sherds are heavily tempered with gravel, which was used "to open up the fine-textured [North Devon] clay to enable it to dry right through, and to reduce the time and temperature needed for firing" (Grant 1983:40). The single North Devon Plain ware (i.e., untempered) sherd recovered has a distinctly smooth, fine-grained body. "The materials of which [North Devon] ware is composed cost little or nothing," according to John Watkins (1792:79), an 18th-century local historian. The pottery's strength and heat-resistant qualities were as appealing as its inexpensiveness.

While a comprehensive survey of the North Devon industry has yet to be attempted, it has been established that by the 17th century, the main North Devon pottery centers of Barnstaple and Bideford manufactured a large quantity of plain and gravel-tempered coarse wares for everyday use. Much of the ware, like others, was lead glazed by this time, with large jugs and jars, dairy wares, baking dishes, chafing dishes, and dripping pans common types (Grant 1983:55). The North Devon potteries, too, made tableware, such as dishes, plates, and cups, some of which were decorated with a clay slip. (The Building 4/5 analyzed assemblage includes a slip-decorated North Devon ware sherd. It is discussed below with the other slipware recovered.)

North Devon ware was well made and efficiently marketed, both in England and in North America. Not only did many of the first immigrants in the northern colonies hail from Devon, providing the connections that laid the foundations of colonial trade in the 17th century, but the west of England in general was well placed to benefit from development of this trade. English shipping records, colonists' probate inventories, and archaeological finds of North Devon wares dating from as early as the 1630s reveal that many New England settlements had trading relations with Bideford or Barnstaple (Grant 1983:114-125). These potteries provided the developing rural New World settlements with the earthenware needed for household and dairy use. Port records reveal that Devon merchants also made contact, often via New England, with the English islands in the Caribbean, such as Barbados and Antigua, where earthenware was presumably needed for the developing sugar industry. Although household utilitarian vessels must also have been required, there was little dairying to create the large demand which was the basis of the trade elsewhere (Grant 1983:126). Indeed, archaeological excavations yield far fewer North Devon ware sherds from English Caribbean sites than are recovered from sister settlements further north. According to historian Alison Grant (1983:127), Jamaica, in particular, due to its location, was seldom a first destination for English ships going directly from North Devon: "Only one vessel, the Jerimiah of Barnstaple, [in 1695] took [North Devon] earthenware direct to Jamaica in the seventeen years for which port books survive for the period." However, as Grant (1983:127) continues, a ship called the Port Royall, which brought sugar to Bideford in 1688, is likely to have been built for trade with Jamaica. It seems reasonable to suggest, given the evidence not only at Building 4/5 and the rest of the Port Royal site excavated by Hamilton but also from Mayes' (1972:81) excavation of the city's old naval dockyard site, which yielded fragments of bowls and milk pans, that practical coarse earthen pottery was part of this exchange.

The North Devon industry peaked around the 1680s, and this is indicated by the small number of sherds found at Port Royal. The Atlantic trade had decreased considerably by the end of the 17th century, "partly through competition with other ports, and finally because the risks of wartime trading in the 1690s could scarcely be justified" (Grant 1983:125). It did not provide the returns to the merchants to warrant its continuity. As the demand for finer wares increased in the 18th century, the industry at Bideford and Barnstaple began a long and slow decline.

Border Ware. More refined than North Devon ware, 'Border ware' denotes the various products of the extensive potteries that developed along the borders between the counties of Hampshire and Surrey, southwest of London (Orton and Pearce 1984:35-36; Pearce 1992). The location of this region, "where agriculture was hindered by poor soils, but where clay deposits and woodland were available" (Crossley 1990:248), undoubtedly influenced the growth of the Border ware industry. Its proximity to London was an obvious additional advantage. Border ware circulated over much of England in the 16th and 17th centuries. The term includes wares with an off-white to buff fabric, and with yellow, green, olive, or brown glaze; and wares with a reddish colored, slightly rough fabric, and with clear, green to olive or brown glaze (Orton and Pearce 1984:35-36; Pearce 1992).

Border ware sherds (n=24) account for 11.2 percent of the analyzed English coarse ware at Building 4/5. These sherds represent an estimated minimum of seven vessels, over half (n=4) of which are the red, sturdier variety (see Figure 4.2*d-f* for examples). Interestingly, these numbers do not concur with other Border ware finds in well-dated late 17th-century contexts in London. There, the proportion of Red Border ware is minimal compared with the quantities of White Border ware (Orton and Pearce 1984:Figure 30; Pearce 1992:97-101). As far as the London market was concerned, White Border ware was considered the most desirable; Red Border wares had a more limited local distribution.

Border ware evolved from a flourishing and long tradition of potting that dates to as early as the mid 13th century (Pearce 1992:89). Its most direct predecessor, thin-walled, untempered, green-glazed tableware, known collectively as Tudor Green, was introduced into the Surrey industry from the Continent in the late 15th century (Crossley 1990:248; Pearce 1992:89). Border ware 'proper' appeared in the late 16th century. Cheaper and easier to make and less fragile than Tudor Green, Border ware was more extensive in its use and appeal, and the industry flourished throughout the 17th century (Pearce 1992:5-6).

The major output of the Border potteries from the late 16th century was for kitchen and general household use, with "quantity rather than quality the main concern" (Pearce 1992:84, 102). While many forms were lead glazed only on interior surfaces, some vessels, such as chamber pots, porringers, and mugs, were glazed on both surfaces. Yellow glaze is predominant, but green to olive and brown glazes were also used (Pearce 1992:85, 94). Bowls, sometimes with handles, mugs, porringers, tripod pipkins (cooking pots with three short stubby feet and one long rod handle), and chamber pots were among the most common forms used throughout the 17th century (Pearce 1992:97). All but the mug and bowl forms were identified in the analyzed sample from Building 4/5.

The Border ware industry peaked towards the end of the 17th century, when it was one of the major suppliers of everyday utilitarian pottery to London and the south of England (Pearce 1992:102). As with the North Devon industry, it appears that the increasing popularity of decorated tableware was largely responsible for the industry's decline in the first half of the 18th century.

Continental European Coarse Ware at Building 4/5

Iberian sherds (n=48, 12.1 percent of the analyzed coarse ware), identified on the basis of fabric color, surface treatment, and form, comprise the majority of the analyzed Continental European coarse ware sherds (n=49). Taken together, these sherds represent at least five vessels, or 9.6 percent of the estimated minimum number of coarse ware vessels taken to be associated with Building 4/5 prior to 1692 (Figure 4.3). A sixth vessel, represented by a basal sherd from a 'costrel,' or oval-shaped flask, is possibly Italian in origin (see Figure 4.3*a*). This manufacturing source is, however, uncertain, and the sherd has been recorded in the 'lead-glazed unidentified' category in the coarse ware typology (see Appendix B).

Of the Iberian sherds, an attached rim and basal fragment of one large storage jar, several unattached body sherds of (possibly) another, several neck and shoulder sherds from a partially reconstructed ewer, and remains of two olive jars are represented. The sherds from the storage jars were identified as of Iberian manufacture on the basis of Deagan's (1983, 1987:36) 'Spanish Storage Jar' type description, i.e., "unglazed coarse-earthenware sherds that fall within the range of associated Olive Jar paste but have elements of form that eliminate them from being classified as Olive Jars." The form represented by the rim and base is similar to the Spanish *bacín* (basin), with its straight sides, everted wide-mouthed rim, and flat base. Both it and the body sherds from the second form have thicker walls than olive jars (20+ mm vs. 10-15 mm), and have a light yellowish brown (10YR 6/4) to reddish brown (5YR 5/4) fabric that in areas appears dark gray.

FIGURE 4.3. Examples of Contintental European coarse ware recovered from the 17th-century context at Building 4/5. *a*, Italian (?) costrel basal sherd; *b*, reconstructed neck and shoulder from an Iberian ewer; *c*, neck and mouth of an Iberian olive jar

The partially reconstructed white, unglazed, two-handled ewer is, perhaps, one of the most interesting finds recovered from Layer 3 at the building complex (see Figure 4.3*b*). The form, which is Middle Eastern in origin, appears throughout the Moorish period on the Iberian Peninsula, and similar ewers have been recovered from 17th-century American sites that were in contact with Spanish shipping in the 16th and 17th centuries (Lister and Lister 1987:13, 19, 27).

Iberian olive jar sherds (n=26), representing at least two vessels, are, unsurprisingly, the most prominent of Building 4/5's analyzed Continental European coarse ware inventory (see Figure 4.3*c*). Identification of this vessel form was based upon form and fabric texture (relatively fine-grained) and color (ranges from a very pale brown [10YR 8/3, /4 to 7/3, /4] to a light yellowish brown [10YR 6/4]). A white to pinkish white (5YR 8/2) film is evident on most sherd exteriors, and a characteristic green glaze is evident on the interior of some fragments. (Marken [1994:42], citing James [1988:51], notes that this white film may be a chemical change attributed to the firing process.)

Olive jars are, perhaps, the most common of the New World Iberian ceramics. They have been recovered from sites that date from the late 15th century to well into the 18th century. Particularly abundant are finds in 16th- and 17th-century Spanish colonial terrestrial and shipwreck sites in the Americas (see, e.g., Marken 1994:41-138). The basic shape of the vessel is very old in the Mediterranean region, dating back to about 5000 B.C., and it has no real counterpart in English ceramics (Goggin 1960:5; Fairbanks 1973:143, 147). The exact provenance of these vessels has not yet been determined. Goggin (1960:5) suggests that Andalusia (Cádiz and Seville) was a major source of production; Fairbanks (1973:144) suggests that it is also possible that eventually they were made in the New World, at the numerous Spanish kilns known to have been established. It is also possible that the Portuguese made similar vessels.

John Goggin (1960) wrote what remains the definitive study of the Iberian olive jar form. This small-mouthed, sturdy jar was used primarily to ship and store all manner of goods, particularly liquids and foods (Goggin 1960:6; Fairbanks 1972:142; Deagan 1987:31). Wine, olive oil, vinegar, olives beans, chick peas, and lard are mentioned in colonial bills of lading as being shipped in *botijas*, or short-necked jugs (Fairbanks 1973:144; Lister and Lister 1987:133); excavations have revealed that once at the port of disembarkation, the jars often served as general storage containers for corn, flour, and other products.[1] Perhaps those found at Building 4/5 were used to store water. The slightly porous jars would allow the water to cool by evaporation, no doubt a welcome aspect in the warm Caribbean with no artificial refrigeration available.

Goggin (1960:8-21) established three styles and several types among the olive jars recovered from New World sites. The typology is based upon vessel form, as well as on the presence or absence of handles, surface treatment, and vessel dimensions. The work by Goggin aided greatly in identification of the Port Royal examples. It is, therefore, appropriate to outline his classificatory system, which is also illustrated in Figure 4.4.

The first of the three olive jar styles is known as the Early Style and dates from about 1500-1580. The form at this time is characterized as a "medium-sized globular vessel with a small flaring, or collared, mouth, having a loop handle on each side" (Goggin 1960:8). The body, which was made of a sandy fabric, was thrown on a wheel in two vertical halves; the neck and handles were added

[1] Soap was found in olive jars recovered from St. Augustine in Florida (Fairbanks 1973:144), and some 18th-century types have been found filled with pitch (James 1985:29, 41). It has been proposed that the Spanish chose to use such vessels for transport containers rather than the casks and barrels that were so important and abundant in northern European shipping because the Iberian Peninsula, unlike England and the Low Countries, had relatively scarce timber (Fairbanks 1973:143; Deagan 1987:31). However, Marken (1994:62) notes that in addition to the large quantity of olive jar fragments recovered from the Spanish *Atocha* (1622) shipwreck, a considerable number of barrel hoops was also found, suggesting "that in 1622, at least, a great deal of storage was still accomplished in casks."

later.[2] A thin white film is usually present on exterior surfaces of these jars; a green lead glaze is often present on the interior. The Early Style olive jar height ranges from ca. 21-26 cm; a cork or wooden stopper would have sealed its mouth (Goggin 1960:10; Fairbanks 1973:145; Deagan 1987:33).

The change to the Middle Style form in the late 16th century may represent merely a change in taste or a change in sources of supply from which New World specimens were drawn. It has basically the same fabric as used in the Early Style but differs both in shape and in method of manufacture (Fairbanks 1973:145-146). Further, Middle Style forms were more widely distributed and are more common than the earlier form due to the spread of Spanish New World settlement, as well as to increased trade (Fairbanks 1973:146-147).

Goggin (1960:12-13) distinguished three types among the slightly varied Middle Style shapes, all of which are somewhat elongated in form. The nearly spherical shape seen in Early Style jars is noticeably absent, handles are never present, and the flaring mouth of the Early Style is replaced by a thick ring neck. A thin white exterior film is common on Middle Style jars, and glazing can occur on either the interior or exterior surfaces. Middle Style olive jars date from ca. 1580-1780. Heights range from ca. 23-55 cm (Goggin 1960:13-14; Fairbanks 1973:145-146; Deagan 1987:33-34).

The Late Style olive jar, which dates from the last half of the 18th century and persists virtually until the present, is characterized by a new, finer fabric, almost without temper. Shapes vary, with perhaps a greater number of specimens with pointed bases, which would have allowed them to be suspended in a wooden ring stand or in a net sling (Goggin 1960:18; Fairbanks 1973:147; Deagan 1987:35). Marks of various kinds appear on both Late and Middle Styles. These may be impressed, incised, or engraved, are usually rather stylized, and may appear on either the thickened ring mouth or on the shoulder. Fairbanks (1973:146) suggests that all seem to be owners' marks, perhaps the marks of merchants identifying their trade goods, rather than potters' marks, as suggested by Goggin (1960:15). Indeed, as Marken (1994:76) points out, if the marks do denote ownership, it is reasonable to assume that the jars were recycled and, perhaps, used primarily as shipping containers. This would also support the assumption that the jars were intended for specific contents, shipped in fairly consistent volumes. However, so little is known about the manufacture of these jars that this is uncertain.

As expected, both of the olive jars identified from the analyzed sherds from Building 4/5 probably belong to Goggin's Middle Style. The complete rim of one of the jars has the characteristic thick ring neck covered with the white film (see Figure 4.3*c*). Fragments of the other jar show vestiges of this same exterior film on a fine, untempered fabric.

That Continental European wares should be recovered from Building 4/5 is not surprising, considering Port Royal's location in the center of the Spanish Caribbean and its renowned reputation as a bustling maritime trading center. Olive jars, in particular, are quite a common find in port towns of the period. Diagnostic sherds, for example, have been identified in some quantity in southern England and Wales. They have also been found in Holland, as well as in Virginia, Maryland, and Massachusetts (Watkins 1973:192).

Fairbanks (1973:143) accounts for the presence of olive jars in these and other regions as being the result of "sporadic trade from Spanish areas." Watkins (1973:192), however, suggests that this trade was regular, and that it was carried on primarily in English ships: "English trade laws would have limited the importation of olive jars in the English colonies to English ships, even if we allow for more than occasional illicit colonial contacts by foreign vessels."

Fairbanks (1973:147) notes that the olive jar "represents a distribution system somewhat foreign to northern Europe, especially to England." Watkins (1973:193), however, disagrees, and argues that not only was it "adopted by the English," but it was "used both for transport and secondary use functions."

Watkins (1973:193) corroborates his argument with evidence from British port records, which show numerous listings of both imports and exports of "Portugal oyle." Of specific interest to this study, in 1678, the ship *Nicholas* sailed from Plymouth, England, to Jamaica, carrying "200 jarrs with 1 ton of oyle" assigned to one merchant, and "400 jars of oyle" to another. In the same year, "40 jars oyle" went from Plymouth to Virginia. Several years later, on 2 January 1682, the ship *Concord* arrived in Plymouth from Lisbon, with "640 jars Portugal oyle" (cited in Watkins 1973:193).

Thus it would seem that the English were picking up, certainly at Lisbon, and perhaps other Iberian ports, whole cargoes of jars filled with olive oil and shipping them first to England, as prescribed, then transhipping them to the New World colonies, including Jamaica. While it is Portuguese, rather than Spanish, jars that are specifically being referred to above, it is not known what the difference was, if any, between the two countries. As Watkins (1973:193-194) notes, too, it reminds us that England in the 17th century had better trade relations

[2] Marken (1994:107-108) doubts such a two-piece construction, since there was no evidence of midsection or shoulder joints in olive jars collected from several Spanish shipwrecks. "A two-piece construction is also unlikely due to the wholesale crudity experienced in the midsections of the vessels, which would have made it difficult to match the two halves. If the jars were thrown in one process, the use of a mold for the part of the jar is possible. After forming the shoulders on the mold, the midsection would be pulled upwards by the potter's hands on either side of the vessel walls."

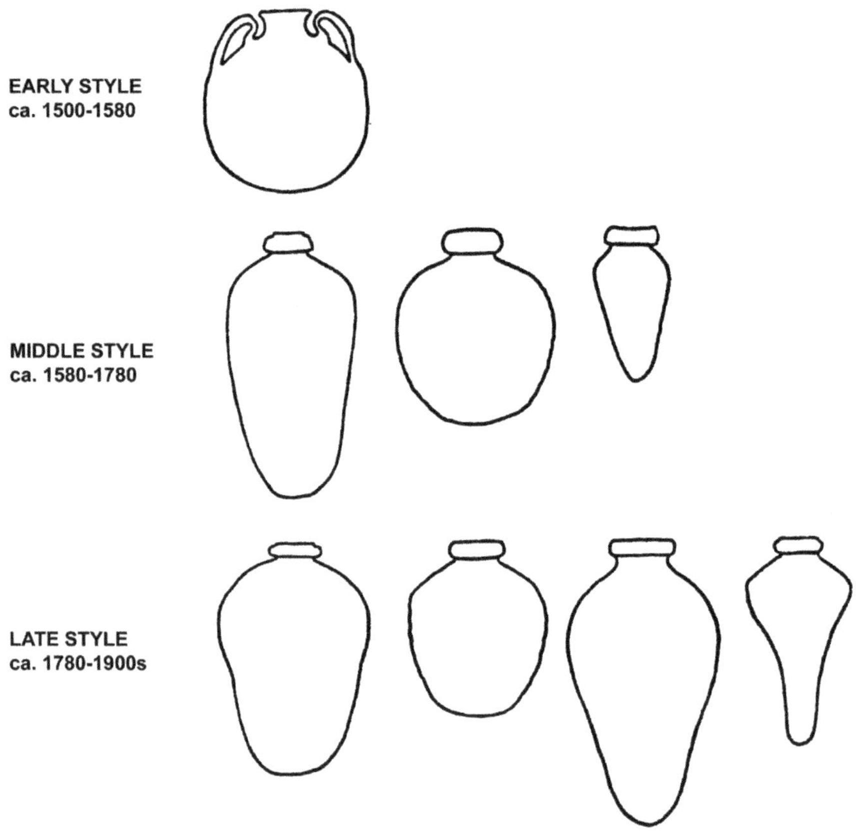

FIGURE 4.4. Comparative vessel shapes of Early, Middle, and Late Style Iberian olive jars (scales vary) (after Goggin 1960:28, Figure 9)

with the Portuguese than the Spanish as a result of the former's independence from Spain. It may be that many of these jars and their contents were deposited at Lisbon by Spanish ships where they were then picked up by the English.

In the case of Port Royal, indirect trade via England was likely not the only route by which these jars reached the city. It is known, for example, that many of Port Royal's merchants traded directly and extensively with Spanish colonial settlements throughout the Caribbean. Spanish shipwreck salvage is another possible scenario, given that olive jar forms found on a 1691 Spanish fleet wrecked on the reefs north of Jamaica (see Hoyt 1984) are exactly like the whole jars and scattered sherds recovered from Port Royal.

African-Jamaican Coarse Ware at Building 4/5

Local slave-made coarse ware appears in significant quantities in the Building 4/5 Layer 3 assemblage, accounting for 125 sherds, or 31.6 percent of the total coarse ware analyzed. Sherds were identified on the basis of mode of production (hand-coiling), form (generally globular bowls), as well as other attributes (fabric composition and color, minimal decoration). Taken together, the sherds represent an estimated minimum of 14 vessels, which were by African slaves.

These wares should be considered African-Jamaican. In Jamaica, they are known as Yabba.

The term Yabba has been traced to the West African Akan word *ayawa*, which means 'earthenware vessel or dish' (Mathewson 1972:55). More recently, McDonald (1993:108) has suggested that it may be traced to the African Igbo word *oba* ('calabash' or 'pot'). As noted above and earlier in this chapter, Yabba is hand-coiled or hand-modeled and inconsistent in fabric texture (Armstrong 1990:147). Fabric colors range from brown to red to reddish brown. Darkened cores and color banding within the fabric probably are the result of the use of the open-hearth technique, which was a common West African tradition. Mineralogical inclusions indicate locally occurring minerals not found in European wares (Armstrong 1990:147, 151; Bratton 1992:3-4). Distinctive European influences are apparent, however, in the use of lead glazes, flat bottoms, and double handles (Mathewson 1973:29). Yabba is still made today for sale in local markets and in some of Jamaica's tourist centers, although it is now predominantly wheel-thrown (Mayes 1972:103).

Excavated Yabba pottery, which is found throughout Jamaica in contexts dating from the late 17th to the late 18th centuries, was at first thought to have been made by the local Arawak Indians. Their recurrent presence in slave contexts on the island's European plantations was likely a result, it was argued, of trade or exchange between the Indians and the enslaved Africans (see Noël Hume 1962). Subsequent research, however, has shown that it appears more likely that the slaves themselves manufactured this pottery, and Yabba is now recognized as part of a wider African-based ceramic tradition, which developed within some slave communities of the Caribbean area (see Handler 1963; Mathewson 1972, 1973; Armstrong 1985, 1990; Wheaton and Garrow 1985; Heath 1991; Ferguson 1992).

The manufacture of Yabba as a distinct industry was first recognized by Duncan Mathewson in the early 1970s. Mathewson (1972:55-56, 1973:28-29), who analyzed over 2000 coarse ware sherds recovered from Old King's House in Spanish Town, the 18th-century Jamaican residence of Governor Thomas Lynch, suggested that many of the pottery fragments reflected African ceramic traditions with European-type elements added. On the basis of vessel form, method of manufacture, and probable vessel function, he proposed that the pottery was "an on-going Jamaican crafting enterprise . . . introduced by transplanted slaves of probable West African origin" (Mathewson 1973:28). They "represent the first tangible evidence of a relatively well-developed craft tradition within the slave population (Mathewson 1973:28).

Also in the early 1970s, during excavations at the 18th-century naval dockyard in Port Royal, Philip Mayes (1972:103) identified local coarse wares as being "recognizably African in origin." Later excavations (Armstrong 1985, 1990:146-158) of the 18th-century slave settlement at Drax Hall plantation on Jamaica's north coast similarly yielded Yabba sherds. While there is certainly diversity among the Yabba from Old King's House, the old naval dockyard, Drax Hall, and the more recently excavated Port Royal assemblages, the defining criteria for the ware—coarse fabric and color differentiation, hand-coiling, and globular forms—are remarkably homogeneous.

The majority of Yabba vessels recovered from Jamaican plantation sites is found in slave occupation areas (kitchens, living quarters) and are associated with food consumption and preparation. Armstrong (1990:274) argues that such forms (primarily bowls and cooking pots) are reminiscent of West African dietary patterns, which emphasize shared, or communal, food preparation and eating practices. Materials that were probably used along with these ceramic forms include the calabash and wooden vessels, both of which are (usually) perishable non-survivors in the archaeological record (Phillippo 1970[1843]:217; Armstrong 1990:153). (As noted in Chapter III, a calabash gourd dipper was recovered from Building 5; several wooden artifacts have been recovered from across the Port Royal site.)

In the case of Port Royal, a crowded and bustling city, there were no slave quarters as such. Rather, the African slave population was interspersed among hundreds of households. To find Yabba at Port Royal, then, does not necessarily indicate a distinct and separate slave residence, since research has shown that locally made ceramics were often used by European colonists (see Orser and Fagan 1995:223-235).

Specifically, Building 4/5's Yabba vessels, as notionally constructed from the analyzed sherds, were recovered from the yard and hearth areas at the back of the complex (most noticeably from Yards 4A and 4B). While bowls are the most prominent form (n=8), cooking pots (n=4) and storage pots (n=2) were also recognized. Bowls were identified on the basis of form (open, unrestricted rim) and the presence of an interior dark green to olive (2.5GY 6/8 - 7.5Y 8/4) 'sugary-textured' lead glaze (after Armstrong 1990:153). In most cases, the glaze is patchy and rarely covers the entire interior surface of the bowl sherds. Cooking pots were also identified on the basis of form (simple, restricted rim). Other distinguishing attributes include the general absence of surface treatment, such as glazing, and evidence for fire use and sooting on sherd exteriors (after Armstrong 1990:147). Storage vessels were distinguished from cooking pots on the basis of thicker walls and absence of external smudging from fire use. One cooking pot and one storage pot, which perhaps functioned as a water jar, have been reconstructed (Figure 4.5).

Yabba, as with most coarse utilitarian ware, is rarely decorated, and only a few of the analyzed sherds (n=8) exhibit this form of surface treatment (Figure 4.6). The stamped designs, as illustrated in Figures 4.6*a* and *b*, were produced by impressing the wet clay with a stamping die. It has been suggested that in Africa, carved wood, carved bone, and gold weights were among the items that often served as dies (Meyers 1999:211). The incised parallel grooves seen on other examples were most likely crafted with a blunt stick or other crude tool (Meyers 1999:212).

Each of the two stamped bowls in Building 4/5 (see Figure 4.6*a*, *b*) also has a small applied and stamped handle, while the handle on one of the storage vessels (see Figure 4.6*c*) is embossed (i.e., the decorative element is raised above the outer surface of the vessel by pinching the wet clay with the thumb and index finger). These handles appear to not have served any practical function.

The low percentage of decorated Yabba in Layer 3 at Building 4/5 parallels its incidence across the Port Royal

FIGURE 4.5. Examples of African-Jamaican coarse ware (Yabba) recovered from the 17th-century context at Building 4/5. *a*, reconstructed cooking pot (interior glaze); *b*, reconstructed storage pot (unglazed) (white areas are plaster-of-Paris)

site. Of the 899 sherds recovered from all three layers by Hamilton's team, less than 5 percent (n=28) exhibit this surface treatment (Meyers 1999:209). Yabba decoration, like Yabba vessel form, appears to be firmly established in West African ceramic traditions. (It might be of interest to note that even fewer decorated Yabba examples were recovered from the 18th-century plantation site of Drax Hall, indicating that stamped and incised Yabba is almost exclusively a 17th-century trait.)

Coarse Ware Summary
The total number of coarse ware sherds recovered from Layer 3 at Building 4/5 (i.e., the analyzed sherds) is 396. The total number of coarse ware sherds recovered from all of the other buildings (Layer 3 only) is 1048. This gives a combined Layer 3 total of 1444 coarse ware sherds for the portion of the Port Royal site excavated by Hamilton's team from 1981-1990. The analyzed assemblage from Building 4/5 thus comprises 38.7 percent of this total. It may be of interest to note that the larger count across the site breaks down as follows: English Redware totals 482; North Devon ware totals 50; Border ware totals 22; Continental European ware totals 233 (over half of which are olive jar fragments); and Yabba totals 423. The remaining 234 sherds are catalogued as miscellaneous/unidentified.

SLIPWARE
Technology and Development
Slipware is a form of decorative lead-glazed earthenware. Vessels are ornamented with a colored 'slip,' which is painted over a white slip. (Slip is loose clay and water mixed together into a creamy consistency.) The vessels are then covered in a lead glaze and fired at temperatures similar to that for coarse ware. Like coarse ware, too, slipware is fired only once.

A decorative slip on pottery can be applied using several methods, each making highly individualized, free-flowing patterns. According to Dr. Plot (1686:123), who observed English slipware potters in the late 17th century, "they Slip or paint [their vessels] with their severall sorts of Slip, according as they designe their work, when the first Slip is dry, laying on the others at their leasure, the Orange Slip makeing the ground, and the white and red, the paint; which two colours they break with a wire brush, much after the manner they doe when they marble paper."

Slip was also often 'trailed' from a fine nozzle attached to a small container. When filled with slip and tilted, the flow of the mixture from the nozzle could be controlled by covering/uncovering an air vent with the thumb (Cooper 1968:12; Draper 1984:15; Wondrausch 1986:21). 'Combing' or 'feathering' the slip employed slip-trailed lines, usually in parallel orientation, on a contrasting slip ground. The surfaces were then 'combed' using a knife or similar pointed-tip tool to create patterns of alternating peaks and troughs (Draper 1984:15; Wondrausch 1986:27).

'Marbling,' also called 'joggling,' is a slipware decorative technique whereby slip-trailed lines are applied over a contrasting ground, and the vessel is then moved around causing a pattern to emerge. Applying brush strokes of clay slip is known as slip painting, while 'impressing' is a technique that uses a notched stick pressed into the thick trailed slip to produce a raised and patterned effect (Draper 1984:15; Wondrausch 1986:28). A rather unique type of decoration, common in the medieval period in Italy, France, and Germany, and then later in the southwest of England, is *sgraffito*, named after the Italian word for 'scratched.' Dried clay vessels were dipped into a colored slip, whereupon a pattern was then scratched through the surface to reveal the contrasting color of the clay body beneath (Draper 1984:20; Wondrausch 1986:30).

FIGURE 4.6. Examples of decorated Yabba recovered from the 17th-century context at Building 4/5. *a, b,* rim sherds from two stamped bowls, each with a small and applied stamped handle; *c,* rim sherd from a storage pot with an embossed handle

The use of slip as decorative technique has been known from earliest times. It appears to have originated in the Far East, where fragments of red-slipped pottery, thought to be 5000 years old, have been found in Japan (Cooper 1968:6). In the West, examples of white slip decoration date from 2000 B.C., in the Minoan culture on the island of Crete. With their famous Black- and Red-Figure vases, the Greeks perfected the craft several centuries later (Cooper 1968:6). From around 200 B.C. to A.D. 200, potters in China are known to have painted their wares with a combed/feathered slip decoration. By the 7th century A.D., the Chinese repertoire included brushed and incised marbleized patterns carefully applied through white slip washes (Cooper 1968:7-8; Eden and Eden 1999:13).

Beginning in the 15th century, Italian potters made white-slipped dishes incised with various patterns, often with brush strokes of colorful mineral pigments. Portugal, Spain, and France developed a slip tradition in the 16th century; by the early 1600s, Dutch and German potters were using varieties of the slip technique (Wondrausch 1986:77-106; Eden and Eden 1999:14-17). Contemporary with these wares was a specifically English development, known as Cistercian ware, which used as decoration contrasting-colored clays covered with a black glaze. This ware has often been regarded as a primitive kind of slipware, developing into Blackware, which was widespread in north-central England in the mid 17th century. Blackware is characterized by a red body covered with a black or very dark brown iron-stained lead glaze (Crossley 1996:245-247; Eden and Eden 1999:9).

By the mid 17th century, slipware manufacture was well established in many centers in England, and it is in England that the slip technique reached a height of skill and excellence that it never attained elsewhere. The oldest, and perhaps most vibrant, of its folk pottery traditions, England's slip-painted wares show an unrivaled freedom of expression and imagination. A certain cheerful character emanates from these sturdy forms, with their warm and glossy colors. Their decoration is made even more charming with the often-inscribed wishes of good cheer or invitations to drink and be merry. The delightful effects produced are even more impressive when the daunting circumstances of their manufacture are considered: in cold, ill-protected structures with fire hazards and appalling transport conditions for raw materials and finished wares (Wondrausch 1986:7). The potteries of Burslem in Staffordshire and of Wrotham in Kent were the principal centers of English slipware in the 17th century. The ware was also made at Harlow, near London, and in the Devonshire villages of Bideford, Barnstaple, and Fremington.

Slipware at Building 4/5

In the Building 4/5 analyzed ceramic assemblage, the slipware group can be generally included in a non-utilitarian, or tableware, category, and vessels can be defined by their use in food and beverage consumption (eating, drinking, serving). As was noted with some coarse ware vessels, certain slipware forms overlap the tableware and utilitarian categories (e.g., a pudding/pastry pan was recovered that may have been used both as a serving piece and a baking dish). This aspect of vessel form and function is discussed more fully in Chapter VI.

As shown in Figure 4.7, most of the analyzed slipware was found in the front rooms of the building complex. Presumably, these are parts of vessels that were in active use at the time of the earthquake. Sherds recovered from the yards may be from vessels that were also used in these areas or were destroyed and then displaced during the disaster. It is also possible that at least some of the fragments in these areas, as well as those in the alley, were intentionally dumped there as trash, the vessels of which they were a part having broken and become useless during the life of the building (see South 1977:47-80). The slipware distribution pattern is discussed more fully in Chapter VI.

A total of 80 sherds, only 6.9 percent of Building 4/5's analyzed sherd count, was identified as slipware. (This can be compared to the 96 slipware sherds recovered by Hamilton from all of the other excavated buildings combined [Layer 3 only].) These 80 sherds represent an estimated minimum of 11 vessels, or 7.6 percent of the total estimated minimum number of vessels taken to be associated with the building prior to 1692. Unlike the coarse ware group, which is represented by wares from a wide variety of sources, the slipware at Building 4/5 is predominantly English, identified on the basis of fabric color and style of decoration. Only one vessel, a costrel, is likely of Italian provenance. A white-slipped, reddish brown (5YR 5/4) 'flatware' sherd, recovered from the alley to the east of Building 4 is possibly from North Holland.

This relative paucity of slipware at Building 4/5 and the Port Royal site as a whole is not surprising when it is considered that in the late 17th century, much of it was still a folk- rather than a mass-produced industry. Not only was slipware not marketed on the same scale as other wares at this time (English slipware potters, in particular, only began to enlarge their markets post-1700, a direct response, in part, to the ever-increasing imports into the country of foreign wares), but it was overshadowed in popularity by tin-glazed earthenware.

English Slipware at Building 4/5

A total of 65 sherds, or 81.2 percent of Building 4/5's analyzed slipware sherds, is of English manufacture. Taken together, the sherds represent at least nine vessels. All but one of the sherds likely are from the Staffordshire potteries. A body sherd, possibly from a bowl, is believed to be from North Devon.

Staffordshire Slipware. That most of the slipware in Building 4/5 is of Staffordshire provenance is not surprising. The area in the late 17th and early 18th centuries was well known for its pottery industry. Indeed, potters of all kinds of wares worked in the Staffordshire region, attracted to it for its abundance of clays, wood and coal for firing, and for the source of lead ore nearby that was used to make lead glaze (Wondrausch 1986:43; Eden and Eden 1999:10). The hallmark of the Staffordshire industry is probably its 'presentation' dishes, made solely for display and usually commissioned individually by some well-known personage. Many of these dishes, which can measure as much as 50 cm in diameter, are, in fact, extant in museum collections; few,

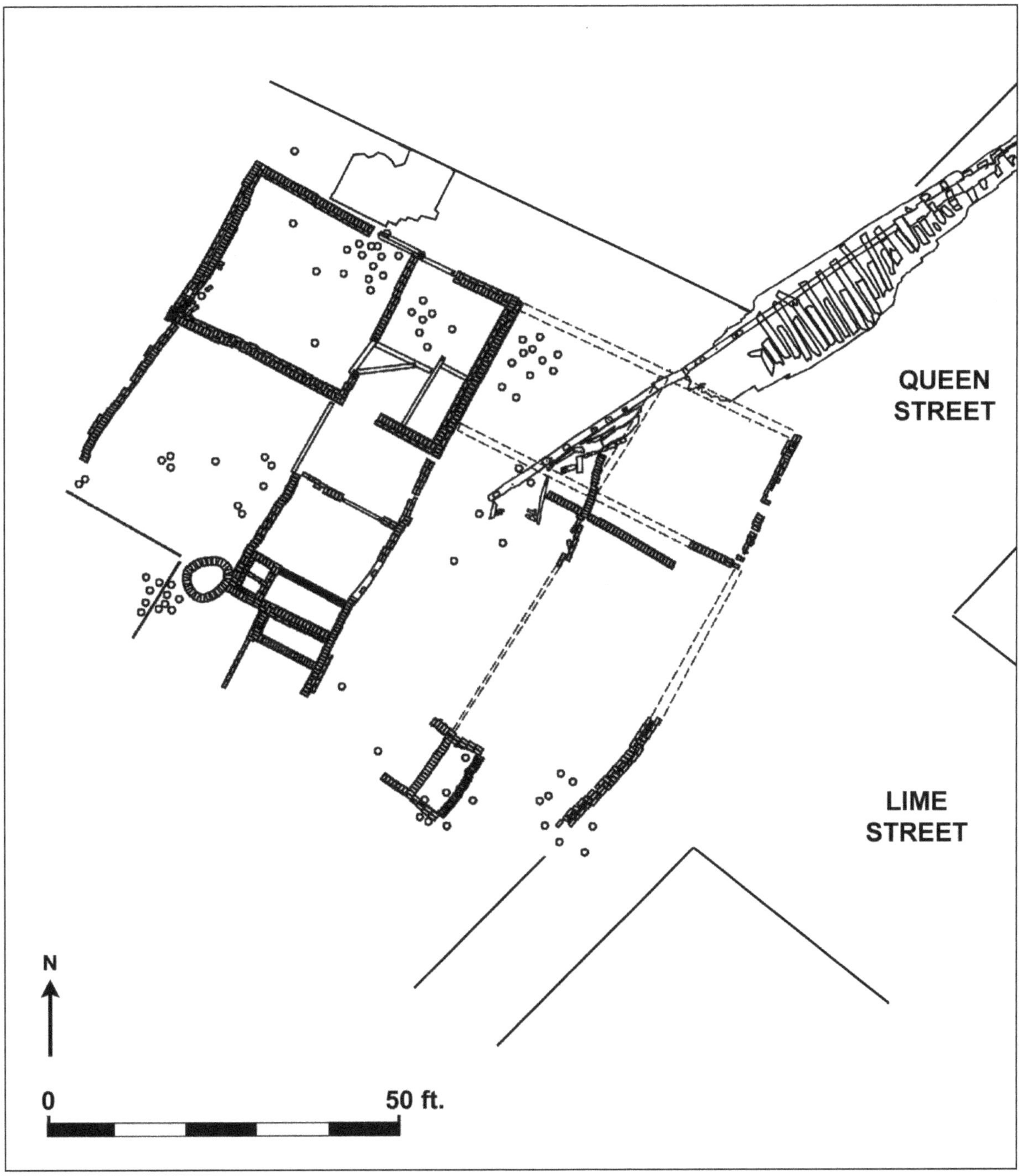

FIGURE 4.7. Distribution of slipware ceramic sherds in the 17th-century context at Building 4/5

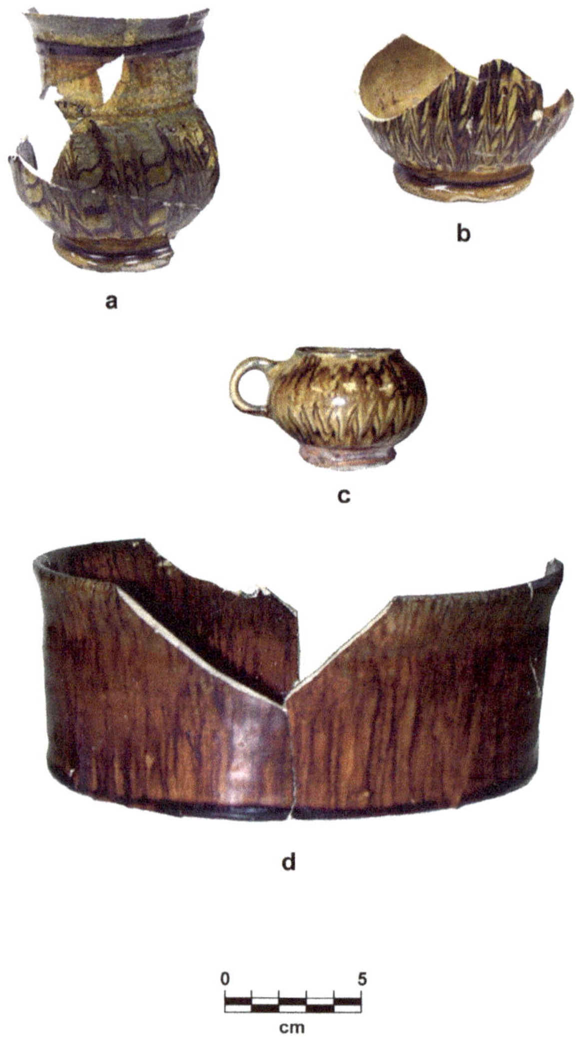

FIGURE 4.8. Examples of English slipware recovered from the 17th-century context at Building 4/5. ***a-c*****, Staffordshire cups;** ***d*****, Staffordshire Mottled ware pudding/pastry pan**

if any, have been uncovered through archaeological excavation.

The more elaborate of the Staffordshire products are often inscribed with the name of the potter, Thomas Toft being the most important in the last quarter of the 17th century. Most of the known Toft pieces are the large, ornamental dishes noted above; usually, they are decorated with monarchist themes to commemorate the 1660 restoration of Charles II to the English Crown. Coats of arms and figures of royalty are also seen in the dishes made by Toft's brother, Ralph, and later, in the works of his sons, Thomas II and James. Ralph Simpson and George and William Taylor, the elder Toft's contemporaries, also tended to employ the same royalist designs (see Cooper 1968; Wondrausch 1986).

As spectacular as they are, these pieces formed only a small part of the Staffordshire slipware industry. The production of more useful domestic wares took up most of the potters' time, and it these pieces that were recovered from Building 4/5. Sherds representing four cups, a bowl, a saucer (likely used for serving condiments), and a porringer were identified by their fabric color (buff to pale yellow) and style of decorative treatment (dark reddish brown [5YR 2.5/2] trailed and combed/feathered slip) (Figure 4.8*a-c*). A reconstructed oval serving vessel that may have been used as a pudding/pastry pan was also recognized as of Staffordshire provenance (see Figure 4.8*d*). That it may have been used in cooking is evidenced by its solid form and more simplified slip decoration (known as Staffordshire Mottled ware).

North Devon Sgraffito. As noted in the discussion on North Devon coarse ware, vast quantities of Devonshire vessels, among them decorated tableware, were exported to the New World colonies throughout the 17th century. Much of this decorated ware employed the slip technique of sgraffito (see above), which was, in fact, used by the Devonshire industry until well into the 19th century. Sgraffito ware was brought to Italy from the eastern Mediterranean in the Middle Ages.

From there, it spread to northern Europe, where it was taken up particularly in France, at the large pottery center of Beauvais. It is from France that the sgraffito technique is believed to have arrived at the North Devon potteries (Grant 1983:2; Wondrausch 1986:57). The analyzed assemblage at Building 4/5 includes a single North Devon Sgraffito ware sherd, identified by its light reddish brown (2.5YR 6/4) fabric, which is coated on the interior with a white slip through which a geometric pattern is incised (Figure 4.9). The curvature of this small body sherd suggests that it may have been part of a small bowl or possibly a deep plate.

FIGURE 4.9. North Devon Sgraffito slipware sherd recovered from the 17th-century context at Building 4/5

Continental European Slipware at Building 4/5
North Italian sherds (n=14), identified on the basis of fabric color, style of decoration, and form, comprise all but one of the analyzed Continental European slipware sherds. The sherds have been cross-mended to form a single, near-complete costrel/flask, which has a white/cream marbled design slipped over a soft light red (10R 6/8) exterior-glazed fabric. On each side of the vessel, attached at the shoulder, a small loop handle is present, each molded in the shape of what looks like a lion's head. A cord or leather thong passed through each of the handles would have allowed the costrel to be suspended and/or carried. Specifically, this vessel type can be attributed to Pisa, and it has been found on sites in London (Orton and Pearce 1984:52), Santo Domingo and Mexico City (Lister and Lister 1976a:33-34), St. Augustine in Florida (Deagan 1987:47), and at the old naval dockyard site in Port Royal, which was excavated by Mayes (1972:82).

The second Continental European slipware vessel is, as noted earlier, represented by a small body sherd, which is possibly from North Holland and which is possibly part of a saucer. The identification of provenance was based upon its red (2.5YR 4/ -5/6) to reddish brown (5YR 5/4) sandy fabric and remnants of a white slip design.

As was noted in the discussion on Continental European coarse ware, the presence of non-English ceramics is not unexpected at the international city of Port Royal. Their comparatively limited numbers is similarly not surprising in an English colonial port of the late 17th century when it is considered that the distribution of material goods in the New World settlements was largely determined by official commercial policies. Trading acts and duties imposed by England in the mid 17th century curtailed attempts by foreign vessels to conduct trade with England's colonies. (But this did not stop English ships from going to these ports directly and trading. The foreign goods were then brought back to English ports.)

England sought not only to increase her market share of trade but also to ensure that domestic manufacturers had a guaranteed market in the colonies.

Slipware Summary

The total number of slipware sherds recovered from Layer 3 at Building 4/5 (i.e., the analyzed sherds) is 80. The total number of slipware sherds recovered from all of the other buildings (Layer 3 only) is 96. This gives a combined Layer 3 total of 176 slipware sherds for the portion of the Port Royal site excavated by Hamilton's team from 1981-1990. The analyzed assemblage from Building 4/5 thus comprises 45.4 percent of this total. It may be of interest to note that the larger count across the site breaks down as follows: Staffordshire slipware totals 151, North Devon Sgraffito slipware totals 6, and Continental European slipware totals 15. The remaining 4 sherds are of the type known as 'Reversed Slip' decoration, i.e., they are covered in a solid dark reddish brown (5YR 2.5/2) slip, which itself covers a white-slipped buff to pale yellow fabric. All 4 sherds appear to be the remains of mugs.

TIN-GLAZED EARTHENWARE

Technology and Development

Tin-glazed wares were the luxury product of the earthenware industry up to the introduction of creamware in 1762.[3] The pottery is recognized by its thick, white lead-tin oxide glaze, which is often painted over with blue or polychrome designs. Like the common lead glaze, tin glaze provided a waterproof layer to the porous earthenware body. Its other purpose, however, was to mask the clay fabric and provide a light background for decoration. Tin-glazed earthenware was a successful painter's pottery, which took as its inspiration designs from Chinese porcelain, a more expensive and refined ware that was not widespread in the West until well into the 18th century.

Tin-glazed earthenware was much more complicated to produce than common coarse ware or even slipware. Vessels were made of a mixture of clays in carefully measured quantities, and they required to be twice-fired, once before and once after the glaze was applied. Even the glaze itself was difficult to produce, and those responsible for the recipes relied heavily on inherited experience, much of which was kept secret (Caiger-Smith 1973:203).

A certain amount of the clay used to make tin-glazed earthenware is buff/white. The red clay commonly used for most earthenware products was long recognized to be 'fat' or plastic, which tended to make wares shrink and spoil their glazes upon firing. The buff/white clay, along with some blue, were 'lean' or aplastic clays, which counteracted these tendencies (Weatherill and Edwards 1971:173; Edwards 1974:19; Britton 1990:65; Archer 1997:13). The whitish colored clay also made the body of the vessel whiter, so that it did not contrast too sharply with the white of the glaze (Weatherill and Edwards 1971:173; Archer 1997:13).

Tin-glazed wares were fired once before they were glazed, at temperatures of about 900°-1000°C. This 'biscuit' firing made the vessel stronger for dipping in the glaze and for decorating. The transparent lead glaze was made white by the addition of lead-tin oxide, which was a mixture of white sand, potash, lead, tin, some copper filings, and a little cobalt blue (Weatherill and Edwards 1971:174; Caiger-Smith 1973:208; Weatherill 1986:59; Britton 1987:11). "These ingredients were melted in the fire-chamber of a kiln during firing and the resultant 'frit' [a partly fused mass] was ground to a powder and mixed with water to make the glaze" (Archer 1997:17).

If a tin-glazed vessel was to be painted (often, they were simply left plain white), this was done only once the glaze coating was dry. The pigments used to paint tin-glazed earthenware were obtained from various minerals: blue from cobalt, yellow from antimony, and orange from iron rust. The green found on some vessels of the early 17th century was based on copper (by about 1695, green generally had been abandoned for blue and yellow). Manganese produced a wide variety of shades, from a near black through a full purple to lilac, depending upon the dilution of the pigment. Iron oxide was used to produce red (Ray 1968:88-90; Noël Hume 1970:106; Caiger-Smith 1973:208-211; Archer 1997:19).

Some of the designs seen on tin-glazed ceramics are likely to have been produced on paper first, with prints from pattern books much used as sources. 'Pounce paper,' which carried a pricked-through outline of the drawing to be reproduced on the vessel, would be placed over the unfired glaze, and powdered charcoal would be deposited through a small muslin bag by lightly patting the bag on the paper. The painter would then follow the grains of charcoal with a fine brush to put on a strong dark blue or purple outline, known as 'trek' (Caiger-Smith 1973:131; Britton 1982:18; Archer 1997:18-19). After they were painted, the vessels were given a second firing at a higher temperature (950°-1050°C) to fix the colors and fuse the glaze, creating an even hard white surface (Weatherill and Edwards 1971:174; Caiger-Smith 1973:208; Weatherill 1986:58; Britton 1987:11). Indeed, the success of the glaze was due to its very dense and hard surface. As with most ceramics, however, the tin-glazed vessel also had its drawbacks: the soft clay provided little support for such a hard tin-glaze shell and thus some degree of chipping and crawling was common. It was also difficult to erase a brushstroke after it was applied, and some crudeness of decoration is thus seen in some pieces (Price 1978:45, 74; Draper 1984:25; Archer

[3] Creamware, a refined earthenware, was developed by Josiah Wedgwood, one of the most famous English potters and ceramic innovators of the 18th century. It can be identified by its cream-colored fabric and yellowish green hue of its glaze. Decoration includes molded rims, sprigged motifs, and hand-painted and transfer-printed decoration. Creamware was relatively cheap and so was affordable to the masses, and it became extremely popular from ca. 1770-1800. It was the first widely available ceramic with a hard-fired surface that allowed one to cut food on its surface without chipping the glaze.

1997:18).

Development of Tin Glazing in Spain and Italy
The discovery of tin oxide as an opaquing material for lead glazes first occurred in the ancient Near East as early as 1000 B.C. It fell into disuse, however, and was not rediscovered until about A.D. 700, in Persia. During the Muslim conquests, the technique of tin glazing spread along the North African coast into Moorish Spain, where it arrived around the end of the 12th/beginning of the 13th century (Fairbanks 1973:148; Lister and Lister 1976b:2; Deagan 1987:26). Tin glazing was introduced into Italy about the same time (Fairbanks 1973:148; Caiger-Smith 1973:83; Wilcoxen 1987:57). From there, it moved through France, Germany, and Holland, finally reaching England in the 17th century.

Majolica is the name given to the tin-glazed wares of Spain and Italy. The early tradition in Spain developed independently from that of Italy, influenced as it was by Islamic motifs (e.g., palmettes, spirals, arabesques) and forms (e.g., ewers and *albarelo* [drug jars]) (see Caiger-Smith 1973:66-69). Many such traits, however, disappeared from the Spanish potteries with the fall of Muslim Seville in the Reconquest of 1248. Workshops were seriously damaged, and the Muslim potters were expelled and fled to Grenada (Lister and Lister 1976b:1-2; Deagan 1987:26). The fall of the Muslim kingdom of Grenada in 1492, and the expulsions of Jews and the Muslims from Spain thereafter, similarly affected the tin glaze potteries on the Iberian Peninsula (Lister and Lister 1976b:6; Deagan 1987:26).
By the beginning of the 16th century, the presence of a new foreign influence, in the shape of Italian majolica, was apparent in the Spanish potteries. The "energized atmosphere of Seville, gateway to the New World and all its riches," was particularly attractive to the Italian potters, who had by this time a well-earned reputation for nomadic wanderings (Lister and Lister 1978:6-7). As Caiger-Smith (1973:103) notes, "where their work found a good reception, the [Italian] potters themselves soon followed." Indeed, it was not long before the Italian influence dominated all of Spain's ceramic industries. By the mid 16th century, Spanish versions of Italian wares were being produced not only in Seville but in the developing centers of Talavera de la Reina in Castille and in Catalonia (Lister and Lister 1976b:7, 1978:7). By this time too, much Italianate-Spanish majolica was being exported to Spain's colonies in the New World (Lister and Lister 1982:57-60; Deagan 1987:61).

Development of Tin Glazing in Holland
Holland's tin-glazed earthenware industry developed around the mid 16th century (Church 1911:20; Draper 1984:25-26; Rackham 1987:1:166; Wilcoxen 1987:53). The city of Delft, in particular, harbored many of the potteries, since not only was it a port city (the Dutch East India Company had one of their import offices there), but it was already a center for painters and engravers, who possessed many of the basic skills needed to paint pottery (Caiger-Smith 1973:131).

During most of the 16th century, Dutch tin-glazed designs, usually floral and fruit motifs, were painted either in cobalt blue or in the majolica colors of purple, green, yellow, and orange (Caiger-Smith 1973:127; Rackham 1987:1:310; Wilcoxen 1987:57-58, 62). The earliest of these Dutch wares were glazed with the tin oxide mixture on only the painting surface, with backs of plates and dishes and undersides of other forms given the cheaper, transparent lead glaze (Britton 1987:11; Wilcoxen 1987:60). Not until the early 17th century did Dutch potters coat their vessels completely in the white tin glaze.[4] Also adopted at this time was the use of kwaart, a transparent glaze that was applied after the painted tin-glazed vessel had gone through its second firing. In a third firing, the kwaart and the tin glaze melted, with the former remaining as a semi-permanent film that "gave depth to the finished surface . . . deepening and smoothing the blue pigment" used in the painting (Caiger-Smith 1973:129; Wilcoxen 1987:60).

From about the mid 1620s, the Chinese influence appears on some Dutch tin-glazed wares (most of the tin-glazed ware made for domestic use was left plain white). The most common of the decorative motifs borrowed from the East were the blue-and-white painted landscapes, flowers, plants, waterfalls, arched bridges, and Chinese pavilions. Employing these designs, which were copied from the popular but expensive imported porcelain, was, in fact, a canny tactic used by the Dutch tin glaze potters to create a cheaper version of this fashionable and exotic ware (Wilcoxen 1987:58, 61). Fueled by the decline at this time of the Chinese porcelain factories and the increasing demands for fine yet modest tableware, this 'imitative' product was an immediate and resounding success and continued in popularity throughout Europe until the end of the 17th century.

Development of Tin Glazing in England
The English tin glaze potters also borrowed motifs from Chinese porcelain. It also would seem that until about the middle of the 17th century, all of the tin glaze potters working in England were either Flemish or Dutch (Caiger-Smith 1973:161). It is known, for example, that one of the earliest English potteries producing tin-glazed earthenware was set up near Aldgate in London in 1571 by Dutch potters Jacob Jansen and Jaspar Andries (Noël Hume 1970:105; Quinn 1971:62; Britton 1987:19). Excavations at a 'pot-works' set up by Dutch immigrants Edmund Bradshaw and Hugh Cressey in Montague Close

[4] Charlotte Wilcoxen (1987:54), an authority on Dutch tin-glazed earthenware, employs the words 'majolica' and 'fayence' to distinguish between the early Dutch wares with the tin glaze on only the obverse, and its successor, the more sophisticated ware, with a tin glaze on both surfaces. As was noted earlier, 'majolica' is more commonly used for Spanish and Italian tin-glazed earthenware, whereas 'faience' is often applied to the French variation (in Germany, it is called 'fayence'). 'Delftware' is the name usually given to Dutch tin-glazed earthenware, whereas 'delftware' (lower case 'd') is synonymous with its English variation.

in Southwark in 1613 reveal that tin-glazed earthenware was probably made there as well (Quinn 1971:68; Edwards 1974:8-10; Dawson 1978).

Indisputable evidence for English manufacture of tin-glazed earthenware comes with Christian Wilhelm, who arrived in London in 1605, and who was documented as living in Southwark as a 'gally-pot maker' in 1618 (Tait 1960, 1961; Caiger-Smith 1973:164). Wilhelm founded a tin glaze pottery at Pickleherring Quay in Southwark in the 1620s (see Britton 1990, 1993) and was given a monopoly for 14 years "for the makinge of earthen gally potte and dishes called by the name of gallyware And all kinde or sorte of bottells of all Colo[r]s basons & ewers salte dishes of all sorte drinking potte pavinge tyles Apothecaries & Comfittmakers potte of all sorte & all kinds of earthen works" (cited in Mountford and Celoria 1968:2; see also Quinn 1971:71). Unsurprisingly, the early wares out of Southwark were similar in decoration to the Dutch type. It would appear also that England was indebted to the Netherlands for the words 'gallypot' and 'gallypotmaker,' as tin-glazed earthenware and the potters were then known, for in the mid 16th century, makers of the ware in Antwerp are practically all described in the records as *geleyers* or *geleyerspotbakkers* (Rackham 1926:29, 97; Britton 1987:22). It has been also suggested that the name is in reference to the Venetian *gallées*, which first brought painted pottery, doubtless Italian majolica, to northern Europe, and from which the tin glaze potters were inspired.

In the course of the 17th century, the alternative term 'white' or 'whitt' was used to describe English tin-glazed wares, since the tin glaze which it covered produced a white appearance that was radically different from any other pottery made before the 18th century. This brilliant glaze and the brightly painted colors were, in fact, in large part responsible for the considerable popularity of this ware (Archer 1997:3). Indeed, tin-glazed earthenware was the first white ware and the first painted colored pottery to be made in England (Draper 1984:25). By 1698, English potteries making tin-glazed earthenware were described by an owner as 'White Earthenware-houses.' Tin-glazed wares were, by this time, described as 'China' or 'Whit Chiny' (Weatherill and Edwards 1971:172; Archer 1997:3). (Although perhaps the most familiar, the term 'delftware' was not used to describe English tin-glazed earthenware before the 18th century [Noël Hume 1970:106].)

By the end of the 17th century, English tin glaze potteries were responding to the same changes of taste and fashion as were felt all over Europe. The introduction of tea, coffee, and chocolate drinking led to a demand for receptacles which did not easily transmit heat, and since pewter, the material most directly in competition with tin-glazed ware, was not suitable for this purpose, it was met by the pottery trade. The change in drinking habits also prompted the manufacture of different vessel forms, such as cups and teapots. Vessel decoration, too, became looser and more flowing, as the English pottery painters began to develop their own style, which according to Caiger-Smith (1973:161), was "ingenious, direct, sometimes eccentric." The painted designs on the ware by this time were not directly copied from Chinese porcelain as they were in the Dutch provinces, but were, instead, modified to fit in with the English setting.

Tin-Glazed Earthenware at Building 4/5

Tin-glazed earthenware sherds were recovered in great quantity from the 17th-century context at Building 4/5. Like the slipware group, this ware was used primarily at table, with most vessels defined by their use in food and beverage consumption (eating, drinking, serving). Certain forms are distinguished by their health- and hygiene-related roles (drug jars and ointment pots, chamber pots and washbasins); others served a 'decorative' purpose (e.g., a flower vase, a large ornamental dish). In fact, the fragility of tin-glazed wares led them often to become items, which fulfilled a social rather than a utilitarian role. Many of the cups, bowls, and plates identified in the assemblage may not, in fact, have served any other purpose than for display on a wall or a dresser, as part of the decoration of a room. This aspect of vessel form and function is discussed more fully in Chapter VI.

As shown in Figure 4.10, many of the analyzed tin-glazed earthenware sherds were found in Rooms 2 and 3. As noted in the discussion of the slipware distribution, most, if not all, of these fragments are likely parts of vessels that were in active use at the time of the earthquake. A significant number of sherds were also recovered from the yards, particularly Yard 5. The array of materials found in the brick-paved Yard 5 point to it being used, at least in part, as an activity area and not simply as a place where refuse was dumped (a common occurrence in both England and the American colonies in the 17th centuries). It is possible, then, that some of the tin-glazed fragments found in this location may be parts of vessels that were also being used there. The fragments found in Yards 4A and 4B, both of which had sand floors, are more likely to have been intentionally dumped in these areas as trash, the vessels of which they were a part having broken and become useless during the life of the building. Again, as noted above, some tin-glazed fragments in all three yards may be from vessels that were displaced during the disaster. The tin-glazed earthenware distribution pattern is discussed more fully in Chapter VI.

With a total of 588 sherds, tin-glazed earthenware, at 50.9 percent of Building 4/5's analyzed sherd count, represents the largest group of ceramics in the assemblage. (This can be compared to the 507 tin-glazed earthenware sherds recovered by Hamilton from all of the other excavated buildings combined [Layer 3 only].) These 588 sherds represent a minimum of 59 vessels, or 41.0 percent of the total minimum number of vessels taken to be associated with the building prior to 1692. This is the largest number of vessels in a single ware in

FIGURE 4.10. Distribution of tin-glazed earthenware ceramic sherds in the 17th-century context at Building 4/5

and around the structure. English and Dutch wares, taken together, comprise the majority of the tin glaze assemblage, at 51 vessels. The source of the English type is probably London, the source of the Dutch type is unknown. Mediterranean and Mexican majolica are also represented. Their numbers in comparison to the Anglo-Dutch type are minimal, but they have, for the most part, retained their glaze better.

English and Dutch Tin-Glazed Earthenware at Building 4/5

It is almost impossible to distinguish between English and Dutch tin-glazed wares, since clay was often imported to be mixed with local clays (Weatherill and Edwards 1971:173; Orton and Pearce 1984:37). For the purposes of this study, therefore, English and Dutch finds recovered from Layer 3 at Building 4/5 are considered together.

The Anglo-Dutch tin-glazed sherds were recognized by their soft fabric, buff to pinkish cream in color, the slightly opaque white glaze, often with a pale blue tint, and hand-painted decorative motifs in either cobalt blue or polychrome (blue in various shades, yellow, green, and purple). Birds, insects, flowers, and geometric designs are evident. Chinese figures in pastoral landscapes are also a common theme. Some of the fragments (n=134) are undecorated, the wares to which they belonged likely left a plain white. Others (n=248) are too fragmentary, with most of the glaze either crazed or completely flaked off, to make further identification possible.

That so much tin-glazed earthenware should be found in a 17th-century building is, perhaps, not surprising given that the ware was the most popular and attractive type of decorated domestic pottery then readily available. The variety of identified forms represented is equally as impressive as the numbers: decorated and undecorated plates (n=7); at least one large painted dish; a blue-rimmed saucer; plain white porringers (n=2); communal drinking pots (n=5); plain and decorated cups (n=9); at least two bowls; and part of a bottle handle. Small (some intact) ointment pots (n=11) and larger (possible drug) storage jars (n=5) were also found, as were the remains of two plain white chamber pots and two washbasins. A partially intact plain white salt stand, a flower vase, and the decorated lid of an unknown vessel were also recovered. It is probable that even more vessels are, in fact, represented, but due to the softness of the ware, many of the finds are, as noted above, fragmentary, and some difficulty was experienced in attributing some of the forms represented. All of the identified vessels are discussed further in Chapter VI. Figure 4.11 shows some examples.

Mediterranean Tin-Glazed Earthenware at Building 4/5

Seven sherds (1.2 percent of the analyzed tin-glazed sherd count) are almost certainly of Mediterranean provenance (Spain, Italy, and Portugal). Identification was based upon the slightly more refined fabric than is evident in the Anglo-Dutch types, and the higher quality glaze, which has remained almost pristine even after centuries of burial. Taken together, the Mediterranean tin-glazed (majolica) sherds represent at least five vessels: three plates, one saucer, and a large bowl.

Two of the decorated plates, the saucer, and the bowl were identified as of either Spanish or Italianate-Spanish manufacture (Figure 4.12*a-c*). Differentiating these types is difficult, since as noted earlier, beginning in the 16th century, many Italian potters worked in Spain. Lister and Lister (1976b:37, 1978:12, 1982:59) suggest that, in general, the former can be distinguished by its creamy yellow, or sometimes rosy yellow, fabric color and by a less refined and precise design execution. Spanish tin-glazed earthenware was fired using cockspurs (small pieces of fired clay used to separate vessels in the kiln), which leave marks on the obverse of forms. The Italian wares, in contrast, have saggar scars on the reverse (Lister and Lister 1976b:37, 1978:12, 1982:59; Deagan 1987:61).

The third plate, represented by uniquely decorated body sherd, is certainly of Portuguese provenance (see Figure 4.12*d*). Its well-preserved decoration, a Chinese-inspired daisy motif, leaves no doubt as to either its origin or its date, since the motif was popular with Portuguese potters in the first part of the 17th century (Calado and Bart 1987:76; Monteiro 1994). Its chalky-white (10YR 8/3) temper-free body and excellent condition of its thick white glaze is also consistent with Portuguese tin-glazed wares of this period. Although few Portuguese sherds have been recovered from the Port Royal site, many have been excavated from 17th-century ports in England (Noël Hume 1977:95-96; Allan and Barber 1992). Portuguese ceramics, sometimes in large numbers, have also been found throughout New England (see Pendery 1999). Their presence in these locations, as previously noted in the discussion of Iberian olive jars, relates directly to the trade between England and Portugal and thence to England's North American colonies during a period of Anglo-Portuguese alliance after Portugal's separation from Spain in 1640. Much of this trade was in wine. It seems reasonable to suggest that tin-glazed ceramics, the newest and most fashionable pottery to come out of Europe, accompanied these shipments.

The relative paucity of Mediterranean tin-glazed earthenware forms is to be expected when it is considered that by the late 17th century, the English and Dutch monopolized the tin glaze industry. Moreover, according to Goggin (1968:215), by 1650-1660, Spanish tin-glazed earthenware, in particular, "had become unimportant in the Caribbean area because of the rapidly expanding Mexican majolica industry." Deagan (1987:21), writing 20 years later, reiterates that "the 17th century was a period of economic decline for Spain [whereby] no significant Spanish shipping took place in the Caribbean after 1630." Many of the decorative patterns are characteristic of a pre-1655 date. The original vessels of

FIGURE 4.11. Examples of English/Dutch tin-glazed earthenware recovered from the 17th-century context at Building 4/5. *a*, reconstructed rim from a decorated Dutch dish; *b*, rim and partial body from a decorated English apothecary pot (or ginger jar); *c*, reconstructed blue-glazed English drinking pot; *d*, near-intact decorated Dutch (?) cup; *e*, intact decorated English apothecary pot; *f*, reconstructed undecorated English chamber pot (with most of the tin glaze flaked off); *g*, reconstructed undecorated English caudle cup

FIGURE 4.12. Examples of Mediterranean and Mexican majolica recovered from the 17th-century context at Building 4/5. *a*, body sherd from a Spanish bowl; *b-c*, rim sherds from Spanish/Italianate-Spanish plates; *d*, rim sherd from a Portuguese plate; *e-i*, remains of Mexico City ware plates; *j*, Puebla ware sherd

which these sherds are a part may thus have been brought to Port Royal by Portuguese Jewish immigrants.

Mexican Tin-Glazed Earthenware at Building 4/5
Ten of the analyzed sherds, representing at least three decorated plates, were identified as of Mexican provenance. Mexican majolica was widely distributed in the Caribbean region by the second half of the 17th century, so it is not unusual to find some at Port Royal. Specific to Building 4/5, Mexico City wares and a single sherd from Puebla have been identified on the basis of fabric color (buff on earlier wares; red [2.5YR 4/ -5/6] on later forms) and a glossy, heavy, cream-colored glaze (Goggin 1968:155; Lister and Lister 1978:12-15).

One of the Mexico City wares is represented by a small body sherd with cobalt blue bands that radiate upwards from the center to what would have been the plate's rim (see Figure 12.4*e*). It has the typical glossy glaze on both surfaces; its light tan to cream fabric suggests a mid 17th-century date. The second Mexico City ware plate, represented by several unattached sherds, is possibly a fine-grade type (see Figure 12.4*f-i*). Its dull blue coloring and floral and dot motif pattern on a red fabric is characteristic of San Luís Blue-on-White (Goggin 1968:154; Lister and Lister 1978:14, 1982:18). That the glaze is heavily crazed and has pooled on the surfaces in some places is also typical of San Luís Blue-on-White (Lister and Lister 1978:14; Deagan 1987:71).

The Puebla ware sherd (see Figure 12.4*j*) has a geometric pattern of dark blue bands, which encircle an intricate lacelike design of narrow black lines on a creamy white ground. Its glaze retains the brilliance that is characteristic of 17th-century Puebla pottery of fine quality. Usually this Puebla pattern has a scalloped border in a black lace design (Goggin 1968:173-182; Wilcoxen 1992a:29), but the border treatment on this example cannot be determined from the small fragment.

There may be more Mexican majolica in the analyzed assemblage. As already noted, some of the tin glaze finds are too small to allow pattern identification. This, combined with the loss of color and glaze in many cases, as well as the esoteric nature of Mexican pottery of this period in general, makes it difficult to classify these wares with certainty (Wilcoxen 1992a:29). That Mexican wares should even be present at an English colony may be attributable to the trade that is known to have been conducted between Jamaica and the various Spanish New World ports.

Tin-Glazed Earthenware Summary
The total number of tin-glazed earthenware sherds recovered from Layer 3 at Building 4/5 (i.e., the analyzed sherds) is 588. The total number of tin-glazed earthenware sherds recovered from all of the other buildings (Layer 3 only) is 507. This gives a combined Layer 3 total of 1095 tin-glazed earthenware sherds for the portion of the Port Royal site excavated by Hamilton's team from 1981-1990. The analyzed assemblage from Building 4/5 thus comprises 53.7 percent of this total. It may be of interest to note that the larger count across the site breaks down as follows: Anglo-Dutch sherds total 888 (over half of which appear to be from undecorated vessels), Mediterranean sherds total 32, and Mexican sherds total 12. The remaining 163 tin-glazed earthenware sherds are catalogued as 'unidentified' due to their fragmentary nature. On the basis of their fabric color, it is probable that most are of either English or Dutch provenance.

STONEWARE
Technology and Development
Stoneware, a very hard, vitrified ceramic, is fired at temperatures of about 1200°-1350°C. These temperatures are high enough to achieve partial fusion of the stoneware fabric, which is made of highly plastic, low-iron clays. Stonewares may be unglazed, may have an applied lead glaze or, more commonly, a salt glaze (Oswald et al. 1982:15; Draper 1984:33; Horne 1985:4; Rice 1987:6).

Salt-glazed stoneware is harder and more durable than earthenware and is characterized by a shiny coating, with a slightly pitted, or 'orange peel,' texture. The glaze can vary in color and thickness depending upon the nature of the clay(s), impurities in the clay(s), temperature of the kiln, and whether the stoneware vessel was covered in a colored slip or thin wash before it was fired (Noël Hume 1958:439-441; Oswald et al. 1982:15; Draper 1984:33). The contrast between dark and light brown stoneware, for example, is achieved by dipping the vessel in varying concentrations of a brown, iron-rich wash; a stoneware vessel can appear white by dipping it in a white slip. When salt glazing alone is used, the color of stonewares varies from pale buff through gray to dark brown and even purple. Unlike lead or tin glaze, salt glaze does not run or crackle. Its texture is not slippery to the grip, making vessels easier to hold (Elliot 1986:86).

Stoneware vessels are often embellished with applied clay motifs in a technique known as sprigging or sprig-molding, whereby a mold-made decorative clay ornament is applied to the unfired body of the vessel. Occasionally, during firing, the applied details lift from the surface and are said to have 'sprung.' Sprigged details, such as grapes and vine leaves, always stand out from the vessel in greater relief (Mountford 1971:14; Draper 1984:35-36). Stoneware decoration can also be rouletted, stamped, or incised.

The processes of stoneware manufacture have become increasingly clear from excavations and documentary sources. A letter written by a Sir John Lowther describes the methods as related to him in a meeting with English stoneware potter John Dwight in 1697/8:

> tis made of white Clay or Rather of

> Tobacco pipe Clay, that is brought out of Dorsetshire [which is] Mixed with a fine smale sort of sand [and] . . . water and wrought to a plyable termper. . . . After it's through Dryed it is Leaded [this is either a misinterpretation on the part of Lowther or a deliberate obscurity by Dwight] and sett in the Furness, which Furness is built a bout Eight feet square, Arched at the top like a valt with holes to let out the smooke; in this Furness it is sett or plased a Closae as maybe one peece by another up to the very top, the smal ware being put in Cases [saggars], but the bigger without, under the Furness is the Fier place built the full bigness of the Furnesse Arched at the top with Brick, with holes, through which the flame Assends into the Furness (cited in Weatherill and Edwards 1971:165).

This letter is referring, in particular, to the production of white stoneware, which Dwight succeeded in making (only in small quantities) in the last years of the 17th century. As the letter indicates, fine white clay at this time was already being used to make clay tobacco pipes. The first attempts were designed to obscure the usual light gray stoneware fabric by dipping it into a white clay slip. The earliest of these slip-dipped wares, which were usually tavern mugs, have a brown wash over the top half of the vessels. These were soon followed by vessels on which white dominated (Draper 1984:36). White stoneware, in the sense of the whole vessel being made from white clay, did not develop until around 1720, in Staffordshire (Mountford 1971:35-47; Draper 1984:37).

Development of Stoneware in Germany

The conditions necessary for stoneware, i.e., easily prepared clay, an abundance of wood for fuel, and knowledge of high-firing kilns, were achieved in northern Syria during the 3rd millennium B.C., and most successfully in China in the mid 2nd millennium B.C. Stoneware development occurred independently in Europe, beginning approximately in the mid 12th century, in the Rhineland region of Germany (Gaimster 1997:34). In fact, there are differing accounts as to the beginning of production of a true stoneware fabric in Germany. Some researchers (e.g., Beckmann 1974; Stephan 1983; Gaimster and Hook 1995:69) suggest a fully fused stoneware body was achieved in the early 14th century, in the town of Siegburg. Others (e.g., Troy 1977:15; Beebe 1980:126; Elliot 1986:86) note this 'early Siegburg' type should be considered only a proto-stoneware, with real stoneware not appearing until the 15th century.

Similar dissension prevails as to the advent of salt-glazed stoneware. It is agreed that it was first manufactured in Germany, but the discovery of the technique has been dated from as early as the 12th century (see Church 1911:17) to as late as the end of the 14th century/early 15th century (see Mountford 1971:17; Troy 1977:15; Beebe 1980:125; Horne 1985:4; Gaimster and Hook 1995:69). There is as much speculation as to why potters began the rather unnatural practice of throwing salt into their kilns to produce glazed wares. One theory suggests that salt was used to keep red clay slips in suspension, and that the sodium was sufficient to produce a faint glaze during firing. No experiments show that this is the case, however, and no solutions have been advanced as to why salt was thrown into the kiln itself (Troy 1977:12). Another suggestion is that wood from old salt-impregnated fish-storage or sauerkraut barrels was used to fire the kilns (Troy 1977:14). It seems feasible that if such fuel was employed during the last stages of firing, potters might well have concluded from the results that salt had played a part in whatever differences were apparent in the fired objects.

The potteries of Raeren and Frechen, together with those of Siegburg and Cologne, are among the most famous centers for stoneware manufacture in the northwestern Rhineland. The high quality of the wares was attributable in part to the geography of the Rhine Valley, which met all of the criteria necessary for building and sustaining potteries, with its "clean, very fine, fusible clays, inexhaustible wood fuel and sea salt" (Elliot 1986:85). The salt, which as noted above was used for the glaze, was undoubtedly acquired through the Low Countries, where a prosperous salt industry flourished, stimulated by the nearby herring plant, which needed salt for preserving the fish (Gaimster 1997:33).

It is usually relatively easy to differentiate between the wares from each of these Rhenish centers. This is generally the case even with fragments recovered from archaeological sites. Siegburg ware, for example, which was produced through the 1630s, is characterized by a light gray to near-white fabric, which by the mid 16th century, was sometimes also salt glazed. Cobalt-blue painted decoration was introduced ca. 1570 (Beebe 1980:126-127; Crossley 1990:262; Gaimster 1997:166). Cologne ware, in contrast, is distinguished by an even, brown glaze and often has sprigged acanthus leaves. Cologne in the 1520s was, in fact, responsible for producing some of the most technically and artistically accomplished salt-glazed stoneware in Germany. Its glory, however, was cut short by the enormous pollution problems caused by the salt-glazing process, which resulted in the city imposing strict laws to minimize firings. Some of the smarter potters moved to other areas, (e.g., Frechen); those too stubborn to leave were heavily fined, had their wares confiscated and kilns destroyed, and were sometimes even imprisoned (Troy 1977:15; Beebe 1980:133; Crossley 1990:263; Gaimster 1997:47-48, 191-193). The Raeren industry, whose height of production was in the early 16th century, and whose most popular export at the time was a small, dark gray salt-glazed drinking jug, similarly faded early, towards the end of the century. Its decline was also a result of the departure of its potters, this time fleeing to escape the Thirty Years War (1618-1648) (Beebe 1980:136-138; Crossley 1990:262-263; Gaimster 1997:224-225). Frechen ware was the most widely traded of all German stoneware and developed more or less as a result of the decline in the Cologne industry. Indeed, the Frechen industry developed at a remarkable rate, which it maintained throughout most of the 17th century. While many of the same techniques employed at Cologne were continued at Frechen, such as the use of even, rich brown glazes, the ware is particularly noted for its distinctive brown and gray 'mottling' on vessel surfaces.

The Westerwald region of Germany, known as Kannenbäckerland (country of stoneware potters), was established as a competitive stoneware-producing center by potters, largely from Raeren and Siegburg, who fled to the region to escape the unsettled conditions of Europe's Thirty Years War. The region, in fact, contains the only Rhenish stoneware potteries to have survived to the present, although on a much reduced scale. Westerwald stoneware is noted for its attractive grayish blue fabric color and its cobalt blue and manganese purple molded decorations. Straight-sided mugs and jugs are typical 17th-century forms, with those made in the early years of production carefully and beautifully decorated. As the Westerwald stoneware industry grew, the workmanship deteriorated. By the beginning of the 18th century, the once finely molded applied motifs had become simple decorations hurriedly stamped or incised into the vessel surface (Noël Hume 1967:349-353, 1970:281; Reineking von Bock 1986:65-70; Wilcoxen 1987:73-75).

German stoneware was one of the earliest European domestic products to make an impact on an international scale. Unlike contemporary earthenware, stoneware vessels fulfilled innumerable functions due to their physical properties. Possessed of a robust, highly durable, impervious body and a stain- and odor-free surface, stoneware was particularly suited to the transportation, storage, drinking, and decanting of liquids, as well as to preservation, pharmaceutical, and sanitary purposes (Elliot 1986:86; Gaimster 1997:117). Moreover, the sprigged decoration enabled the ware to compete with other pottery in the dining sphere, such as tin-glazed earthenware and the more expensive glass and metalware (e.g., pewter, silver, and gold).

Development of Stoneware in England

The enormous commercial success of the Rhenish stoneware trade, as surviving port books and customs accounts demonstrate (see Haselgrove 1992), provided the stimulus for several attempts to manufacture a stoneware body in England. In the late 16th century, William Simpson, a member of the Merchant Taylors Company, petitioned for an import monopoly on "all the pottes made at Cullein [Cologne] called drinking stone pottes"(cited in Oswald et al. 1982:21; see also Quinn 1971:66). There is no evidence, however, that Simpson either knew how to make stoneware or made any attempt to learn. His petition for a patent must have been unsuccessful, for in 1593, the courtier Henry Noell was granted a 15-year monopoly to "provide and buy in att and every the partes beyonde the Seas all manner of stone pottes earthen pottes stone bottles and earthen bottles" and to sell them anywhere in England and Wales (cited in Oswald et al. 1982:21; see also Quinn 1971:66). Again, however, there is no evidence that Noell actually made any stoneware.

After another unsuccessful attempt in 1614, when Thomas Browne, Tobie Steward, and Nicholas Burghley were granted a 21-year patent to make "all manner of stone potte stone Jugges and stone bottelle not heretofore usuallie made . . . within this Realme" (cited in Oswald et al. 1982:22; see also Quinn 1971:73), Thomas Rous and Abraham Cullen, two wealthy Dutch merchants, were granted a 14-year patent by Charles I in 1626 (Mountford and Celoria 1968:1; Quinn 1971:73). Rous and Cullen could well have been the first to succeed in manufacturing stoneware in England, since they arranged to bring over a potter from Frechen. But there is no evidence that this actually happened, and, in fact, other records show that Rous was still importing large quantities of German stoneware in 1639 (see Quinn 1971:75). Crossley (1990:266) makes a case for associating Rous and Cullen with a small stoneware kiln excavated at Woolwich-on-the-Thames, downstream from central London, and dated to the 1640s to 1660s (Pryor and Blockley 1978). The excavation showed a limited range of well-made mugs and Rhenish-type bottles, likely made by immigrant potters, using clays and molds brought from Frechen (clays were tested using neutron activation analysis, see Gaimster and Hook 1995; Hook 1997). Another kiln at the same site produced some white stoneware mugs and bowls with decoration in cobalt-blue of a type known to have been made in Cologne in the mid 17th century.

The recovery of stoneware sherds with a 'WK' monogram and the date of 1672 shows that Captain William Killigrew, whose patent application is dated 30 April 1672, made a market stoneware in England using Rhenish potters (Oswald et al. 1982:23; Horne 1985:4; Haselgrove 1990:154-156). The search for a successful English substitute for the robust Rhenish product, however, only came with John Dwight of Fulham, who, also in April 1672, was granted a 12-year patent for the sole right of production of English stoneware (Mountford and Celoria 1968:11-12; Price 1978: 52; Grigsby 1990:38).

Indeed, the story of John Dwight and the founding of the first commercially successful English stoneware factory at Fulham in London has been told many times from both the documentary and production viewpoints (see, e.g., Blacker 1922; Bimson 1961a, 1961b; Edwards 1974; Oswald et al. 1982). The majority of his early wares replicated the standard Rhenish types, but he also made original designs. Bottles and mugs were the products most in demand by the public, and it was these that provided the financial success that allowed Dwight to continue his experiments, which eventually led to the reliable manufacture of a fully vitrified stoneware, with or without glaze, in the mid 1670s (Oswald et al. 1982:16, 28-30; Draper 1984:33; Horne 1985:7; Grigsby 1990:37-38). In 1684, Dwight was issued a second patent for the same product, as well as for the making of 'Marbled Porcellane Vessels' and 'the opacous redd and dark colored Porcellane or China.'[5] Dwight died in 1703, but his pottery was continued by his family (Price 1978:52-53; Oswald et al. 1982:27-28).

By the end of the 17th century, the foreign monopoly on stoneware was broken, and the production method of Dwight's stoneware manufacture was common knowledge in England. The first London rivals were in Southwark, the Bear Garden and Gravel Lane potteries being established in the 1690s (Edwards 1974:16-18). Talbot's pottery in Lambeth followed soon thereafter. In Nottingham, James Morley was making a smooth glossy brown stoneware by 1696. Staffordshire Brown stoneware, which ranges from a sandy light yellow color to a lustrous dark brown, similar to Nottingham wares, was made from the late 17th century to the 1730s (Noël Hume 1970:114; Troy 1977:37; Oswald et al. 1982:102-138; Horne 1985:31). (Often, Staffordshire stoneware vessels have a light yellow-colored glaze on their lower body and a dark brown glaze on their shoulders and neck.)

Indeed, the Staffordshire stoneware industry took over from London in the early 18th century (Mountford 1971:35-47; Crossley 1990:267). White-clay stoneware, made in quantity there from the middle of the century, was particularly popular, since the only comparable ceramics available were imported porcelain (white but very expensive) and tin-glazed earthenware (almost white but very heavy and prone to chipping). As archaeologist/ceramic historian Jo Draper (1984:37-38) notes: "White stoneware was ideal for the growing middle-class market for newly fashionable tea wares, and for table pottery generally. . . Its shapes relate more to contemporary silver than to the traditional shapes of . . . earthenware."

Stoneware at Building 4/5

The functional versatility of stoneware, as noted above, is directly related to its unique combination of technical, physical, and decorative properties, which allowed it to fulfil both utilitarian and social roles. One cannot, therefore, separate the analyzed stoneware assemblage at Building 4/5 into 'utilitarian' and 'tableware' categories. While most of the identified vessels can be defined by their use in food and beverage storage (i.e, utilitarian), some of these (e.g., bottles) were also likely used in beverage consumption (i.e., as serving vessels, or tableware). Other recovered forms (mugs) should also be considered as tableware.

As shown in Figure 4.13, the distribution of the analyzed sherds supports the interpretation of the multi-functional purpose of stoneware vessels. A concentration of sherds is evident in Building 4's yards, particularly in Yard 4A. Some pieces were also found in the alley to the east of the structure. As was noted above in the distribution patterns both of slipware and tin-glazed earthenware, it is possible that some of the stoneware sherds recovered from these areas were displaced there during the disaster (perhaps from Rooms 4A and/or 4B). It is equally possible that they belonged to vessels that had been either originally used in these back areas or dumped there as trash as new vessels took their place in the day-to-day life of the building.

A total of 76 sherds, only 6.6 percent of the analyzed sherd count, was identified as stoneware. (This can be compared to the 417 stoneware sherds recovered by Hamilton from all of the other excavated buildings combined [Layer 3 only].) These 76 sherds represent an estimated minimum of 16 vessels, or 11.1 percent of the total minimum number of vessels taken to be associated with the building complex prior to 1692. German wares comprise over half of the assemblage. English types, most of which look to be from London and which are possibly intrusive from upper layers, are represented in minimal numbers. Two thin-walled stoneware costrels of possibly either Dutch or Southeast Asian provenance were also identified, based upon their similarity of form (specifically, short, high-arched handles) to stoneware vessels recovered from wrecked 17th- and 18th-century Dutch East Indiamen (see Green 1977:1:Figure GT 913; Ingelman-Sundberg 1978:74, 146, 197; Van der Pijl-Ketel 1982:223-242). An unglazed storage pot of unknown provenance (possibly Iberian) has also been recorded. Represented by a base and body sherd, it is covered on the exterior with a white slip wash.

It would be surprising if German stoneware was not found in relative abundance at Port Royal considering that by the 17th century it was being exported on a massive scale, first by ships of the Dutch East India and Dutch West India Companies and then later by the English merchant adventurers (Noël Hume 1970:276-285; Elliot 1986:88; Gaimster 1997:98). That English stoneware is almost absent from the 17th-century context is similarly to be expected, as it was not produced in quantity until after 1700.

German Stoneware at Building 4/5

Some 52 sherds, or 68.4 percent, of Building 4/5's analyzed stoneware sherds, are almost certainly of German provenance. Taken together, the sherds represent at least nine vessels, two of which are intact.

[5] This red 'porcelain' was actually a fine red unglazed stoneware, which was introduced to England in the 1690s by John Philip Elers, a Dutch immigrant. Elers ware (as it came to be known) was made to imitate the so-called Chinese red porcelain being then imported in large quantities in the forms of teapots (Troy 1977:37; Price 1978:57; Draper 1984:35).

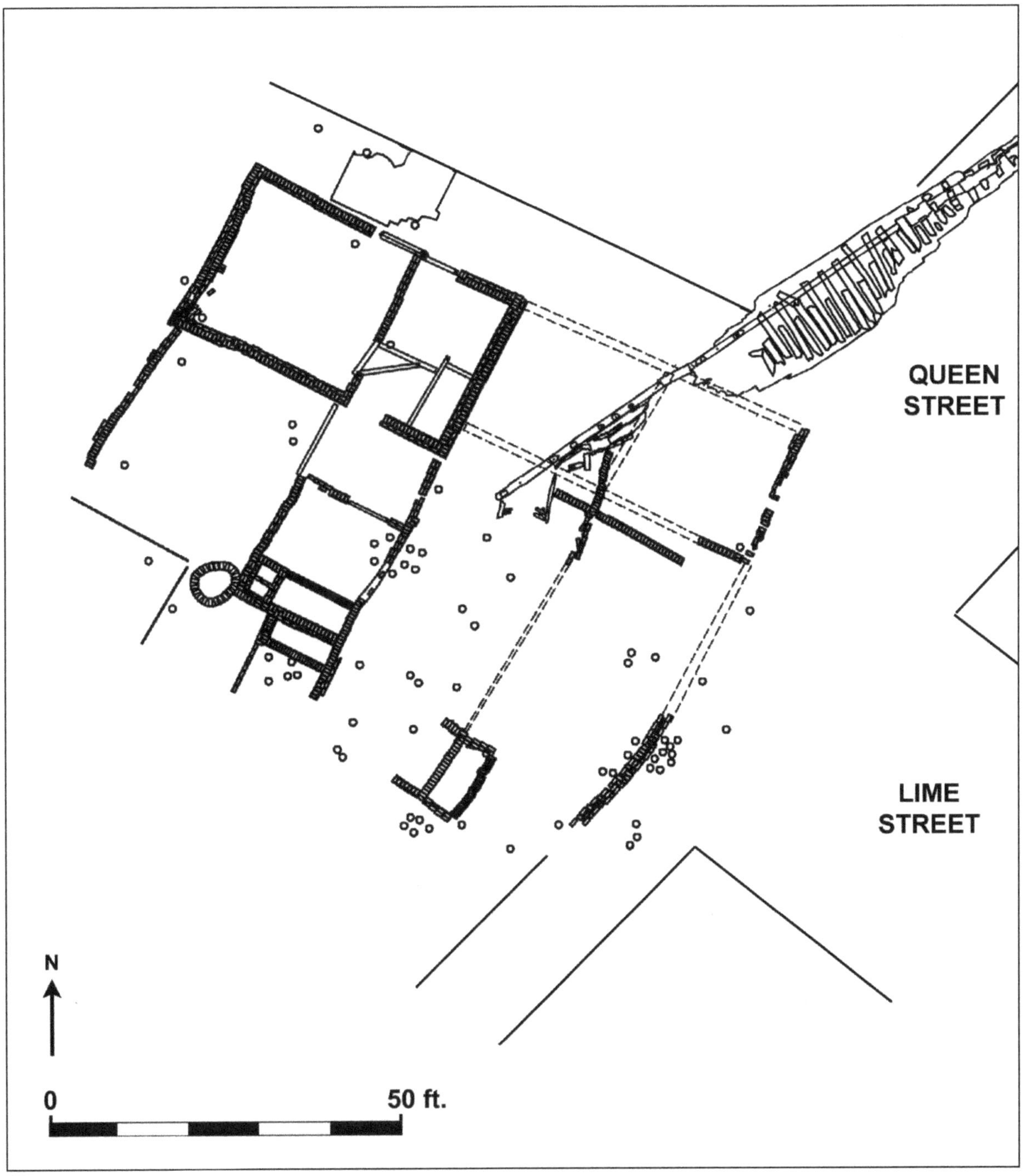

FIGURE 4.13. Distribution of stoneware ceramic sherds in the 17th-century context at Building 4/5

Types from the northwestern Rhineland (Raeren and Frechen) and the Westerwald region to the south were recognized in the assemblage. Identification was based upon fabric composition and color, as well as on vessel form and decorative treatment. Five additional sherds are also possibly of German provenance, based upon similar criteria. These latter fragments are, however, too small to identify with certainty. As such, they were recorded in the appropriate 'unidentified' category in the stoneware typology used for this study (see Appendix B).

Raeren and Frechen ware. The majority (n=46) of the identified German sherds belong to at least six vessels that were likely manufactured in the Rhenish potteries of Raeren, near Aachen, and Frechen, near Cologne. Some of the smaller fragments in this group (n=4) could not be distinguished further, since wares from both of these centers often have a similar fabric (light gray to light brown in color) and a similar brown-colored iron oxide wash over which an even salt glaze was applied. A specifically Raeren provenance was more positively attributed to a select few sherds in the group (n=9), primarily based upon their distinctive blue- to gray-colored glaze. Beginning in the 17th century, the Raeren industry developed and used such a glaze on many of its products (Noël Hume 1967:349; Beebe 1980:136; Crossley 1990:262). The remaining sherds (n=33, over half of the total German assemblage) were attributed specifically to Frechen, primarily on the basis of a distinctive brown-colored mottling on sherd exteriors. This mottling, which is often referred to as 'tiger flecking,' is due to the presence in the Frechen clay of iron salts (Noël Hume 1958:439, 1970:55; Elliot 1986:85). A specific type of plastic ornament evident on some Rhenish sherds also pointed to a Frechen provenance.

Bottles of various sizes comprise all of the recognized Raeren/Frechen forms recovered from Layer 3 at the building complex. Two of the most interesting types were recovered intact and are in extremely good condition (Figure 4.14*a*, *b*), a testament to not only the durability and strength of the product but also to the almost unparalleled preservation conditions of the Port Royal site. Both bottles, which are very small (between 10 and 12 cm tall), are identical in shape and in method of manufacture. It is, in fact, the latter detail that confirmed their German provenance: each has concentric string- or wire-marks on its base, a feature typically found on German bottles, jugs, and mugs from the 17th century (Wilcoxen 1992a:37; Gaimster 1997:35). Such marks reveal that the vessels were removed from the wheel while it was still moving, which, in turn, points to mass production, whereby "it was uneconomic to slow the wheel down before removing the pot and replacing it with a fresh ball of clay" (Gaimster 1997:35). Other supporting evidence of a German origin for the bottles is the configuration on the handle termination of each, and the thumb impressions immediately below each handle's lower end (Wilcoxen 1992a:37).

While plain (i.e., no sprigged decoration) small bottles such as these are not well documented from terrestrial sites, and no others have as yet been recovered from Port Royal, they are quite a typical find in 17th- and 18th-century Dutch and English shipwrecks (see Andersen 1974:93, Figure 3; Holman 1975:255-256, Figure 3; Barber 1977; Green 1977:1:105-109; Owen 1988:289, Figure 4). Recently, a similar bottle (both in size and color) to one of the Port Royal specimens (see Figure 4.14*b*) was found on the 17th-century French ship, *La Belle*, which sank in Matagorda Bay, Texas, in 1686 (Hamilton 2001, pers. comm.). Texas A&M archaeological conservator Dr. Helen Dewolf suggests that it is possible that both, if not all three bottles, came out of the same pottery.

It is possible that these small bottles were used to store medicines or ointments, since similarly sized Rhenish bottles found on both English and Dutch ships contained residues of pharmaceutical substances (see Andersen 1974:92-93, Figure 3; Rule 1982). Indeed, the brown-mottled specimen was, in fact, found sealed with a cork and contained the remnants of its liquid contents, which included phytoliths (plant silica crystals) and pollens that have yet to be identified.

The remains of at least four much larger Rhenish bottles were also found in the building complex and its surrounding areas (see Figure 4.14*c-f*). All appear to be of a particular type made first in the potteries of Cologne, then carried on in Frechen from at least the mid 16th century. Indeed, this bottle type, known in Germany as the *Bartmannkrug* (bearded-man bottle) and in Britain and North America variously as the Bellarmine, Graybeard, or Long-Beard, can be considered the German stoneware industry's equivalent of the Iberian olive jar in terms of popularity as an exported product. It has been found in profusion across Europe, along the West African Coast, in the East and West Indies, and throughout the North American continent (Noël Hume 1958:439, 1982; Thwaite 1973:255; Beebe 1980:133; Elliot 1986:86-89; Gaimster 1997:99-103). Fragments and often whole vessels have also been recovered from 16th-century Spanish shipwrecks (Arnold and Weddle 1978:262, Figure 45) and from numerous 17th-century Dutch and English wrecks (see, e.g., Forster and Higgs 1973:297-298, Figures 6-8; Stanbury 1974; Sténuit 1974:239-243; Green 1977:1:95-104, 110-142; Holman 1975:255-256; Barber 1977; Van der Pijl-Ketel 1982:246-247; Lessman 1997). While only broken pieces remain of Building 4/5's *Bartmannkrüge*, a near-intact vessel was found by Hamilton in Building 1. Diagnostic *Bartmannkrüge* sherds were also found by Mayes (1972:74-75) in the 17th-century components of Port Royal's old naval dockyard. Brown (1996:170) describes several fragments recovered from a 17th-century tavern in New Street, Port Royal, which was excavated by Anthony Priddy in the 1970s.

FIGURE 4.14. Examples of Raeren/Frechen salt-glazed stoneware recovered from the 17th-century context at Building 4/5. *a-b*, small intact bottles; *c-d*, neck and mouth fragments from large bottles known as *Bartmannkrüge*, or Bellarmines; *e-f*, fragments of molded medallions from *Bartmannkrüge*

The Rhenish *Bartmannkrug* is a full, round-bodied vessel with a short, cordoned neck and a single vertical strap handle. Its name derives from an unusual decorative motif—a bearded face or mask—which is often found sprig-molded on to the vessel's neck. A floral or heraldic medallion is also usually molded on to the vessel's body (Figure 4.15). The bottles were made originally for shipping Rhine wine, but like the olive jar, the form also served as a household container for any number of purposes, such as for storing vinegar or oil. The vessels were also used as decanters for ale or for the distinctive wheat beer called 'mum' (from the German *Mumme*), a popular beverage in Europe in the late 17th century.

Since the mid 19th century, much research has been devoted to the Bartmann bottle (see Chaffers 1850, 1946[1856]:45-49; Marryat 1850; Holmes 1951; Bardenheuer 1960; Reineking von Bock 1966; Thwaite

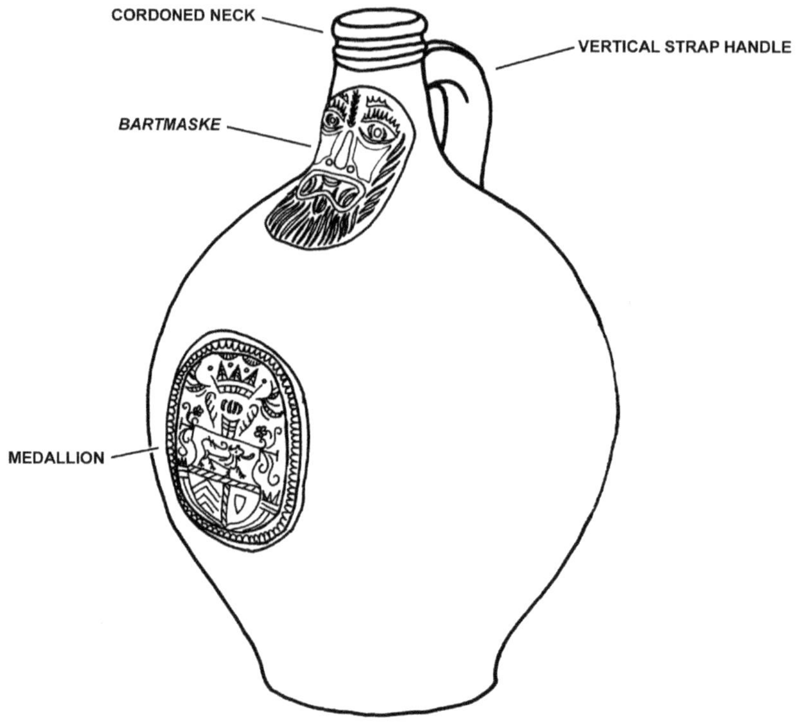

FIGURE 4.15. A Rhenish *Bartmannkrug*, showing typical stylistic characteristics

1973). Most significantly for archaeology, it has been used, with mixed results, as a chronological tool (see Holmes 1951; Thwaite 1973). It is argued that both the bottle shape and characteristics of the mask gradually changed in style throughout Bartmannkrug production. It is possible, however, that the different styles of the masks merely denote the work of individual potters (see Haselgrove and Van Loo 1998:46).

The origin of the mask (known as the *Bartmaske*) has been similarly debated. It possibly originated from the popular tradition of the Wild Man, a mythic creature believed to live in the remote alpine regions of Germany and Austria, and who features prominently in northern European folklore of the 14th-16th centuries (see Husband 1980). Popular tradition, however, attributes it to the Roman Catholic theologian Cardinal Roberto Bellarmino (1542 1621) in an attempt to caricature the hated zealous opponent of Protestantism in the Low Countries and northern Germany (Chaffers 1946[1856]:46-47; Holmes 1951:175; Noël Hume 1958:439-440; Thwaite 1973:258). While the connection is incorrect, since Bellarmino was just a child when the earliest dated examples of Bartmann bottles were made, the term 'Bellarmine' has, perhaps, become the most common name for these Rhenish vessels.

It appeared in English popular literature from the early 17th century as references to bearded bottles in, for example, Ben Jonson's works, "The Gypsies Metamorphos'd" (1613) and "Bartholomew Fair" (1614) indicate. It is also mentioned in William Cartwright's comedy, "The Ordinary" (1634), where a drunken curate describes his opponent as "Like a larger jug, which some men call a Bellarmine, but we a conscience" (cited in Holmes 1951:177; Noël Hume 1958:440; Thwaite 1973:260-261; Beebe 1980:134-135).

Cardinal Bellarmino is not the only historical figure likened to the bearded faces. The Duke of Alba, another opponent of Protestantism, was also compared to the sometimes grotesque mask. In 1677, Dr. Plot mentions the stone "d'Alva Bottles, Jugs, Noggins . . . made only in Germany" (cited in Noël Hume 1970:283; see also Beebe 1980:135). The diarist John Evelyn, in his 1697 "Numismata," also linked the bottles to the duke, "of whom there are a thousand pictures showing a malicious stern and merciless aspect, fringed with a prolix and squalid beard, which draws down his meagre and hollow cheeks, emblems of his disposition" (cited in Beebe

1980:135).

Evelyn's description is inappropriate for the *Bartmaske* fragment recovered from Building 4/5, since the face, of which only the eyes, brows, and part of the nose remains, appears almost worried or sad (see Figure 4.14*d*). Its design is carelessly applied. It is certainly a far cry from the masks of the mid 16th century, which were "handsome, even noble . . . well defined" (Beebe 1980:133). Its expression is more like that of the *Bartmaske* of later years, whose poor design and execution led to it being completed abandoned as a decorative motif by the end of the 17th century (Noël Hume 1958:439-441, 1970:57; Thwaite 1973:257-258).

Building 4/5's other possible *Bartmannkrüge* are represented by a neck and rim fragment (see Figure 4.14*c*) and by two molded medallions, one of which is in the shape of an oval rosette, a design or variation of it commonly represented in 17th-century contexts (see Figure 4.14*e*). The second, more complex, heraldic medallion was recovered from the brick-paved sidewalk in front of Building 5, and, therefore, is possibly part of a bottle that was used either in Building 5's upper story or in Building 8, which is located across the street, north of the complex (see Figure 4.14*f*).

Westerwald ware. Very few Westerwald sherds (n=6) were recovered from the 17th-century context at Building 4/5, but all show the typical Westerwald fabric and cobalt blue decorative motifs. Manganese purple, introduced into the Westerwald palette around the 1660s (Noël Hume 1970:281), is not apparent on any fragments found in the building's occupation layer.

Despite the small number of sherds, the diversity of their patterns (stamped, incised, and sprigged circles, hearts, and rosettes) indicates that at least three distinct vessels are represented. A mug, the glaze and colors of which are still pristine, is partially reconstructed; the other two (represented only by body sherds) are here also catalogued as mugs (Figure 4.16).

Such paucity of Westerwald stoneware in Layer 3 at Building 4/5 may be thought surprising, given that such vessels appear in England from the mid 17th century (Oswald et al. 1982:18; Draper 1984:33) and occur, sometimes in quantity, on contemporary English settlements in North America (see Gaimster 1997:100-104 for examples). Westerwald wares have also been found on wrecked ships of the period (see, e.g., Stanbury 1974; Green 1977:1:95-146; Lessman 1997). On the other hand, this decorated stoneware only came into prominence as reasonably priced tableware around the 1670s (Haselgrove and Van Loo 1998:49). It may have arrived at Port Royal as the personal cargo of a settler who had access to the German ceramic market at this time. Alternatively, it may have arrived, likely via English ports, as a traded commodity, with much more yet to be recovered from other parts of the site.

Rhenish stoneware received its first serious competition in the middle of the 17th century with the growing enforcement of the English Navigation Acts after the 1650s and the development of the English stoneware and glass industries. The fashion for chocolate, coffee, and tea drinking also brought new table and dining customs, and with it, as already has been shown, the development of tin-glazed earthenware. The increasing cargoes of Chinese porcelain (to be discussed in the following section) were also a contributing force in the industry's decline. However, Rhenish stoneware was still the ware of choice in taverns, in the form of mugs and handled jugs/bottles, where beer was still a popular beverage.

English Stoneware at Building 4/5

A total of only five sherds, or 6.6 percent of the analyzed stoneware sherds, is almost certainly of English provenance. Taken together, the sherds represent an estimated minimum of four vessels. Identification of the ware type was primarily based upon fabric color and surface treatment.

FIGURE 4.16. Westerwald stoneware recovered from the 17th-century context at Building 4/5

The English stoneware vessel forms are a (possible) heavy-duty bottle or jug (represented by a large body sherd), a small, sturdy storage pot or bottle (an inkpot?) (represented by a base and body sherd), and two mugs (each represented by rim sherds).

The bottle/jug may be Crouch ware, given its dark gray fabric and reddish brown (5YR 5/4) salt-glazed wash. Crouch ware was a cheaper alternative to white salt-glazed stoneware products, and according to contemporary historian Simeon Shaw, was first made in Staffordshire about 1690 (cited in Mountford 1971:21-22).

The storage pot is glazed on its lower exterior body with a buff to light yellowish brown (10YR 6/4) glaze, with a band of dark reddish brown (5YR 2.5/2) glaze faintly evident on its shoulder. It may be of Staffordshire provenance.

Both of the mug rims are covered in a white slip. One (represented by two cross-mended sherds and illustrated in Figure 4.17*a*) is edged with a thin band of dark reddish brown (5YR 2.5/2) slip; the other is covered on the exterior with freckled brown wash (see Figure 4.17*b*). It is possible that both of these vessels are intrusive from upper layers at the site, since both are similar to early 18th-century Staffordshire types illustrated in Mountford (1971:Plates 15, 52). London potter John Dwight, however, was also making white-slipped wares, usually tavern mugs of this sort, in the late 17th century (Draper 1984:36).

Stoneware Summary

The total number of stoneware sherds recovered from Layer 3 at Building 4/5 (i.e., the analyzed sherds) is 76. The total number of stoneware sherds recovered from all of the other buildings (Layer 3 only) is 417. This gives a combined Layer 3 total of 493 stoneware sherds for the portion of the Port Royal site excavated by Hamilton's team from 1981-1990. The Building 4/5 analyzed assemblage thus comprises 15.4 percent of this total. It may be of interest to note that the larger count across the site breaks down as follows: Raeren/Frechen ware totals 325 (94 sherds of which are Bartmannkrüge fragments), Westerwald ware totals 66, and English stoneware totals at least 23 (a further 33 sherds are covered in a brown wash and are also possibly English). The remaining 46 sherds (including the white-slipped [possibly Iberian] basal sherd noted above) are catalogued as miscellaneous/unidentified.

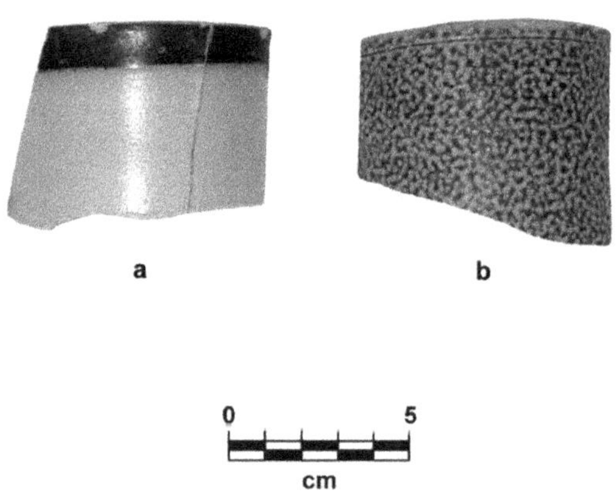

FIGURE 4.17. English white-slipped stoneware mug rims recovered from the 17th-century context at Building 4/5

PORCELAIN

Technology and Development

Porcelain is a thin, hard, non-porous ceramic. It is often called china or chinaware, since it was first made in China, and is the pinnacle of the potter's art in terms of technical accomplishment. Unlike earthenware and stoneware, which use clay only, porcelain is composed of two related materials, namely kaolin clay and petuntse. Kaolin (often called china clay) is a white-firing, highly refractory clay derived from ground feldspar, granite, and pegmatite. Petuntse (often called china stone) consists of feldspar and silica. The kaolin-petuntse mix is fired at very high temperatures that range from about 1280°-1400°C or more. In the process, the petuntse melts and forms a colorless glass, which fuses to the kaolin. The kaolin is tolerant to high temperatures and so the clay object holds its shape during firing. A (purely aesthetic) glaze, which often covers the unfired body is also made from petuntse (Hobson 1948:46, 114-115; Draper 1984:53; du Boulay 1984:9-10; Mudge 1986:26, 249; Rice 1987:6).

There are three main types of porcelain: (a) hard-paste porcelain, (b) soft-paste porcelain, and (c) bone china. Hard-paste porcelain, or 'true' porcelain, is that which is described above and has always been the model and ideal of porcelain makers. It can be distinguished from the other types by its pale gray to off-white, very hard and compact fabric, high-gloss glazed translucency, and melodious ring when tapped (Hobson 1948:19, Noël Hume 1970:258; Mudge 1986:249). In true porcelain, it is impossible to differentiate the body from the glaze.

Soft-paste porcelain, sometimes called artificial porcelain or *pâte tendre*, was developed in Europe in an attempt to imitate the Chinese ware. Experimenters, beginning in the 16th century in Italy, used a wide variety of material in their efforts to produce a substance that was hard, white, and translucent, but they were hindered (until the early 1700s) by a lack of a suitably plastic and white-firing kaolin (Tait 1962:12; Mudge 1986:249). Instead, they developed a soft-paste type, using a combination of ground glass and various clays. Soft-paste porcelain, so named because it required a lower or 'softer' firing temperature (ca. 1200°C), is usually covered with a lead-based glaze, which is brittle and liable to crack (Weatherill 1986:248). Further, the body, being more porous than hard-paste, is easily stained (Mudge 1986:250). The ware did have its merits—the surface is creamy in tone, and the colors used to decorate it merge with the glaze to produce a soft, silky, pleasing effect (Draper 1984:53; Mudge 1986:249-250; Weatherill 1986:249)—but it eventually fell out of favor when a hard-paste type was developed in about 1710, in Meissen, near Dresden, Germany (Tait 1962:25; du Boulay 1984:9).

Bone china is a late 18th-century English innovation (developed in Staffordshire by Josiah Spode) in which calcined ox bones are added to the hard-paste clay to provide the desired translucency (Tait 1962:89; Draper 1984:53; Rice 1987:6). Although not as hard as true porcelain, bone china is more durable and less susceptible to chipping and cracking (Mudge 1986:249). It is still made today, almost exclusively in England.

All three types of porcelain can be either wheel-turned or mold-made. If decorated, motifs may be incised or carved on to the unfired body; molded or slip-applied relief designs are also common. Porcelain is usually painted, and this can be done in several ways. Colored glazes are common on early Chinese wares. Painting under the glaze, usually in cobalt blue, was popular both in China and Europe. Paints applied over the glaze are commonly called enamels (which require a second firing to make them more permanent), and a large palette of colors—varied greens, yellow, purple-brown, and reds—was perfected in China at an early period (Hobson 1948:49-50). Painted decoration in Europe differs greatly from that in China: Chinese designs usually have clearly defined outlines, while European artists tend to blend colors together. Further, European decoration is used for its artistic value, while the Chinese ornament is filled with symbolic meaning. Indeed, the whole matter of Chinese ceramic decoration—the repeated combinations of certain designs, the curious symbols, the strange and mythical creatures, and the numerous figure scenes—while of absorbing interest, is a topic unto itself and beyond the scope of this study. For more information, the interested reader is directed to early works by Bushell (1910), Hobson (1915, 1923), and Honey (1927, 1944); more modern works include Lion-Goldschmidt (1978) and Rawson (1984).

Porcelain developed in China, but its beginnings are obscure. Historical texts are vague, and there is no single Chinese word for porcelain as distinct from other kinds of pottery (Hobson 1915:1:140-142, 1948:19). While some of the opaque-glazed stoneware ceramics made in the Han dynasty (206 B.C.-A.D. 220) have been called 'porcellaneous stoneware' or 'protoporcelain,' the technique of mixing kaolin and petuntse and firing them at sufficiently high temperatures to make a vitrified white substance probably did not develop until early in the T'ang period (A.D. 618-906) (Hobson 1948:4-5, 12; du Boulay 1984:19; Rice 1987:16).

Certainly, high-fired, gray-green glazed (but non-translucent) porcelains are well known in China from the 9th and 10th centuries, when many were exported to the Islamic world, where they were highly prized (Hobson 1948:19; du Boulay 1984:13; Mudge 1986:13). But it appears that it was not until the middle of the Sung dynasty (A.D. 960-1279) that these wares were refined to the white, hard, translucent, resonant type we know today (Hobson 1915:1:149, 1948:36; du Boulay 1984:13, 103). Indeed, this 'true' porcelain was compared "by lyrical Chinese poets to jade, snow, and lotus leaves" (Rice 1987:16). It is known among Chinese scholars as *qingbai* (pale blue or bluish white) or *yingqing* (misty blue) in reference to the tone of its glaze (Hobson 1948:36; du Boulay 1984:59, 103; Mudge 1986:75).

Chinese *qingbai* porcelains are most closely identified with the 'imperial kilns' at Jingdezhen (formerly Chin-tê Chên, or Ch'ang-nan Chên) in northern China, which, at the beginning of the 11th century, became "the metropolis of the ceramic world" when Sung emperor Chên Tsung decreed that its kilns should produce wares for the imperial capital (Hobson 1915:1:152). The glazes and decoration made at the royal factories were at first intended to reproduce natural colors, and thus many of the types from the early period are monochromatic. Celadon, a sea-green glaze, was especially popular in the 12th century and was called by various names, based upon its shade and tone or its pattern of crackles (Hobson 1948:21, 23-26). Molded or applied relief or incised decorative motifs are also common on *qingbai*, and much attention at this time was paid to vessel forms, the fashion favoring the reproduction of bronze and other metal artifacts (Hobson 1948:19; du Boulay 1984:103).

The last Sung emperor was overthrown by the Mongols in 1279, making way for the Yuan dynasty (1279-1368) of Qubilai Qan. Contacts with the Near East flourished during and after this period, and there was considerable interchange between the Chinese and Islamic ceramic arts. Perhaps the most significant Islamic contribution to the porcelain industry was the imported cobalt pigment used for underglaze blue decoration (the local Chinese cobalt was inclined to be dull and grayish). The result of this trade laid the foundation for the Chinese blue-and-white decorative tradition that has continued to the present (Hobson 1948:48; du Boulay 1984:13, 20; Mudge 1986:15-17; Rice 1987:17).

The porcelains of the native Ming dynasty (1368-1644) are more refined in both potting and glazing than those of the Yuan. The bodies are also much whiter, the blues deeper in tone. Indeed, the Ming period is associated with new experiments in color (although blue-and-white was by far the largest group): underglaze copper-red was perfected (it was introduced in the Sung); a yellow glaze made from antimony was developed; and green, brown, and purple glazes appeared, slightly later, after 1500 (Hobson 1915:2:98-101; du Boulay 1984:151-152, 189; Mudge 1986:15-17). Ming porcelains are also often decorated in two colors, the most popular being the combination of underglaze blue and yellow enamel. Polychrome (three- and five-color) enameled wares are common from the 16th century (Hobson 1948:49-52; du Boulay 1984:152, 189; Mudge 1986:18, 63).

The Ming is also known for the production, in the mid 17th century, of *blanc de chine*, "a pure white porcelain, sometimes with a faint smoky ivory, very glassy surface" (du Boulay 1984:153). Both it and other white porcelain types are either left undecorated, molded in low relief, appliquéd, carved, or incised (Hobson 1948:51-52; du Boulay 1984:151). *An hua* (veiled or secret decoration) is another well-known Ming white ware innovation. In this process, a fine, lightly incised decoration or an almost invisible relief slip molding is applied to the unfired, unglazed body in such a way that the decoration is very faint. On the finest of these wares, the decoration can only be discerned by holding the piece up to a strong light (Hobson 1915:2:5-6, 1948:47-48; du Boulay 1984:151). Throughout the Ming dynasty, the dragon and the phoenix were the most popular decorative motifs on ceramics. Other animals, plant forms, and human figures in gardens and interior settings were often used as decor for blue-and-white wares.

The peak of Chinese ceramic production was reached in the reigns of K'ang Hsi (1662-1722), Yung Chêng (1723-1735), and Ch'ien Lung (1736-1795) of the Ch'ing, or Manchu, dynasty. In this period, improvement was seen in almost all ceramic types, especially the blue-and-white, which had become a major export item to South East Asia and the Near East (Hobson 1915:2:68, 118, 1948:73-74; Mudge 1986:9). Blue-and-white decorated porcelain also became the favorite of European merchant-mariners, who began to arrive in China in large numbers in the 17th century (Tait 1962:17; du Boulay 1984:251-252; Mudge 1986:20). The Dutch East India Company was the most powerful of these traders throughout the 1600s, and it is the Dutch who were responsible for bringing the ware in quantity to the West (the Portuguese and Spanish had brought limited amounts of porcelain to Europe in the 16th century). Trade with England and France, and later, Denmark and Sweden, was comparatively small until the early 18th century (Mudge 1986:22, 111, 128-134). By about 1750, the European market was saturated with Chinese porcelain, and there was increasing competition from the newly developed home industries.

Porcelain at Building 4/5

Porcelain sherds were not found in any great quantity in Layer 3 at Building 4/5. While the ware should be considered first as an elite special-purpose ceramic used for social display, the vessels recovered from the site can also be defined by their association with the table. Forms recovered include bowls of various sizes, two cups, and the remains of a saucer. While these vessels may have performed their functionally prescribed roles, it is possible that, like tin-glazed forms, some may have never been used.

As shown in Figure 4.18, most of the few analyzed porcelain sherds that were found were associated with Building 5. Some of those on the brick-paved sidewalk directly in front of the structure may be parts of vessels that were originally housed in Building 5's upper story or in Building 8, which is located across the street. The porcelain sherds found in Building 4's yards may have been parts of vessels that were originally located within this smaller building and which were blown out on to these areas as the ship rammed into it. Alternatively, given that some of these sherds were found intermixed with discarded tobacco pipe fragments, it may be that the vessels of which they were originally a part had broken and were thrown away as trash.

With a total of 15 sherds, porcelain comprises only 1.3 percent of Building 4/5's analyzed sherd count, representing the smallest group of ceramics in the Layer 3 assemblage. (This can be compared to the 206 porcelain sherds recovered by Hamilton from all of the other excavated buildings combined [Layer 3 only].) These 15 sherds represent an estimated minimum of six vessels, or 4.2 percent of the total minimum number of vessels taken to be associated with the building prior to 1692. Again, this is the smallest number of vessels of a single ware in the structure. As noted above, some of the porcelain recovered from the front of Building 5 may, in fact, have originated in Building 8, "a commercial establishment specializing in the sale of fine-quality tableware" (Dewolf 1998:144). As Dewolf (1998:121) comments in her study of Chinese porcelain at Port Royal: "when reconstruction of various artifacts was completed, it was determined that many of these [porcelain] sherds [at the front of Building 4/5] were, in fact, originally from Building 8." Some of these displaced sherds may have been dumped by the post-earthquake tidal wave as it washed over the site. Others were retrieved from the Building 8 excavation dredge piles that had been piled on top of the sidewalk.

Unlike the other ceramic wares so far discussed in this study, which are represented by types from a wide variety of sources, all of the analyzed porcelain is of Chinese provenance. Specifically, the ware should be defined as 'export porcelain,' since the types that were traded to the West were markedly different in character (Europeanization of designs and forms) from those made for the domestic Chinese market. Sherds from Building 4/5 were identified on the basis of fabric composition and color (hard, vitreous, and white in color), glaze (shiny, transparent, with a slight pale blue tint) and, in some cases, on style and color of decoration (cobalt blue 'chinoiserie' motifs). Some forms are undecorated.

All of the Chinese export porcelain recovered from the Port Royal site is discussed and illustrated in Dewolf (1998). Much of the following is taken from that analysis. Specific to the analyzed assemblage at Building 4/5, the remains of two decorated bowls, two undecorated *blanc de chine* cups, a decorated tea bowl, and a undecorated white saucer were found.

The base and some body fragments of one of the bowls is illustrated in Figure 4.19*a*. It is decorated on the interior with a blue-and-white floral pattern, and although not evident on the photograph, it has on its exterior surfaces the characteristic brown wash of Batavia ware. (Batavia ware is so named for the Dutch port in Indonesia, now called Djakarta, where much of it was transhipped to the West.) The second, slightly smaller, bowl (see Figure 4.19*b*) has a blue-and-white floral motif centrally placed

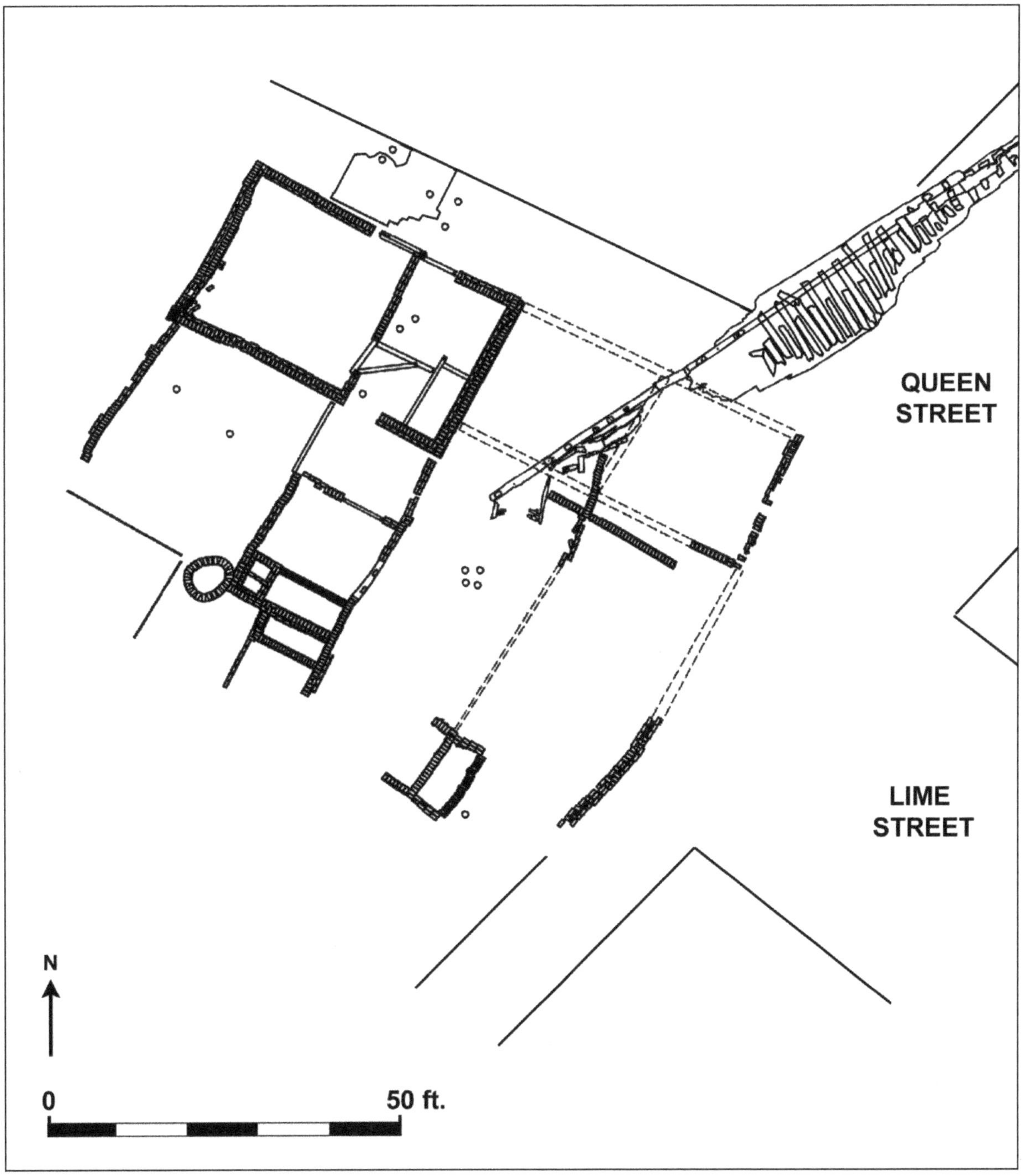

FIGURE 4.18. Distribution of Chinese export porcelain sherds in the 17th-century context at Building 4/5

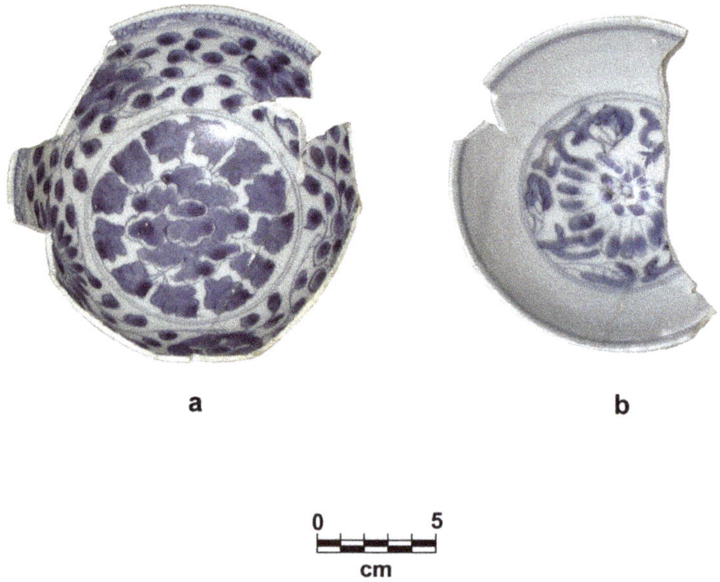

FIGURE 4.19. Examples of Chinese export porcelain recovered from the 17th-century context at Building 4/5. *a*, reconstructed Batavia ware bowl, showng an underglaze blue-and-white floral design on its interior surface; *b*, reconstructed underglaze blue-and-white bowl, showing a floral motif in the interior center

on the interior; a blue-and-white floral pattern covers the exterior. One of the *blanc de chine* cups was recovered intact, while the other is represented by fragments. The decorated (Batavia ware) tea bowl is represented by two basal fragments. The undecorated white saucer is represented by its base.

Chinese export porcelain was a relatively expensive tableware in the 17th and 18th centuries. To find it at Port Royal is significant, since "the Chinese civil war made porcelain more or less unobtainable to the west between about the late 1640s and the early 1680s" (Curtis 1988:28). This, coupled with the ban during the same period on English merchant adventurers from using Chinese ports, because "they apparently behaved so badly," provides important clues to the social background of those who lived in the port (Noël Hume 1970:259). Not only could the occupants of Building 4/5 obviously afford to acquire material goods purely for status value, they had access to markets not yet widely available to the majority of Europeans.

Porcelain Summary

The total number of Chinese export porcelain sherds recovered from Layer 3 at Building 4/5 (i.e., the analyzed sherds) is 15. The total number of Chinese export porcelain sherds recovered from all of the other buildings (Layer 3 only) is 206. This gives a combined Layer 3 total of 221 Chinese export porcelain sherds for the portion of the Port Royal site excavated by Hamilton's team from 1981-1990. The Building 4/5 analyzed assemblage thus comprises 6.8 percent of this total. It may be of interest to note that note that over half of the larger count across the site (n=125) was recovered from the area around Building 8.

SUMMARY

The 1155 sherds (representing 144 vessels) recovered from Layer 3 at Building 4/5 and detailed above are itemized in Tables 4.1 and 4.2, respectively. These tables present the breakdown of the analyzed assemblage into its component wares and types. It can be seen that tin-glazed earthenware is the most common, comprising half of the sherds and about 40 percent of the vessels. Most are, unsurprisingly, of English and Dutch provenance, the former likely being more prevalent. That Mediterranean and Mexican examples should also be found speaks to the international trading activities that are well documented for Port Royal.

Tables 4.1 and 4.2 also show that coarse ware is prevalent, comprising approximately one-third of the total number of sherds recovered and slightly over one-third of the total estimated minimum number of vessels. Most of the coarse ware is English, of types common in the 17th century. Again, given the Anglo heritage of Port Royal, this is not surprising. That Continental European coarse ware is also present again speaks to the location of Port

Royal, at the center of Atlantic trading routes.

Stoneware, slipware, and porcelain are present in Building 4/5 in far fewer amounts. Stoneware in the 17th century was relatively expensive compared to many other ceramics; this may account in part for its paucity at the site. Alternatively, the Building 4/5 occupants may have simply preferred non-ceramic materials (as noted in Chapter III, for
example, barrels and wooden casks are just some of the other storage ware that has been recovered from the site). The lack of slipware is, as noted earlier in this chapter, probably due in large part to the fact that slipware was not marketed on the same scale as other wares at this time. Further, it was not as popular for tableware as was tin-glazed earthenware.

As to the porcelain: that this ware should even be present in the assemblage again—like the tin-glazed earthenware—gives some indication as to the lifestyle enjoyed by at least some of the site's occupants, since porcelain in the late 17th century was first and foremost an ornamental ware, often displayed as merely a statement of status.

In short then, Building 4/5's Layer 3 ceramic assemblage includes a variety of ceramic wares and types from a variety of places, mostly England, and to a lesser extent Continental Europe, with a good proportion being decorative quality tableware. This would seem to confirm, then, what has already been suggested with regards to the standard of living of at least some of the building's occupants, namely that (1) they were aware of the latest fashions in decorative household ware and were able to acquire it, (2) they appear to have taken advantage of the wide commercial contacts afforded by their location in a large international port, and (3) one could also suggest that they enjoyed reasonable social contacts.

Since the author has also briefly compared the ceramic inventory of the complex with that from the other buildings excavated by Hamilton's team (Layer 3 only), it may be worthwhile to see how the figures play out. It should be borne in mind, however, that the original data base in relation to the other buildings is provisional at the time of writing for the reasons given in Chapter III (see page 19). Nevertheless, even the provisional figures produce some intriguing results, as shown in Table 4.3.

The author's attention has been confined to five ceramic wares: coarse ware, slipware, tin-glazed earthenware, stoneware, and Chinese export porcelain. Across the Port Royal site (Layer 3 only), including Building 4/5, the sherds of these wares total 3429. A total of 33.7 percent of these (n=1155) were retrieved from Building 4/5. Examining the numbers further, we see that Building 4/5 contained 27.4 percent of all of the coarse ware, 45.4 percent of all the slipware, 53.7 percent of all the tin-glazed earthenware, 15.4 percent of all the stoneware, and 6.8 percent of all the porcelain.

As a whole, probably the most interesting finding from this comparison is that more tin-glazed earthenware was found in Building 4/5 than in all of the other buildings combined, and that a large percentage of the slipware was also found in the complex. In contrast, coarse ware, stoneware, and porcelain are disproportionately higher in the other buildings. That so much decorated fine earthenware was, indeed, found in the complex again indicates that the structure (or at least part of it, namely Building 5) housed residents who had the means to acquire and use attractive and relatively expensive tableware, whether it be for their own private use, or as a way to attract customers to the commercial hospitality outfit that was probably working out of it, in Room 1. That proportionately less coarse ware and stoneware was found may be simply due to the fact that, as already noted, other materials were being used in the complex for cooking and storage purposes, as revealed by excavation. As to the porcelain: the total count at the site is somewhat misleading, since almost half of the sherds (n=125) were found in four 10-x-10-ft. excavation squares comprising Building 8. If these sherds are discounted, Chinese export porcelain at Building 4/5 increases to 15.6 percent.

TABLE 4.1. Ceramic Wares and Types by Analyzed Sherd Counts (n) and Relative Percentages (%) from the 17th-Century Context at Building 4/5

Coarse Ware	n	Slipware	n	Tin-Glazed Earthenware	n	Stoneware	n	Porcelain	n
English:		English:		English / Dutch	571	German:		Chinese	15
Redware	153	Staffordshire	64	Mediterranean:		Raeren/Frechen	46		
North Devon ware	38	North Devon	1	Spanish/Italian	6	Westerwald	6		
Border ware	24	Continental European:		Portuguese	1	English	5		
Continental European:		Italian	14	Mexican:		Dutch/SE Asian	18		
Iberian	48	North Holland (?)	1	Mexico City ware	9	Unknown (Iberian?)	1		
Italian (?)	1			Puebla ware	1				
African-Jamaican Yabba	125								
Unknown (English?)	7								
Total	**396**		**80**		**588**		**76**		**15**
% Total Analyzed Sherds	**34.4**		**6.9**		**50.9**		**6.6**		**1.3**

TABLE 4.2. Ceramic Wares and Types by Estimated Minimum Number of Vessels (MNV) and Relative Percentages (%), as Represented by the Analyzed Sherds from the 17th-Century Context at Building 4/5

Coarse Ware	MNV	Slipware	MNV	Tin-Glazed Earthenware	MNV	Stoneware	MNV	Porcelain	MNV
English:		English:		English / Dutch	51	German:		Chinese	6
Redware	20	Staffordshire	8	Mediterranean:		Raeren/Frechen	6		
North Devon ware	3	North Devon	1	Spanish/Italian	4	Westerwald	3		
Border ware	7	Continental European:		Portuguese	1	English	4		
Continental European:		Italian	1	Mexican:		Dutch/SE Asian	2		
Iberian	5	North Holland (?)	1	Mexico City ware	2	Unknown (Iberian?)	4		
Italian (?)	1			Puebla ware	1				
African-Jamaican Yabba	14								
Unknown (English?)	2								
Total	**52**		**11**		**59**		**16**		**6**
% Total Analyzed Sherds	**36.1**		**7.6**		**40.9**		**11.1**		**4.2**

TABLE 4.3. Ceramic Wares by Sherd Counts (n) and Percentages (%) from the 17th-Century Context at Building 4/5 and Other Excavated Buildings at Port Royal, Jamaica (Hamilton's Excavations, 1981-1990)

Ceramic Ware	**Building 4/5**		**Other Excavated Buildings**		**Total Port Royal Assemblage**	
	n	% Total	n	% Total	n	% Total
Coarse Ware	396	27.4	1048	72.6	1444	42.1
Slipware	80	45.4	96	54.5	176	5.1
Tin-Glazed Earthenware	588	53.7	507	46.3	1095	32.0
Stoneware	76	15.4	417	84.6	493	14.4
Porcelain	15	6.8	206	93.2	221	6.4
Total	**1155**		**2274**		**3429**	

CHAPTER V

THE DOMESTIC WORLD OF PORT ROYAL IN THE 17TH CENTURY

So that the ceramic vessels recovered from Layer 3 at Building 4/5 can be better understood, this chapter presents a brief look at two aspects of the domestic world of the 17th century. Most ceramic vessels were used within the food domain, and so an emphasis on the culinary sphere is appropriate. Seventeenth-century health and personal hygiene practices are also discussed, since pottery vessels that were made for such purposes occur in Building 4/5's analyzed assemblage.

The diet of Englishmen and women in the early modern period provided the basis for the cuisine of the early Anglo-American colonies, and it is reasonable to assume that, for the most part, much of what was being consumed in England was also being consumed in its Jamaican colony, since many of the island's foodstuffs were imported. Preparation and storage of food and drink in both countries would almost certainly have followed similar patterns. Ideas about personal cleanliness and medicinal treatments would also have coincided.

Fortunately, diet and hygiene changes in 17th-century England can be identified more clearly than for preceding periods. As historian Joan Thirsk (1999:13) notes: "more and more books were being printed at this time, and as food is a subject of never-failing interest, references to it began to appear in literature discussing medicine and health." Fortunately for this study, too, there exists an extensive collection of primary documents pertaining to life in Port Royal in the late 17th century. The writings of Sir Hans Sloane, John Taylor, Richard Blome, Francis Hanson, and Edward Ward all provide valuable data regarding the types of food and drink available and the state of public health in the port.

FOOD PROCESSING (PREPARATION AND COOKING)

"Diet," according to English physician Thomas Muffett (1655), is an orderly course of nourishment "for the preservation recovery or continuance of the health of mankind" (cited in Caton 1999:29), and for most 17th-century households, "obtaining the day's daily meal and gaining nourishment from it was the primary objective of activities carried out within the food domain" (Yentsch 1991:43). This objective did not require elaborate preparation, and "it was usually not practical in a kitchen that was in essence only the enlarged base of the chimney" to prepare complicated meals (Root 1958:41). This is not to say, however, that cooking in early modern times produced flavorless and monotonous dishes, "full of strange compounds . . . . a riddle" (Mead 1931:8). Rather, people of the period were familiar with a wide range of foodstuffs and seasonings and had strong opinions about the flavor and quality of what they ate.

From at least the mid 16th century, in England at least, as long as the population at large "had a roof over their heads and a kitchen fire, [they] were able to eat foods of great variety without spending much money" (Thirsk 1999:13). By the 17th century, this variety incorporated all kinds of meats, barnyard fowl and game birds (including sparrows), as well as all manner of fish. Vegetables, herbs, and fruits, earlier "thought fit only for the poor and for those who chose the monastic life" (Thirsk 1999:16), were now conspicuous at most tables, and people appreciated the various breads that were readily available. Although still relatively expensive and available only to the wealthy, imported 'exotic' goods were widespread by the 17th century: eastern spices were used in many a cook's kitchen, and the "pleasing, delicate taste of cane sugar" (Root and Rochement 1976:83) appealed to everyone's sweet tooth.

It would appear that 17th-century Port Royalists did not want for variety of food either. This is not surprising given that they lived in a thriving merchant town at the center of the Atlantic trade. Visitors to Jamaica note that there were plenty of cattle, sheep, and goats bred in pastures (Blome 1678:10; Hanson 1683:5). Large "stocks of Hogs, Hens, Ducks, Pigeons, Rabbets, Turkeys, and divers kinds of wild Fowl" were also a common sight (Hanson 1683:5). Richard Blome (1678:10-11) writes of the many and varied fish, "the principal sort [being] the Tortoise." Garden herbs and roots, "Pulse and Sallads, and . . . Fruits, some known here as Grapes . . . . Lemons, Oranges, Citrons, Pomgranats, and Musk Mellons" were also available aplenty (Hanson 1683:5). John Taylor describes in detail Port Royal's three daily markets:

> The one an Herb and fruitt Markett held in the hart of ye high street, where stands ye stocks, and market bell, to this Markett is brought plenty of Herbs, fruitts, and fowels fresh every Morning, unless on the Sabath day; The Second market is for fleash, and Turtle, this Market is keept at the Wester end of ye high street, nigh Chocolata hole: over this Markett there are Overseers as Judge Whitle. This Markett is plentifully stored with Beef, Mutton, Hog, Veal, Lamb, Kid, and Tartoise, and the Chief times of Market is in the cold of ye Morning, and Eavenings. The third Market held here is for fish, which is keept on ye Wharf nigh the Wherry bridg: Where is every morning plenty and Variety of excellent fish, verey Cheap (Taylor 1688:253-254).

Some of the fish Taylor notes included salted codfish, mackerel, salmon, and sturgeon, imported from England and North America. Other imported food included beef, pork, bacon, cabbages, peas, onions, garlic, apples, cranberries, biscuits, cheese, and butter (Taylor 1688:261; Pawson and Buisseret 1975:67).

Cooking in Port Royal, as elsewhere in the colonies and in Europe, was usually done over an open fire or over coals in a hearth. But unlike many houses of the time, the Port Royal kitchen, or 'cook room,' was usually located at the rear of the house or in a separate, small brick building at the end of the yard, so that the heat and threat of fire would be at a safe distance from the dwelling.

In households that could afford them, ovens were also used. Made of brick, they were simply built into the side of the fireplace or hearth and heated with a fire. The ashes were then scraped out before the bread or meat was placed inside (Root and Rochemont 1976:77; Hess 1995:19). Heated baking stones were also used to cook food, and there is frequent mention of them in the Port Royal household probate inventories.

Several pieces of cookery equipment were basic to a well-furnished 17th-century kitchen: 'kettles' (broad-mouth cooking pots) for simmering and boiling, skillets for frying and sautéing, a turnspit for roasting, 'pottles' for measuring liquids, and bowls and basins for mixing. "Braising dishes (from the French braise, ember) ingeniously designed with concave lids into which embers could be heaped, would be nested in the ashes to one side of the hearth in such a way that the savory stews contained therein could sit for hours at an imperceptible simmer, virtually untended" (Hess 1995:21). A 'chafing dish of coles' (i.e., a portable grate or brazier mounted on a tripod) would be used "for fussy bits of cooking" that were not feasible in the great heat of the fireplace (Hess 1995:22).

While most of the implements associated with the hearth were, by the 17th century, made in metal, ceramic utilitarian ware was also still used in many households (Deetz 1973:25, 1977:52; Yentsch 1991:31). Glazed ceramic cookware was particularly recommended over metal containers for the preparation of foods with delicate flavors, such as preserved fruits and jellies. Earthen pots also gave an even distribution of heat, and were more reasonably priced than were pots of pewter, iron, or copper (Wheaton 1983:110; Schaefer 1998:21).

FOOD AND BEVERAGE STORAGE

For most 17th-century European households, "it was necessary, or at least prudent to store many foodstuffs, especially for the winter months when the availability of fresh produce declined and prices rose" (Schaefer 1998:36). This would have been also true for the colonists in Jamaica despite its extended growing season, since many of the foods, as noted, were imported from northern Europe and the northeastern American coast. As Schaefer (1998:36) notes: "The piles of wooden barrels, baskets, ceramic and metal pitchers, jugs and other containers, so often shown in paintings of [domestic] interiors were not simply attempts by artists to demonstrate their virtuosity . . . . A major portion of most dwellings was set aside for storage."

Storing provisions required a cool dry space that was both insect- and rodent-proof. The cheapest and most available material in the 17th century was wood, and the wooden casks, hampers, boxes, and barrels listed in numerous Port Royal household probate inventories were no doubt used to store the various consumables available in the city's markets. The wooden containers would have also been used to hold foodstuffs that had been preserved, although as entries in various inventories illustrate—'17 barrells of decayed pork, 7 halfe barrells of pickled pork, 1 bad barrell ditto, 1 goodel badd cocoa' (Jamaica Public Archives [JPA] 1689, 3:320)—food and drink preservation was not always assured. Glazed earthenware pots, while not used for storing salted or pickled foodstuffs, which might have caused the lead glaze to decompose, were valuable as storage vessels because they would provide protection from dirt and damp and animals (Schaefer 1998:36). Vitrified stoneware allows for indefinite storage, and these pots may have been used as an alternative to wood for salted or pickled foodstuffs.

The preservation of food laid in store changed little until the beginning of the 19th century, and methods used were identical or very similar throughout the western world. Meat rubbed with salt, or vegetables sprinkled with salt, were packed into casks "and compressed by means of a board or stone, so that juice flow[ed] out," which not only preserved the foods but also produced a characteristic taste (Horändner 1986:54). Meat and fish were also laid in pickle or brine, a liquid containing salt with admixtures of sugar. Often, pickling (especially of pork) preceded smoking or curing, another method of preservation used exclusively for meat and fish (Horändner 1986:55).

Fruits could be heated and stored in vessels (usually earthen, since metal discolors fruit) for use as preserves and marmalade (see, e.g., Hess 1995:228-268). One interesting method for preserving cherries for tarts was found in a 17th-century family recipe book, which recommended storing the fruit in a barrel of hay placed "under a fether bead where one lyeth continually" (Hess 1995:162). (Interestingly, the French still use hay or straw in this way, although more often in the cellar.)

FOOD CONSUMPTION (EATING AND SERVING)

The period from about 1650 to the beginning of the 19th century was one of fundamental change in popular eating habits. Among these were manners at table, particularly of those who could afford to adopt the new standards of etiquette. The development in the Near East of utensils such as the dining fork, which was introduced to England at the beginning of the 17th century, took over the spearing duties of the knife. Eating and serving vessels, including the adoption of plates for the earlier shallow bowls or wooden trenchers, were added to the stock of basic household equipment (Schaefer 1998:44; Yentsch 1990:37-40, 1991:39-41). The middle classes at the end of the 17th century were also beginning to add to their houses special rooms for dining. In a well-to-do 17th-century English house, meals were served at the table, either in "the main room, called the hall [the descendent of the medieval 'great hall'] . . . or in the second room, newly termed the parlor, after the French word, parler, 'to converse'" (Belden 1983:5). Diners would also set their tables in the upstairs bedroom ('chambers') "and kept their eating and drinking vessels in upstairs or downstairs cupboards, wherever convenience and the season suggested" (Belden 1983:5-6). Port Royalists were no different from their English cousins with regard to this, as evidenced by the probate inventory of Charles Cresso, who kept '6 glass cupps, Dishes Potts and Plates' and '3 brass kettles' in his 'lodging chamber' (JPA 1688, 3:217), or Jobe Newberry, whose 'chamber above staires' contained '6 pottengers [porringers], 13 spoons, and a great tankard' (JPA 1689, 3:335).

As a result of this new emphasis on food as social display, which was not restricted to the homes of the wealthy—"rich or poor, a man's goal was a groaning board [loaded table]" (Belden 1983:5)—the number of eating and serving vessels in a household began to increase, and the materials of which they were made became a visual marker of social status. Seventeenth-century genre paintings, such as that illustrated in Figure 5.1, show that with the less affluent, vessels were mostly of wood or earthenware, but some were decorated and even tin-glazed. In contrast, depictions of affluent households show pewter, tin-glazed wares, brass, silver, gold, and porcelain, along with a great variety of specialized vessels, such as salt holders, condiment holders, and display pieces, such as large dishes, or 'chargers' (Figure 5.2).

BEVERAGE CONSUMPTION (DRINKING AND SERVING)

Beverages in the 17th century were as varied as the food. Innumerable alternatives were available to all classes, and every household must have had its own favorite flavors and mixtures. Many involved no expenditure of money but called only for the housewife's effort, interest, and skill in combining herbs or spices to add agreeable flavors to water, ale, and beer (Thirsk 1999:22). In fact, ale and beer were probably the most popular drinks during the 17th century in England and northern Europe, and many manuals, such as those by Thomas Tryon (1691) and William Y-Worth (1692), offered home-brewing tips (see also Markham 1986[1615]:205-208). Certainly, they had been the cheap drink of the masses since the Middle Ages, and only milk and buttermilk, which was strongly "commended to the young, the old, and the sick," offered a little competition (Thirsk 1999:21; see also Schaefer 1998:67). Ale and beer were also considered safer beverages at a time when good drinking water was not assured. Their putative health benefits meant they were often recommended by physicians (Schaefer 1998:67; Thirsk 1999:22).

FIGURE 5.1. 'Prayer Before the Meal,' by Jan Steen, 1660. The Walter Morrison Collection, Gloucester. The Westerwald stoneware jug and the tin-glazed earthenware serving dish contrast sharply with the wooden bread board and plain furnishings of the room.

It is hardly surprising to learn, then, that beer and ale were packed with the rest of their cargo as European settlers began their lives in the New World. Root and Rochement (1976:360-361) note that the first breweries were established by the Dutch in New Amsterdam in the 1620s, using imported raw materials. Hops were planted in New England and Virginia within the next 20 years.

While Jamaica provides neither the soil nor the climate to produce much of any cereal, least of all barley for beer, its colonial inhabitants, according to John Taylor (1688:262), did not want for "good English Bear, and Mum [German wheat beer]." Interestingly, Sir Hans Sloane (1707:28) was not so enamored with Jamaica's imported brews: "the Beer is often sowerish, and the Ale is generally too sweat and heavy, the one too old, and the other not well wrought." Cider, which by the 1640s had become popular among the English gentry, was also noted by Sloane (1707:28) as being spoiled by the Jamaican heat.

It appears, rather, that wine, especially "good Madera Wine both reed and white," which could be bought in Port Royal "att a reasonable rate," was the drink of choice for many (Taylor 1688:261). Madeira was apparently

FIGURE 5.2. 'Easy Come, Easy Go,' by Jan Steen, 1661. Museum Boymans-van Beuningen, Rotterdam. Gold, silver, and pewter jugs and other serving vessels and stemmed glassware are in harmony with the rich surroundings of the affluent household.

favored over the "Canary, Whitwin, Rhenish [and] Clarrat" that were also available (Taylor 1688:262), likely because its heavy sweet flavor accorded with the taste of the times. Further, as Sloane (1707:28) notes, unlike the "French Wines, and all others coming hither, [Madeira wine] keeps better in a hot Place, and expos'd to the Sun, than in a cool Cellar."

Root and Rochemont (1976) give an interesting perspective on the popularity of Madeira wine throughout the Anglo-American colonies:

> In pre-Revolutionary days, when the colonists deeply resented British laws requiring that all European goods going to America should be shipped in British bottoms, the island of Madeira (whose wine was not European merchandise, since Madeira lay off the coast of Africa) was exempt from this restriction. Thus a bottle of Madeira became a symbol of defiance of the King's oppressive interference with the Colonies' right to trade (Root and Rochemont 1976:364-365).

While Port Royalists also enjoyed sherry, port, and various brandies, and a "cold drink called Rumm punch," a particular favorite among the island's planters (Taylor 1688:262), many other beverages were available. We learn from Dr. Sloane (1707:10, 27) that the most common drink was, in fact, water, which if collected from the island's springs and stored "in earthen Jars" to allow the sediments to settle at the bottom, was perfectly good to drink and was great for the "Belly-ach." The new hot drink of chocolate (introduced to England in 1657), continues Sloane (1707:20), was also enjoyed in Jamaica "by all People, at all times." Several Port Royal household probate inventories (e.g., JPA 1688, 3:112; 1689, 3:320; 1693/4, 3:600) have entries of tea and coffee, the other new fashionable drinks. As Thirsk (1999:23) comments: "Drinking these beverages went hand-in-hand with animated conversation in public houses where one could also pick up news of the latest projects in town." By the 1670s and 1680s, in England at least, chocolate, coffee, and tea were being freely drunk at home.

It seems that the initial appeal of these new hot liquids lay not only in their novelty but in the belief that they held medicinal properties. Coffee, for example, is described in an anonymous poem of 1672 as "Ver boon for de stomach, de Cough, de Phisick" (cited in Caton 1999:51). It was also considered a cure for drunken hangovers and was even thought to have the ability to ward off plague or to dispel noxious odors (Ukers 1935:54-55). Chocolate, notes English physician Henry Stubbe (1662:44), "satisfies, and cools the body [and is] not intoxicating in any way the Head," and like coffee, could be sweetened with sugar or vanilla, or mixed with milk, spirits, or a claret solution. (In fact, Dr. Stubbe traveled to Jamaica as physician to Lord Thomas Windsor, governor of the island, in 1662. While there, as he wrote later, he left his own chocolate recipe with a trusted "poor man, Richard Mortimer, in Sun-Alley in East Smith-field," should any other physicians wish to use his recipe [Stubbe 1662:A4-5].) Tea apparently could "hinder the gout and the gravel in the kidneys" (Dufour 1685:40); when prepared with milk and water, "it strengthened the inward parts and removed pains from the bowells" (Garway 1660, cited in Caton 1999:55).

HEALTH AND PERSONAL HYGIENE

Port Royal by the 1680s was no more unhealthy a place than most European large towns of the late 17th century. "There is no such Antipathy betwixt the constitutions of the English, and this clime, for the occasioning Sickness to be Mortal or Contagious more than in other parts," notes Richard Blome (1678:12) in his survey of Jamaica. In their history of Port Royal, Pawson and Buisseret (1975:102) add that the "air was generally accounted very healthy, and children particularly seemed to thrive." This is not to say, however, that the port was free from maladies. John Taylor (1688:268) notes in his journal "Augus [sic], Fevers, Plurisies, Callentuers, Pox Fluxes, Yauws, & the Illiaco Passio (here called ye dri Reyach) [lead poisoning] which deprives many of there strength, and use of Limbs for many years & some for ever." Sir Hans Sloane (1707:122) writes in great detail of 'Colick,' the 'Gout,' consumption, worms, and distemper, the last of which was "usually cur'd by an easie Vomit, or Purgers." Bacillary dysentery and tuberculosis appear to have also been prevalent (Pawson and Buisseret 1975:101, 102).

Much of the reason, of course, for the prevalence of disease at this time lay in the lack of an understanding of the connection between personal cleanliness and good health. The simple act of washing was not generally considered part of one's toilet in the 17th century (water was believed to weaken the organs and leave the pores open to unwholesome air). Cleanliness, instead, was 'created' with cosmetics and perfumes (Vigarello 1988:9-17, 133-141; Schaefer 1998:90).

> Bathing was practiced for its therapeutic benefits or to satisfy curiosity, as at Bath, where Samuel Pepys spent over two hours standing in the hot water in 1668. He still had reservations, remarking, 'methinks it cannot be clean to go so many bodies together in the same water.' Queen Elizabeth bathed once a month, and in between used scented water to wash her face and hands, and it was common practice to perfume gloves, handkerchiefs and other clothing. Hair received little attention. In 1600, William Vaughn's Artificial Directions for Health advised the reader to 'bath your head foure times in the yeere and with hot lie made from ashes' (Schaefer 1998:90).

The concept of public health, or 'collective sanitation,' was also unknown, and most 17th-century cities in Europe lacked appropriate methods for the disposal of waste. "Privies [cesspits] of several types existed to fill human needs, or be filled by them," notes Schaefer (1998:91). Bucket privies, which could be emptied in areas outside the town, were another option. The alternative was the chamber pot, which made its appearance in Europe during the 14th century. The earliest models were made in various metals—tin, pewter, copper, silver, and even gold, as shown in Dutch painter's Michiel van Musscher's 'Doctor Taking a Young Woman's Pulse,' of ca. 1670-1680 (Private Collection, USA) (illustrated in The Philadelphia Museum of Art Catalogue 1984:Plate 124). Those made in earthenware were in common use by the early 15th century (Wright 1960:122; Amis 1968:9-11).

The advantage of the chamber pot was that it could be kept near or under a bed at night (although it must not be assumed that they were only used at night), surely a welcome reprieve from the solitary night-walk to an outside closet (Amis 1968:7; Schaefer 1998:91). Unlike the earlier jugs, jars, urinals, or 'jordans,' the chamber pot's squat, low shape meant it was difficult to knock over. Further, it could be slid around the floor without fear of the contents sloshing over the sides (Wright 1960:120: Amis 1968:7, 9; Janssen 1983:169). Of course, as Schaefer (1998:91) notes, the drawback of the chamber pot was that the smell remained until it was emptied. In his Directions to Servants of 1745, Jonathan Swift complained about ladies who were too lazy to use the outdoor privy; instead, they "keep an odious implement [chamber pot] sometimes in the Bed-Chamber itself, or at least in a dark Closet adjoining . . . which maketh not only the Chamber but even their Cloaths offensive" (cited in Palmer 1973:21). The 'night soil' was simply ejected from an upstairs window, a practice, indeed, still observed in Edinburgh in 1750. Likewise in France: "The chambermaids at the Louvre, like everybody else, soiled the façades by throwing the contents of the night-commodes out of the windows" (Braudel 1973:437-438, cited in Palmer 1973:21). "City streets were common sewers, where pedestrians were inevitably covered in noisome mud" (Schaefer 1998:91).

The street sanitation situation apparently was none too different in colonial Port Royal. Its position, at the end of a long, flat peninsula surrounded by salt water made it impossible for a formal sewage system to be installed. The citizens' idea of an appropriate place for the disposal of waste did not sit well with some of its visitors:

> In the Afternoon, about Four a Clock, they might have the refreshment of a Sea-Breeze, but suffering the Negroes to carry all their Nastiness to Windward of the Town, that the Nauseous Effluvias which arise from their stinking Dunghills, are blown in upon them; thus what they might enjoy as a Blessing, they Ingratefully pervert buy their own ill Management (Ward 1700:15, emphasis original).

THE BUILDING 4/5 COMPLEX

It is worth noting at this point (à propos the foregoing brief discussion) that in relation to food preparation, Building 5 contains the remains of a brick oven built into the brick hearth in Room 4 (the 'cook room'), and that cook-room foundations, each containing a brick hearth, were also located at the end of the yards associated with Building 4. In relation to food storage, a large house like Building 4/5 probably had a buttery, a cellar (at ground level because of the high water table at Port Royal), and a pantry in which pots and pans and other equipment, as well as food and drink, would have been stored. The remains of a privy in the northwest corner of Yard 5 and the recovery of several chamber pots from throughout the complex suggest that its occupants also probably dumped their waste outside the structure.

CHAPTER VI

FORM, FUNCTION, AND DISTRIBUTION OF THE ANALYZED VESSELS

Prior to the second half of the 20th century, archaeologists did not generally interest themselves in how the ceramic vessels they find were actually used. There are several reasons for this: before the 1960s, most archaeological research focused on problems of a culture-historical and chronological nature, where knowledge of vessel function was largely irrelevant. Second, most archaeological ceramic assemblages contain only a few whole or reconstructable vessels (or they are unrepresentative of the original inventory, e.g., unreliable contextual data, sampling strategies), whereas vessel use is most effectively investigated with whole vessels (Hally 1983:3; Rice 1984:246, 1987:208, 232-233). Third, ceramic surfaces may be modified by occurrences unrelated to actual use. Fluvial abrasion, salt erosion, and chemical changes in the depositional environment, for example, seriously undermine the process of inferring vessel use from pottery sherds (Rice 1987:235; see also Schiffer 1987).

Nevertheless, it is the case that interest in ceramic function has grown steadily in recent years to the extent that now many methods to determine function are used by archaeologists. The physical properties of wares are examined, including fabric composition, shape, size, and surface treatment (e.g., glazed vs. unglazed, decorated vs. undecorated) (see Rice 1987:215-232 for a review). Studies of living pottery-using communities focus on "ceramic censuses, or the number of pots in individual households, and on the use life or longevity of the censussed vessels" in an attempt to understand the relationship between material culture and human behavior (Rice 1987:293). Experimental replication "to test the accuracy and reliability of archaeological recovery" is also increasingly employed in ceramic-use research (Skibo 1992:18; see also Coles 1973, 1979). Use-wear analysis seeks to infer use activity by examining changes in the ceramic material (traces of residues left by the contents of vessels and abrasion marks from the use of a utensil, for example) (see Griffiths 1983; Halley 1983; Skibo 1992).

In relation to the possible functions of the ceramic vessels recovered from Building 4/5, this study relies upon (1) the morphological (and decorative) traits of the finds, with site provenance data providing supplementary information; and (2) sources of information independent of the archaeological record. In the first case, each data base sherd was identified to form (where possible) according to a vessel form typology established by Beaudry et al. (1983). Known as the Potomac Typological System (POTS), the scheme assumes that the shape of a vessel is related to its technological or primary function (as opposed to social or ideological functions, see Binford 1962). Of the data base sherds, those identified in the pre-1692 context (i.e., the 'analyzed sherds') were thereafter focused on by the author and grouped into one of several categories according to their assumed domestic or culinary role. These categories, outlined in Chapter III (Table 3.1), include food processing, food/beverage storage, food and beverage consumption, and personal hygiene/health. A sixth 'other' category comprises those vessel forms that could not be identified or whose 'function' was unequivocally ornamental.

In the second case, Jamaican household probate inventories from the late 17th century provided valuable insights into the material culture of the Port Royal community. Contemporary English pottery inventories helped to identify how ceramic vessels might have been used. Catalogues of museum collections were also helpful in this regard. Visual evidence in the form of 17th- and 18th-century European paintings of domestic interiors is also discussed; as a record of the here and now, the quality of social document provided by this genre of art is unequaled.

The author was able to catalogue to form (using POTS) 80 percent (n=924) of the analyzed sherds (n=1155). This represents an estimated minimum of 144 vessels. The following discussion orders the various forms according to the above mentioned functional groups. It also looks at the distribution of the forms throughout the building complex. A summary of the analyzed collection is presented at the end of the chapter.

FOOD PROCESSING VESSEL FORMS

The Building 4/5 analyzed assemblage contains a total of 65 sherds (5.6 percent of the analyzed sherds), which the author has catalogued as belonging to forms used in food processing. These sherds represent an estimated minimum of 11 vessels, or 7.6 percent of the total number of ceramic vessels taken to be associated with the building's pre-1692 context. The vessel forms represented are cooking pots (n=8), a colander, a pipkin, and a pudding/pastry pan, and all except one are made in plain coarse earthenware, or coarse ware. The exception, the slip-decorated pan, has a brown and yellow mottled finish (known as Mottled ware).

The relative proportions of ceramic food processing forms and their respective sherd counts are shown in Table 6.1. Each form and its distribution is now discussed.

Cooking Pots

A 'cooking pot' is here used as a generic term for a vessel, generally globular in shape, that was likely used on the hearth for the preparation of food. Identification of this form assumed vessels used in cooking would show patterns of sooting or blackening on exterior surfaces. Additionally, those surfaces usually show little or no decorative and shape: vessels might be relatively thin-walled (to conduct the heat from the cooking fire more efficiently), and rims would be unrestricted for easy

TABLE 6.1. Ceramic Wares, Forms, and Functional Classifications, as Represented by the Analyzed Sherds (n) and Estimated Minimum Number of Vessels (MNV) from the 17th-Century Context at Building 4/5

Vessel Form and Functional Classification	Coarse Ware		Slipware		Tin-Glazed Earthenware		Stoneware		Porcelain		Total
	n	MNV	n	MNV	n	MNV	n	MNV	n	MNV	MNV
Food Processing											
Cooking Pot	49	8	--	--	--	--	--	--	--	--	8
Colander	1	1	--	--	--	--	--	--	--	--	1
Pipkin	7	1	--	--	--	--	--	--	--	--	1
Pudding/Pastry Pan	--	--	8	1	--	--	--	--	--	--	1
Subtotal	**57**	**10**	**8**	**1**	**--**	**--**	**--**	**--**	**--**	**--**	**11**
Food/Beverage Storage											
Storage Pot	200	23	--	--	--	--	2	2	--	--	25
Bottle	--	-	--	--	1	1	29	7	--	--	8
Subtotal	**200**	**23**	**--**	**--**	**1**	**1**	**31**	**9**	**--**	**--**	**33**
Food Consumption											
Bowl	75	10[a]	5	2	17	2	--	--	10	2	16
Plate	--	--	--	--	53	13	--	--	--	--	13
Porringer	5	2	3	1	25	2	--	--	--	--	5
Saucer	--	--	3	2	5	2	--	--	1	1	5
Dish	--	--	--	--	17	1	--	--	--	--	1
Salt Stand	--	--	--	--	13	1	--	--	--	--	1
Subtotal	**80**	**12**	**11**	**5**	**130**	**21**	**--**	**--**	**11**	**3**	**41**
Beverage Consumption											
Cup	--	--	33	4	80	9[b]	--	--	2	2	15
Drinking Pot	--	--	--	--	40	5	--	--	--	--	5
Mug	--	--	--	--	--	--	15	5	--	--	5
Costrel	1	1	14	1	--	--	15	2	--	--	4
Ewer	7	1	--	--	--	--	--	--	--	--	1
Punch Bowl	--	--	--	--	5	1	--	--	--	--	1
Tea Bowl	--	--	--	--	--	--	--	--	2	1	1
Subtotal	**8**	**2**	**47**	**5**	**125**	**15**	**30**	**7**	**4**	**3**	**32**
Health and Hygiene											
Apothecary Pot[c]	--	--	--	--	39	16	--	--	--	--	16
Chamber Pot	11	4	--	--	40	2	--	--	--	--	6
Washbasin	--	--	--	--	13	2	--	--	--	--	2
Subtotal	**11**	**4**	**--**	**--**	**92**	**20**	**--**	**--**	**--**	**--**	**24**
Other											
Flower Vase	--	--	--	--	75	1	--	--	--	--	1
Unknown Vessel	1	1	--	--	2	1	--	--	--	--	2
Subtotal	**1**	**1**	**--**	**--**	**77**	**2**	**--**	**--**	**--**	**--**	**3**
Unknown Sherds	39	--	14	--	163	--	15	--	--	--	--
Total	**396**	**52**	**80**	**11**	**588**	**59**	**76**	**16**	**15**	**6**	**144**

[a] n=8 Yabba may have been used in preparatory functions
[b] n=3 caudle cups
[c] one of the apothecary pots may be a ginger jar or vase

adding and removing of food (Rice 1987:237-239, Table 7.2). Obviously, these criteria are general guidelines only to help in identification of sherds. Very few of the cooking pots identified in the assemblage have soot deposits, for example. They are thinner bodied only in relation to the analyzed storage pot sherds.

Most of the sherds categorized as belonging to cooking pots are undecorated English Redware, which the author estimated to represent at least four vessels, one of which was recovered intact (Figure 6.1*a*). The remains of four African-slave-made Yabba cooking pots were also identified, based upon attributes discussed in Chapter IV (see Figure 6.1*b-e*).

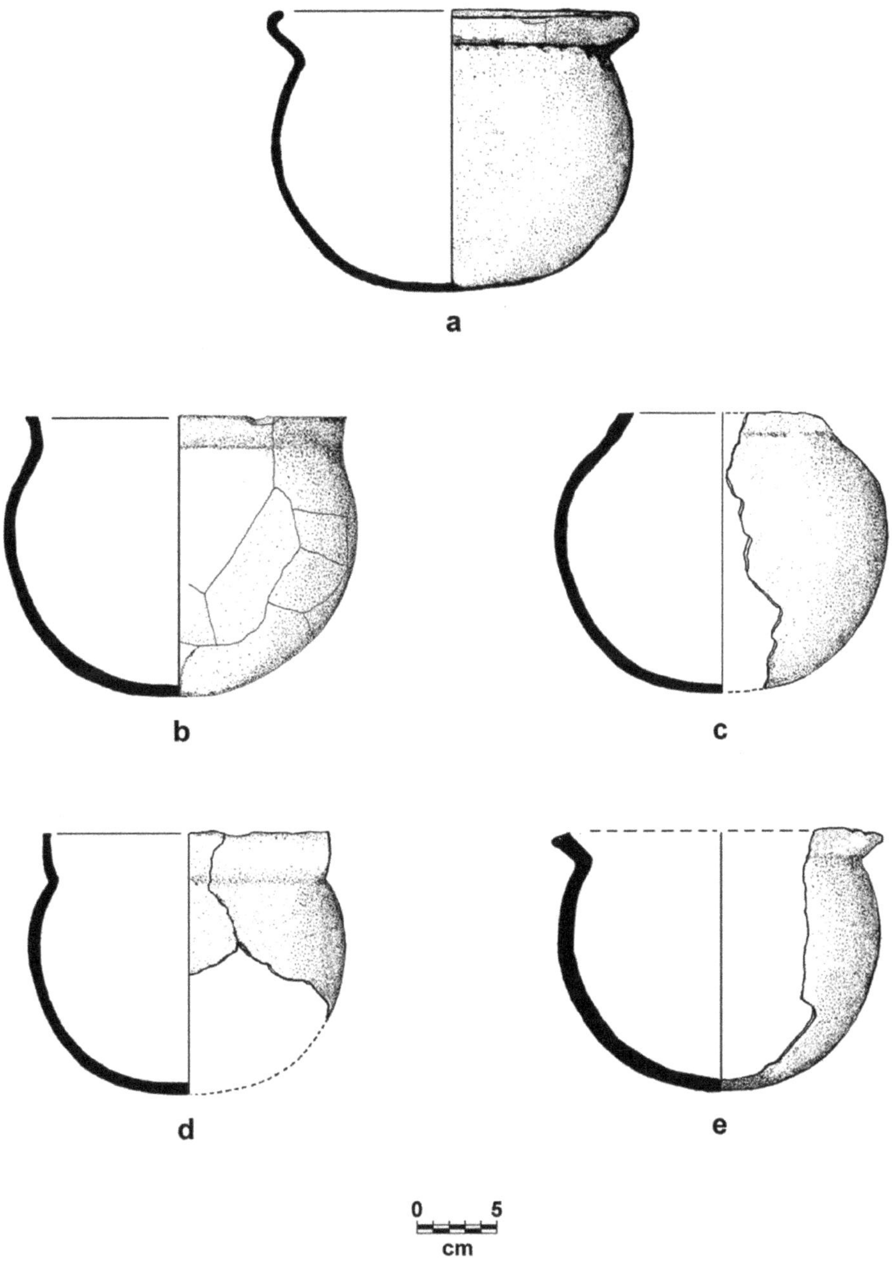

FIGURE 6.1. Examples of cooking pots recovered from the 17th-century context at Building 4/5. *a*, intact English Redware form found in Yard 6; *b*, reconstructed African-Jamaican slave-made (Yabba) form found in Yard 4A; *c-e*, remains of Yabba forms found in Yards 4A and 4B. (Illustration by Madeleine J. Donachie.)

While some of the Redware cooking pot rims have an internal ledge, evidence that they could accommodate the fitting of a lid, no ceramic lids for cooking pots were found. A wooden lid associated with a cast-iron cooking pot was, however, recovered from Yard 5, and it may be that more lids of this material were used and simply have not survived. Indeed, lids at this time were often made of wood. In Pieter de Hooch's 'Women Peeling Apples' (ca.1663, Wallace Collection, London), for example, an iron pot with a wooden lid hangs over the fire. In Frans van Mieris' 'The Peasant Inn,' (1655-1657), illustrated in Figure 6.2, a wooden lid rests on the floor in the lower right of the painting.

Most of the cooking pot fragments were recovered from Yard 4A, in association with the hearth, which would have been enclosed within a kitchen, or 'cook room.' (A cross-mended Redware rim was found in Room 1. A second Redware rim with a handle was found in the alley. The intact Redware specimen was found in Yard 6.) As noted in Chapter V, cook rooms in Port Royal were usually located at the rear of the dwelling, as is the case with Building 5, or in a separate brick structure at the end of the yard, as is the case with Building 4.

It is entirely possible that vessels used for cooking have not been so recognized due to their not being of plain red or brown coarse ware. Also, that a vessel has no carbon (soot) deposits does not necessarily mean that it was not used for cooking. Carbon deposits do not always adhere to glazed surfaces, for example. Steaming food in pots placed in larger cooking vessels would also have left pots soot-free (McCarthy and Brooks 1988:105).

Colander

A rim and body fragment from a dark green (Munsell 2.5Y 5/4-5/6), lead-glazed English Redware colander was found at the entrance to Room 2. The vessel is typical of the inexpensive utilitarian kitchenware found throughout Europe, England, and North America from the 15th through 18th centuries. Randle Holme , in his dictionary of heraldic symbols written in the 1680s, which also gives much information about 17th-century objects and their uses, describes a "cullander" (from the French *couler*, 'to strain') as "Usefull for a cooke . . . having the bottome full of small round holes: in these Herbs, as such like things are washed, whose dirt and filth runs through the holes, leaving them pure and cleane" (Holme 1905 [1688]:14:11). Colanders were also used for draining moisture from fruits to make jams and desserts, and some 17th-century Dutch genre scenes show them containing fish, which required scaling, or shellfish, which needed to be rinsed free of sand and other grit (see, e.g., Figure 6.2). A 17th-century recipe book recommends sifting a boiled, grated hog's liver through a colander to make a liver pudding (Hess 1995:105).

Any or all of these functions may have been undertaken in the example from Building 4/5. The context of the vessel's recovery may, however, indicate that it was originally in Building 8, on the opposite side of the street, and was displaced by the earthquake/tidal wave.

FIGURE 6.2. 'The Peasant Inn,' by Frans van Mieris, 1655-1657. Stedelijk Museum, Leiden. A wooden lid with a strap handle and a glazed earthen colander filled with shellfish are shown in the lower right corner.

Pipkin

A pipkin is a round- to pear-shaped cooking pot, and two general shapes were in common use in the 17th century. The first, a small, bulbous vessel, usually has a rod handle and often three short legs. The second, a larger vessel, more like a cauldron or 'kettle,' generally has two ears and also three legs (Beaudry et al. 1983:34). The legs allowed the pot to be set in the fire. The handle kept the cook's hands out of the fire.

Fragments from a glazed White Border ware pipkin were recovered from Room 2 and Yard 5. The vessel is represented by a cross-mended basal sherd (9.5 cm diameter), which has attached to it the characteristic three short legs, and several cross-mended rim sherds (14 cm diameter) (Figure 6.3). Soot deposits cover the basal fragments, evidence that this vessel was, indeed, used in the fire. The rim sherds reveal that the pipkin was designed to be used with a lid.

Since the pipkin is relatively small, it may have been used for making sauces. Simple sauces were one of the attractions of the then nouvelle cuisine, which became widely popular during the 17th century. Alternatively, its small size and the fact that not only is it White Border ware, which was more expensive than the red variety, but that it is covered in an aesthetically pleasing brightly colored yellow (10YR 6/6) interior lead glaze, suggests that it may have been used at table also, as an individual eating vessel, perhaps as one would use a bowl. Indeed, many 17th-century Dutch paintings show people eating directly from small, pipkin-type vessels. In Nicolas Maes' 'Prayer without End' (Rijksmuseum, Amsterdam), for example, the vessel is set on a wooden disk (to protect the white tablecloth), and the viewer infers that the bottom of the pot may be either sooty from the cooking fire or simply hot. Maes' 'Prayer Before the Meal' (1648, Louvre, Paris) similarly shows a pipkin on a man's lap, with a disk between its base and his pants, protecting him from dirt and/or heat. Gabriël Metsu's 'The Sick Child,' illustrated in Figure 6.4, shows a pipkin-like vessel, with a spoon placed inside it, on the table beside the bed.

FIGURE 6.3. White Border ware pipkin recovered in fragments from Room 2 and Yard 5 in the 17th-century context at Building 4/5

It is interesting that the pipkin fragments were found in Room 2 and in the northeast corner of Yard 5, directly behind the rear wall of Room 1. Room 1 may have been used as a parlor or 'best' room due to the presence of much of the finer wares in the assemblage, the existence of plastered walls and a plastered floor, and the fact that it was located at the front of the building. It is also possible that it was used to serve food and drink to patrons on account of its separate entrance to the street. Room 2, also with its own front entrance, provided access to the rear of Building 5, as well as to an upper floor. An interior doorway connected Room 2 with Room 1. It appears that Room 2 was also used in part for storage (a collection of new clay tobacco pipes were found in the room, and stacks of pewter plates were found under the stairs).

Pudding/Pastry Pan

A single oval-shaped Staffordshire slipware pudding/pastry pan was recovered in fragments from the front of Room 2. It has been almost completely repaired and is in the shape of a modern, small casserole dish (20 cm long x 15 cm wide x 8.5 cm tall) with a slightly everted, rounded rim (i.e., not fitted for a lid), flat base, and nearly vertical sides (see Figure 4.8*d*). The vessel is covered on all surfaces (minus the base) with a dark reddish brown (5YR 3/3) slip splattered with manganese purple. The overlying lead glaze has produced a dense brown speckled effect on a yellow-brown background. Indeed, such decoration is known as Staffordshire Mottled ware, and it was produced in the Staffordshire potteries beginning in the late 17th/early 18th centuries. An almost identical vessel was recovered from excavations in London, where it was dated to ca. 1730-1775 (Thompson et al. 1984:52). The recovery of this form from the 17th-century context at Port Royal suggests an earlier date of production than that conjectured by Thompson et al. (1984).

FIGURE 6.4. 'The Sick Child,' by Gabriël Metsu. Rijksmuseum, Amsterdam. A small earthenware pipkin containing a spoon sites on the table at left.

From 17th-century pottery inventories, it appears that the word 'pan' (usually spelt 'pann') meant much the same as it does today, i.e., referring to a shallow cooking/baking vessel, usually without a cover. Various modifiers were placed before the word, signifying specific uses, such as 'pudding–,' stew–,' pidgeon–,' partridge–,' and 'pheasant–' (see Vaisey and Celoria 1974:25, 26; Britton 1990:72, 76). A cursory examination of 17th-century Jamaican household probate inventories revealed to the author that many people in the colony owned such pans. Madame Judith Freman's 'Cook-Roome,' for example, contained 'two Pye panns and 14 patty panns' (as well as 350 earthen pots) (Jamaica Public Archives [JPA] 1688, 3:105). 'Puding pans' and stew pans are among the stock enumerated for Port Royal butcher Robert Howard (JPA 1689, 3:317), as well as for Samuell Coulson (JPA 1688, 3:248) and Arthur Turner (JPA 1689, 3:307). There apparently was a market for such cooking/baking vessels, as revealed by Port Royal merchant William Robinson's '28 pudding pans' listed at 9d. each (JPA 1687, 3:26).

The fact that the pan is decorated suggests that it was used at table, perhaps as a container from which a pie or pudding was served. As Belden (1983:207) notes, until well into the 19th century, pudding was "a mainstay of the first course of English and American dinners, its currant-studded substantiality lessening appetites for more expensive meats and vegetables." Or, perhaps, the pan was used for a stew. Slip decoration did not always preclude use for cooking, and the vessel is certainly robust enough to stand the heat of an oven.

An economic alternative to ceramic pie or pudding dishes were known as coffins, which were, in fact, used until the early 20th century. The pastry in which the pie was cooked served as the dish and was not meant to be eaten. To serve the cooked filling, the top crust was simply removed and portions scooped out. After the sweet or savory filling was eaten, the coffin could be re-used (Belden 1983:213; Markham 1986 [1615]:98; Hess 1995:83).

FOOD AND BEVERAGE STORAGE VESSEL FORMS

In contrast to food processing forms at Building 4/5, which comprise the smallest functional category of vessels in the analyzed assemblage, ceramic vessels used for storage appear in relative abundance, comprising, in fact, the second largest functional category. A total of 232 sherds (20.1 percent of the analyzed sherds) represent at least 33 vessels, or 22.9 percent of the total number of ceramic vessels identified in the building's pre-1692 context. The vessel forms represented are storage pots (n=25) and bottles (n=8). While the majority of the storage pots are in undecorated coarse ware, all but one of the identified bottles are stoneware (the exception is tin-glazed).

The relative proportions of ceramic storage forms and their respective sherd counts are shown in Table 6.1. Each form and its distribution is now discussed.

Storage Pots

Most of Building 4/5's analyzed storage pots are in fragments. Forms were identified using the following archaeological criteria: storage vessel walls are relatively thick (for strength), and forms are usually not decorated (since spending an inordinate amount of time decorating an ordinary storage pot was not practical). Further, storage forms tend to be restricted in angle, and bases are usually broad and flat (both attributes make the vessel more stable) (Rice 1987:240-241, Figure 7.2). As was noted in the section on cooking pots, obviously, these criteria are general guidelines only to help in identification of sherds. Olive jars, for example, which are also included in this category, are relatively thin-bodied. They served both storage and transfer functions, so thick walls would be a disadvantage because of weight.

The majority of the coarse ware storage pots appear to be English-made, including two relatively small vessels (each represented by a single sherd found in Room 4B), which cannot be positively identified. Undecorated Redware vessels (n=11) are the most prevalent type, and except for a single pot, which was recovered intact, all are represented by rim and basal fragments and diagnostic body sherds (i.e., unique in thickness and/or wall angle). The intact Redware vessel was found, filled with pitch, near the interior door of Room 2 (Figure 6.5*a*, see also Figure 4.2*b*). It has two small holes in the wall, each of which is plugged with a wooden dowel. (Perhaps these holes served to dispense the pitch.) The other Redware examples were concentrated around Rooms 1 and 3 and Yards 4A and 5.

A further two English-made pots, one each in Room 2 and Yard 5, were identified as North Devon Gravel-Tempered ware (see Figure 6.5*e* for one example). Also identified were two intact Red Border ware forms recovered from Yards 4A and 5 (see Figures 6.5*c*, *d*). In fact, one of the Border ware vessels (found in Yard 5) was thought at first to be a mug (see Figure 6.5*d*). On closer inspection, however, its everted, rolled-over rim suggests it would not be a practical vessel for drinking. Rather, the rolled-over rim would have provided a purchase for string to tie down a parchment cover. The vessel is covered on both surfaces with a thickly applied lead glaze, which gives it a warm and pleasing dark brown (5YR 4/2- /6) color.

Non-English storage ware includes the remains of at least four vessels identified as of Iberian provenance, based upon characteristics outlined in Chapter IV. Two are similar to the type known as the 'Spanish Storage Jar' (Deagan (1983, 1987:36) and two are olive jars. Two storage pots locally made by African slaves were also found. One of these Yabba vessels is represented only by a large basal sherd, the other has been almost completely reconstructed (see Figure 4.2*a*). The four Iberian vessels were found in fragments in Yards 5 and 6, in association with the cistern. The Yabba pots were found in Yard 7.

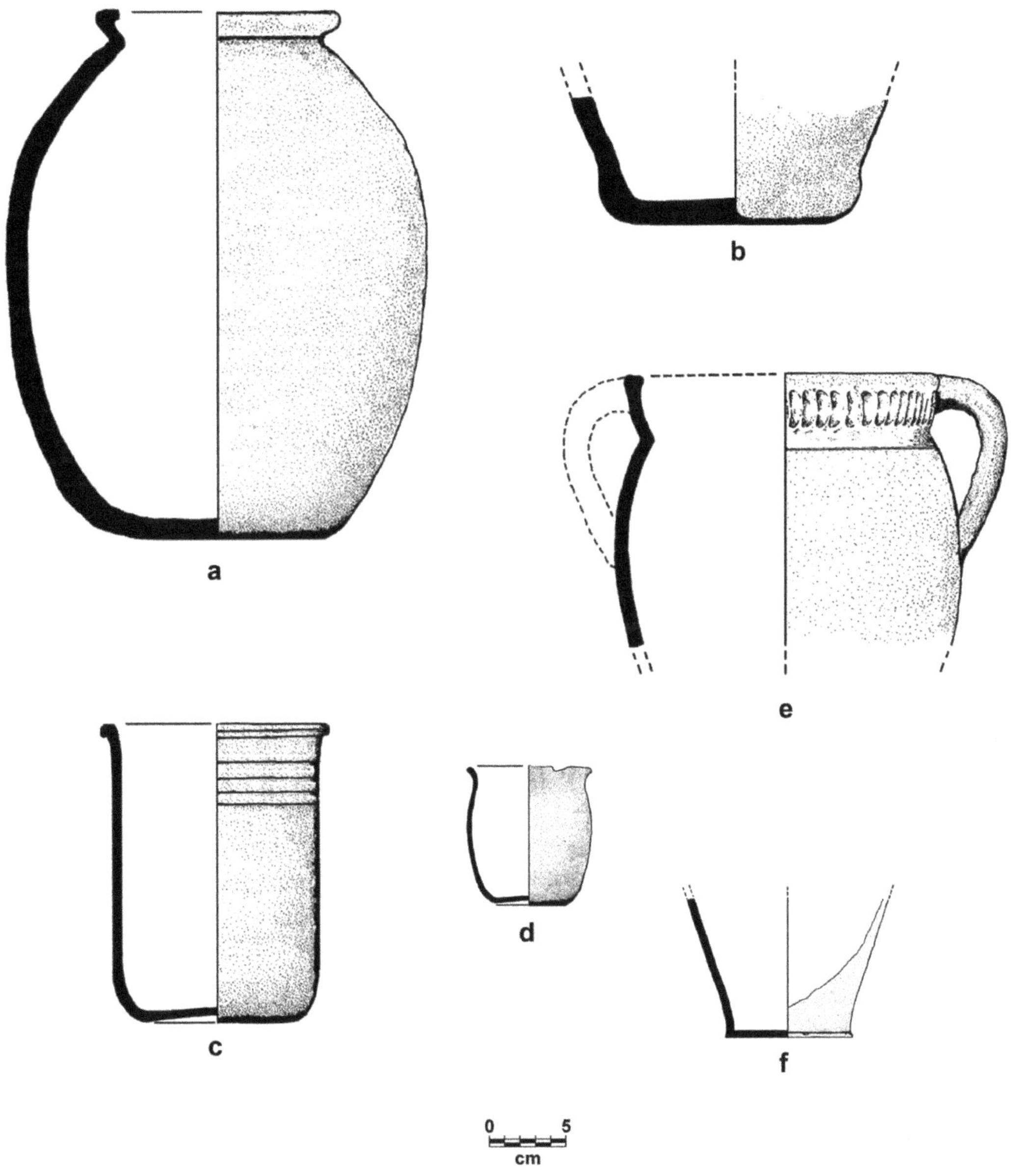

FIGURE 6.5. Examples of storage ware recovered from the 17th-century context at Building 4/5. *a*, intact English Redware form (filled with pitch) found in Room 2; *b*, English Redware basal sherd found in Yard 6; *c-d*, intact Red Border ware forms found in Yards 4A and 5; e, North Devon Gravel-Tempered ware rim and body sherd found in Room 2; *f*, Iberian (?) stoneware basal sherd found in Yard 4A. (Illustration by Madeleine J. Donachie.)

Only two stoneware storage pots were identified in the analyzed assemblage. One, found in Yard 7, is represented by a 10-mm thick base and body sherd, which is covered on the exterior with a light yellowish brown (10YR 6/4) glaze. It is probably English. The other, found near the hearth in Yard 4A, is represented by a thick base and body sherd covered on the exterior with a white slip wash (see Figure 6.5*f*). It is possibly of Iberian provenance. As noted in the previous chapter, the stoneware fabric allows for indefinite storage, and these pots may have been used as an alternative to lead-glazed earthenware (or wood) for salted or pickled foodstuffs. Storing such preserved contents in a lead-glazed earthen pot might cause the glaze to decompose and so poison the contents. That so few are represented in comparison to earthenware forms may be because stoneware was significantly more expensive.

No lids were found associated with the ceramic storage pots. Doubtless, cheap wooden lids could have been used, as with cooking vessels. In a domestic scene by Pieter de Bloot ('Peasant Interior,' ca. 1640), pots are shown open or covered rather casually with a bowl or dish. Alternatively, as noted above, storage pots with flaring rims and constricted necks could also be sealed by placing a piece of parchment or a bladder over the opening and tying it with string.

It has been shown that most of the storage ware was recovered from the yards at the back of the building complex. The remains of some of these vessels werefound associated with the areas around the cook rooms, and this suggests their use for holding various foods and liquids. Many of the larger forms, such as the Yabba jar, the intact English Redware pot, and even the olive jars, were probably used to store water, which had to be shipped to Port Royal from the main part of the island. The porous earthenware fabric of these vessels would allow the water to cool by evaporation, no doubt, as noted in Chapter IV, a welcome aspect in the warm Caribbean. Other uses for these and similar-sized pots may have included the storage of olive oil (oil is frequently mentioned in the Port Royal probate inventories, although it is not clear whether it was used for cooking or lighting), wine or other beverages (e.g., beer), flour/meal, dried legumes, or, indeed, even sugar.[1] Some of the smaller storage pots identified in the assemblage (e.g., the Red Border ware specimen) may have been used to store spices, such as salt, pepper, cinnamon, cloves, nutmeg, or ginger.

[1] The Spanish settlers brought sugar cane to Jamaica in the 16th century. By 1660, when the first shipment was sent to England, sugar was commonly used in medicinal cookery in Europe and was fast becoming an ingredient in prepared foods and drinks (Pawson and Buisseret 1975:63; Mintz 1985:77-147; Caton 1999:30). English physician Thomas Tryon (1634-1703), for example, counsels drinking milk sweetened with "good white sugar" up to three times daily for the sick (cited in Caton 1999:47). Recipes for 'sweetmeats,' such as the "Cheesecacks, Custards, Tartes, &c" noted by John Taylor (1688:261) as being available in Port Royal, were common cookbook items (see Belden 1983; Hess 1995; Longe 1999[1610]).

Bottles

Ceramic bottles in the 17th century were bulbous-bodied vessels with a relatively short, thick, narrow neck and sometimes with a handle (Beaudry et al. 1983:31). Many surviving examples have a series of grooves around the top of the neck to provide a purchase for string, which was used to tie down a parchment/bladder cover. Evidence from several wine merchants' accounts show that by the mid century, corks were also in general use (Berry 1933:4-6, cited in Archer 1997:266). Ceramic bottles were often favored over glass forms (which before about 1650 were not very strong) and were used for storing and serving, as well as individual drinking containers.

The robust, salt-glazed stoneware bottles made in the German Rhineland were particularly popular at this time and, as noted in Chapter IV, were exported in huge quantities throughout Europe and the rest of the world. Indeed, the stoneware bottle was extremely useful for storing all manners of goods, particularly acidic liquids, such as vinegar and wine, which tended to leach the poisonous lead from glazed earthenware containers. Smaller forms often served pharmaceutical purposes, since the fused vitrified fabric of stoneware is resistant to corrosive substances (Elliot 1986:86; Gaimster 1997:117, 123).

Following this, it is not surprising to learn that all but one of the eight analyzed ceramic bottles are made in stoneware. (The eighth bottle is tin-glazed, see below). Six of the stoneware bottles are German; one is English. The German bottles were probably brought to Port Royal either via the English market or by Dutch dealers operating directly with the Rhenish stoneware industries (Noël Hume 1970:276-285; Wilcoxen 1987:73; Gaimster 1997:98).

Two of the German bottles, which are discussed in detail in Chapter IV, were recovered from the building intact (Figure 6.6). As noted in that chapter, both their relatively small size (ca. 12 cm tall) and the fact that one was corked and contained the residue of a foul-smelling liquid may indicate that they may have been used to store liquid medicines. Alternatively, they may have contained a more expensive substance. Miniature German bottles recovered from the *Lastdrager*, a 17th-century Dutch East Indiaman wreck found off the island of Yell, Shetland, were found to contain droplets of mercury (Sténuit 1974, 1977:443-44). Mercury was essential "for producing an amalgam with silver or gold, extracted from crushed low grade ores. When distilled, mercury evaporated, condensed and was recovered, leaving behind extracted pure precious metal" (Elliot 1986:89; see also Green 1977:2:481-483).

Another, and perhaps the most interesting, use of these vessels was in witchcraft. Known as 'witching bottles,' apparently an English invention since there is no evidence of this use on the Continent, they contained

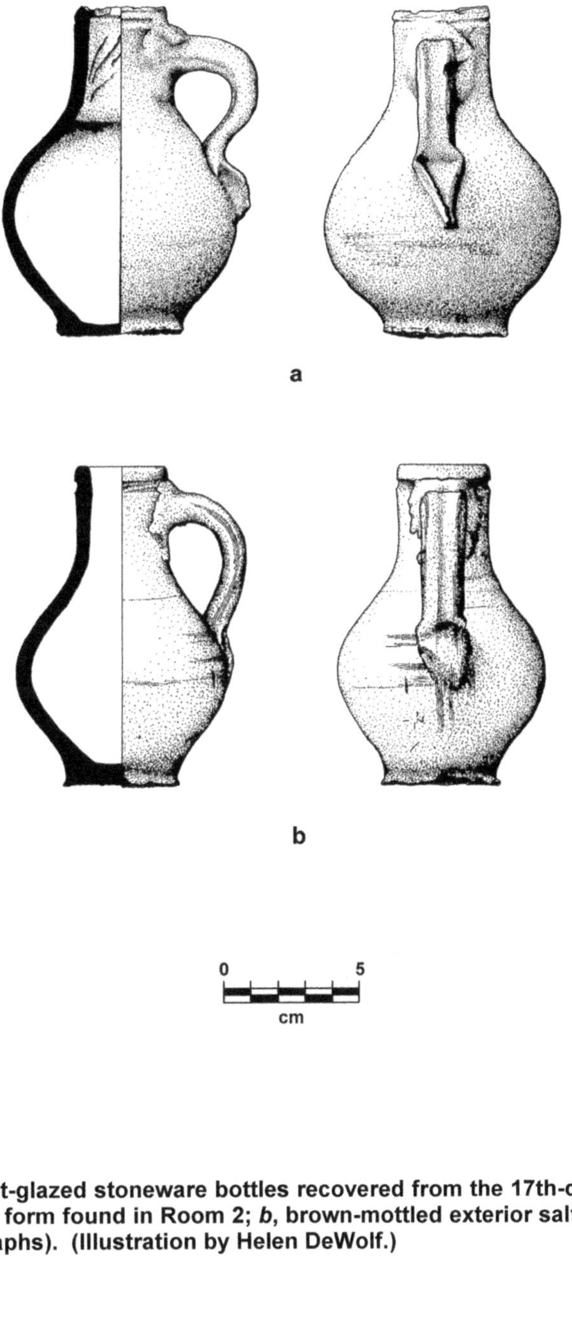

FIGURE 6.6. Intact miniature salt-glazed stoneware bottles recovered from the 17th-century context at Building 4/5. *a*, gray-mottled exterior salt-glazed form found in Room 2; *b*, brown-mottled exterior salt-glazed form found in Yard 5 (see Figure 4.14*a-b* for color photographs). (Illustration by Helen DeWolf.)

such ingredients as fingernail parings, human hair, urine, cloth hearts pierced with pins, pieces of wire, iron nails, thorns, and bent pens, and were stoppered with earth or clay. Witching bottles have been found in London and in East Anglia, sometimes in streams and ditches, but usually under buildings or else suspended from chimneys. It appears that they were devised as counter measures to witchcraft rather than used for malevolent magic (Noël Hume 1958:440-44; Merrifield 1987:163-175). A similar-sized bottle to those at Building 4/5 was found in London, containing "nine brass pins, each carefully bent twice" (Merrifield 1987:165-166).

The four other, larger German bottles recovered from the building are also discussed in Chapter IV. They are of the type known as *Bartmannkrüge*, or more commonly, if erroneously, Bellarmines (see Figure 4.14*c-f*). All are represented by fragments, one of which shows part of the mask to which the bottle form owes it name, and two of which are represented by their molded medallions. The fourth *Bartmannkrug* is represented by a reconstructed mouth and neck section (consisting of about 20 sherds) in a light gray fabric with reddish brown (5YR 5/4) surface mottling.

The distribution of the German bottles and their representative sherds seems to be rather arbitrary (due to their shape, bottles can displace easily). While one of the two intact forms was recovered from Room 2, which may have served in part as a storage area, the other intact form was found in the northeast corner of Yard 5, directly behind Room 1. The *Bartmannkrüge* fragments were recovered from the sidewalk (in front of Room 1), Room 2, and from the alley to the east of Yard 4B. It may be that those bottles in the alley were already broken prior to the earthquake and had been intentionally dumped there as trash. Those recovered from in and around the rooms are more likely to have been in active use. (It may be that the example found on the sidewalk is part of a *Bartmannkrug* that was recovered from Building 8.)

The remains of a dark brown glazed English stoneware bottle were also found in Room 1. So too was part of a rounded, decorated handle from a tin-glazed earthenware (wine?) bottle (Figure 6.7). While the handle fragment is too small to identify the decorative pattern, a distinctive bored hole on its obverse side indicates that a lid (likely pewter or silver) with a thumb piece was originally attached. A similar patterned handle is illustrated in Britton (1982:92, Figure 6.21), where it is dated to London 1690. Other examples of tin-glazed wine bottles are shown in Archer (1997:268-271).

FIGURE 6.7. Remains of a decorated English tin-glazed earthenware (wine?) bottle handle found in Room 1 in the 17th-century context at Building 4/5

FOOD CONSUMPTION VESSEL FORMS

Forms related to the consumption of food comprise the largest functional category of identifiable ceramic vessels in the Building 4/5 assemblage. A total of 232 sherds, or 20.1 percent of the analyzed sherds, were catalogued by the author as belonging to forms used for the eating and serving of food. These sherds represent an estimated minimum of 41 vessels, or 28.5 percent of the total number of ceramic vessels associated with the building's pre-1692 contexts. The vessels forms represented are bowls (n=16), plates (n=13), porringers (n=5), the remains of at least 5 saucers or condiment dishes, 1 large decorated dish, and an almost intact and rather elaborate salt stand. The majority of the forms occur in tin-glazed earthenware, but plain coarse ware, slipware, and Chinese export porcelain vessels were also found.

The relative proportions of ceramic food consumption forms and their respective sherd counts are shown in Table 6.1. Each form and its distribution is now discussed.

Bowls

Like their modern equivalents, bowls may have been used for a wide range of purposes in the 17th century, including that of mixing bowl, drinking vessel, slop basin, serving vessel, or soup bowl. (Soups as an introduction to main meals were beginning to appear in the mid 17th century [Kisbán 1986:7].) Contemporary potters' inventories merely distinguish between different shapes and sizes of bowls, rarely providing a name which describes use (see examples in Wills 1967; Vaisey and Celoria 1974; Britton 1990). Household probate inventories, which rarely even mention ceramics, are similarly non-specific. All of this makes functional classifications of bowls from archaeological sites difficult. In the case of Building 4/5, the locations of these forms—in either the front or back areas of the building complex—and their materials—either coarse or refined ware—provided important information.

Undecorated Coarse Ware Bowls (n=10)

Most of the ceramic bowls identified are coarse earthen Yabba (n=8) and are represented by diagnostic rim and body sherds, some of which show evidence of hand-made impressed and molded decoration (see Figure 4.6). Their presence could possibly indicate an African slave/servant presence in the building complex. More likely though, and that which is noted in Chapter IV, is that this local ware was used by the Anglo/European settlers and African slaves alike. While the distribution of the Yabba bowl sherds is somewhat scattered, the majority of the forms are concentrated around the hearth/cook room areas in Yards 4A and 4B. At least one bowl is represented by fragments in Yard 7, near the hearth.

To find Yabba bowls near the cooking areas suggests that these vessels were used in the preparation of food (i.e., as mixing bowls). They may have been also used as communal eating bowls. The relatively large diameters of the rims identified as belonging to Yabba bowls are between 20 and 25+ cm. Thus either of these two functions (i.e., preparatory and consumption) is plausible.

Other plain coarse ware bowls represented include those made in English Redware (n=2). The sherds from the English forms, like those of Yabba, were recovered from the yards but tended to be concentrated in areas closer to the structures. One of the Redware bowls, which is glazed an olive/dusky red on the interior only, was recovered from Yard 5, directly behind Room 1. The other, which has an olive glaze on both surfaces, is represented by sherds recovered from Yard 4B, directly behind Room 4B. The English Redware bowls are smaller (10-15 cm rim diameter) than the Yabba forms, suggesting they were used as individual eating vessels.

Decorated Fine Ware Bowls (n=6)

Six decorative bowls that may have been used to serve and/or consume food are represented in various fine wares—tin-glazed earthenware (n=2), slipware (n=2), and Chinese export porcelain (n=2)—and most were recovered from areas inside the structure. One of the tin-glazed forms (recovered from Room 4A, Figure 6.8) is represented by several basal sherds, which are decorated on their obverse with a Chinese-inspired foliated landscape scene that includes a Chinese figure. Such scenes were very common on tin-glazed earthenware from around 1680 (Draper 1984:30), and similar patterns are seen on other vessels in the assemblage. The strong turquoise glaze suggests that this bowl was made in the Brislington factories in England (see, e.g., Archer 1997:285, Plate 186). (A similar-colored yet less intense glaze is also found on the products of London and Bristol factories.) The decoration derives from either Chinese Ming Transitional porcelain (i.e., porcelain made after the collapse of the Ming dynasty) or from Japanese versions of the style produced in the late 17th century.

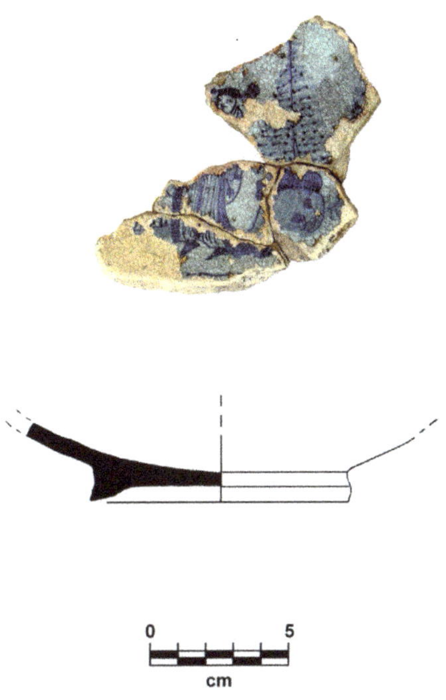

FIGURE 6.8. Remains of a decorated English tin-glazed earthenware bowl recovered from Room 4A in the 17th-century context at Building 4/5

A large majolica bowl, decorated on the interior with a pale blue wave-like design, is represented by a thick (12 mm) light red (2.5YR 6/6 - 10R 6/8) earthenware body sherd (see Figure 4.12*a*). The vessel is probably Spanish, based upon the fabric color and rather crude and carelessly applied design (see Lister and Lister 1976b:37, 1978:12, 1982:59). Perhaps it is part of a deep brimmed plate, or *plato*, a typical Spanish form that has an everted and curled rim (see Deagan 1987:Figure 41). This sherd was recovered from the street directly in front of Building 4.

A North Devon Sgraffito slipware bowl (represented by a body sherd, see Figure 4.9) and a Staffordshire slipware bowl (represented by a near-intact vessel) have also been identified. Interestingly, both pieces were recovered from the yard areas of Building 4—the sgraffito sherd from the back of Yard 4B, next to the hearth, the Staffordshire piece from Yard 4A. It is possible that they were deliberately deposited in these areas as trash. (A large cluster of discarded used clay tobacco pipes was also found in the same area.)

The base of a quite large (15 cm rim diameter) Chinese export porcelain bowl was found in Room 2 (other fragments were found in Yard 5) (Figure 6.9*a*). It has been cross-mended with sherds from Building 8, and it possible that the vessel originated in that building. It is of the type known as Batavia ware, after the Indonesian port from which much Chinese porcelain was exported in the 17th century, and shows a covering on its exterior of the characteristic brown wash of this decorative type. A blue-and-white floral motif covers the interior. All of the sherds that make up a second, smaller, near-intact blue-and-white decorated Chinese export porcelain bowl were recovered from Yard 4A (see Figure 6.9*b*). This vessel may have been displaced to this area in the earthquake. Less likely, but possible, is that it may have broken during the life of the building and dumped in this area as trash.

Both porcelain bowls are slightly bigger and more substantial than tea bowls, so they were probably used as serving/display dishes or simply as decorative ornaments. Alternatively, they may have been used for the cooling and rinsing of wine glasses, as shown in J. F. de Troy's 'Le Déjeuner d'huitres' (1734), where the diners each have a small bowl containing water in which to cool their glasses (Figure 6.10). That quantities of glass stemware were also found in the Building 4/5 complex strengthens this proposal (see McClenaghan 1988).[2]

Charlotte Wilcoxen (1992a:17), a ceramic historian who examined some of the Port Royal sherds, notes that "hard evidence of bowls is surprisingly small in the Port Royal

[2] Much larger circular or oval bowls made especially for this purpose are known as 'monteiths,' several of which were recovered from an excavated 17th-century tavern on New Street, Port Royal [Brown 1996:180-181]. Monteiths have a deeply scalloped rim from which the stems of glasses are hung while their bowls were immersed in water.

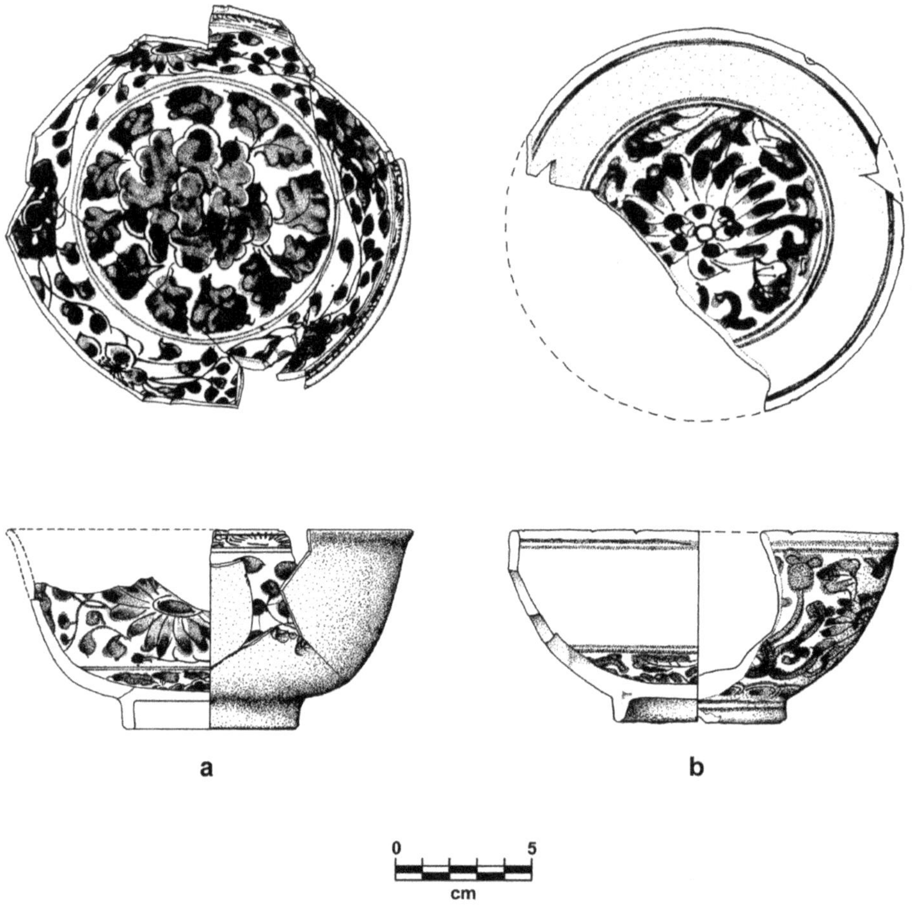

FIGURE 6.9. Reconstructed Chinese export porcelain bowls recovered from the 17th-century context at Building 4/5. *a*, Batavia ware form, showing an underglaze blue floral design on the interior, found in fragments in Room 2 and Yard 5 (and cross-mended with sherds from Building 8); *b*, blue-and-white decorated form found in Yard 4A (see Figure 4.19 for color photographs). (Illustration by Helen Dewolf.)

collection, although reason says that more are represented here than have been identified." The author agrees, but it should be remembered that many bowls, certainly those that were made to be used in kitchen activities, were made of wood, as evidenced by the remains of large wooden bread dough bowl recovered from Yard 4A.

Plates

In the 15th century, thick-sliced, day-old household bread was often used as a plate on which to pile food. These bread trenchers (from the Old Norman French, *trencheor*, 'to cut') were not very practical, however, since they absorbed most of the food's flavor; trenchers made of wood were much more serviceable, and together with pewter plates, soon became common forms on the medieval table (Schaefer 1998:44; Caton 1999:63, 81). Ceramic plates first appear in English archaeological assemblages around the mid 17th century (see Cunningham 1985; Fox and Barton 1986).

Following Beaudry et al. (1983:33), a plate is here defined as an eating vessel, approximately 15 to 25 cm in diameter. In this study, plates are distinguished from dishes by their size (dishes are larger) and function (dishes are serving vessels). In the Building 4/5 analyzed ceramic assemblage, plates are second to bowls as the most common item in the food consumption category (see Table 6.1). Unlike bowls, however, which are represented in various wares, all of the plates identified are made in tin-glazed earthenware. In fact, plates are the second most prevalent of the tin-glazed forms (after apothecary pots) in the entire analyzed assemblage.

Slightly over half of the plates are colorfully decorated (n=7), and it may be that they served to ornament a wall or dresser or mantlepiece. All but one appear to be Mediterranean and Mexican majolica. The exception, represented by a rim sherd recovered from the alley, is possibly of English provenance (Figure 6.11). It shows two narrow blue lines encircling a wide manganese purple border, which is incised with circles and scrolls. The condition of this fragment is excellent, with the colors still vibrant. While the decoration is similar to that found on Wincanton tin-glazed plates from around the 18th century, Wilcoxen (1992a:15) notes that a sherd almost identical to this was recovered from the Adam Thorogood 17th-century site in Virginia. This establishes that this decorative pattern was used prior to 1692.

FIGURE 6.10. Detail from 'Le Déjeuner d'huitres,' by J. F. de Troy, 1734. Musée Condé, Chantilly, Lauros-Giraudon. The diners each have a small bowl containing water in which to cool their glasses.

The undecorated plates in the assemblage are also probably English and were likely used at table on a regular basis. While tin glaze is "generally too delicate to serve as a cutting surface, before the widespread use of the fork, such delicacy was probably not so great a disadvantage" (Schaefer 1998:48). Paintings like Jan Molenaer's (ca. 1610-1668) 'Merry Company in a Inn' (Private Collection, The Netherlands), showing an affluent company at table with individual place settings of tin-glazed plates, supports their hypothesized use as everyday table wares. Indeed, the presence of large numbers of ceramic plates in mid 17th-century Anglo-American households is well noted, both from excavations and analysis of probate inventories (see, e.g., Pendery 1987, 1999; Yentsch 1990:37-40, 1991:39-41). Plates are recorded by Brown (1996:201-204) as comprising the largest percentage of ceramic food consumption forms in an excavated 17th-century Port Royal tavern on New Street.

Both the data from Building 4/5 and these other example, then, seemingly contradict Deetz's (1973:28-29, 1977:56-58) suggestions that in the 17th century the occurrence of plates in proportion to other vessel forms is relatively low, and that plates as table wares (compared to their use as decorative wares) only became common in the 18th century. Deetz believes that shared wooden trenchers continued to be the norm, certainly in New England, throughout the 17th century, and that ceramic plates served primarily as decorative items.

FIGURE 6.11. English (?) decorated tin-glazed earthenware plate rim sherd recovered from the alley in the 17th-century context at Building 4.5

Interestingly, many of the sherds from the plates (both plain and decorated) were recovered from Room 3, with a second concentration in Yard 5. The remaining sherds were found in and around Building 4's yards. Those sherds found in Room 3 were likely in active use at the time of the earthquake. Those sherds in Yards 4A and 4B, both of which have sand floors, are possibly the remains of plates dumped in these areas as trash. It is unlikely that those sherds in Yard 5 are from surface middens, since this yard, being brick-paved, probably served as an extended-use area rather than a place where refuse was dumped. It is always possible, of course, that the plate sherds in Yard 5 were parts of vessels originally housed in Building 5's upper story.

Porringers

A second hollow ware form (following the bowl) recovered from the building complex is the earthenware porringer, which is described by Beaudry et al. (1983:32) as "a vessel usually hemispherical in shape and shallower in relation to its diameter than a cup or a pot. Porringers have at least one and sometimes two handles, either horizontal or vertical." They were used for eating porridge, pottage (stew), gruel, and other semi-liquid foods. Outside the food domain, porringers were quite often used as bleeding bowls and were made in various metals, most commonly pewter and silver.

All of the analyzed earthenware porringers are completely or partially intact. They vary in shape and size, as well as in ware type. Two of the forms are made in glazed Border ware, two are in tin-glazed earthenware, and one, painted with a colored slip, is likely of Staffordshire provenance.

Both of the Border ware porringers were found associated with Building 4. The first (Figure 6.12*a*), recovered intact from Yard 4A, directly behind Room 4A, is short and thick in form (10 cm rim diameter x 6 cm tall) and has a single horizontal handle luted on to the body. Its simple rounded rim and horizontal ribbing around the lower body are common attributes on porringers made in Border ware throughout the 17th century (Pearce 1992:15, 160). Its glaze, appearing reddish brown (2.5YR 5/6) on the deep red fabric, covers both surfaces (the base is unglazed). The second Border ware porringer (Figure 6.12*b*), recovered partially intact from the alley, is made in a white, finer fabric, is covered on the interior with a smooth olive glaze (5Y 4/3), and is slightly larger (12.5 cm rim diameter x 6 cm tall) and thinner walled than the red form. Soot deposits in several places on its exterior surfaces suggest that it was used on the hearth. (That it was found in the alley suggests that it broke while in use and was purposely dumped there as trash.) It, too, has a horizontal, slightly upturned handle, but it is unlikely that it served as a 'handle' in the usual sense, since it is too small to be of any use as a way to hold the vessel. Indeed, many porringer handles are either so small, or attached at such an acute angle (such as in the Red Border ware porringer) that it would have been difficult to hold the vessel by the handle alone. They are more likely to have been used to hang the vessel from a hook.

Both of the tin-glazed porringers (ca. 14 cm rim diameter x 6 cm tall), recovered from Yards 5 and 7, show no evidence of painted decoration and are probably English, since generally tin-glazed porringers from Holland have a slightly different shape and are more varied in design (Wilcoxen 1992a:16; see Britton 1982:82-83; Archer 1997:Figure F.1 for examples of English forms). Like the Border ware porringers, the three-holed horizontal lobed handle on one of the tin-glazed forms (see Figure 6.12*c*) would have allowed it to be hung from a hook for storage or as decoration. While no handles were found attached to the second porringer (not illustrated), a one-holed porringer 'shell' handle was recovered from the vicinity (see Figure 6.12*d*).

Porringer manufacture was considerable in England's tin glaze potteries. The Pickleherring Potteries inventory of 1699, for example, lists 7748 porringers, the second largest category of objects after apothecaries' wares (Britton 1990:89-90, 1993:64). They are further described as 'Spanish,' 'blood,' white,' 'toy,' and 'galley' porringers. Archer (1997:280) notes that 'Spanish' may be a term used to distinguish two-handled vessels, since those from Spain were usually of this form; 'blood' probably means that some were used as bleeding bowls. 'White' and 'toy' are self-explanatory; 'galley' probably refers to a form of simple decoration like that found on gallipots (see below, under 'health/personal hygiene' discussion) (Archer 1997:280). It is possible that both of the tin-glazed porringers recovered from Building4/5 had only one handle, which at one time was thought to indicate that they were used as bleeding bowls. As Archer (1997:280) notes, however, this is not the case, pointing as evidence to a 17th-century woodcut of a family at table, eating from porringers with one handle.

The Staffordshire slipware porringer (10 cm rim diameter x 6.5 cm tall, see Figure 6.12*e*), reconstructed from four fragments, was also recovered from Yard 5. It is decorated on the exterior with dots of dark reddish brown (5YR 2.5/2) slip placed horizontally around the vessel, slightly below the rim, and a single, reddish brown slip-trailed line drawn around the lower body. The pale yellow (2.5Y 8/4) interior glaze appears olive (5Y 5/3) on the vessel's exterior. This marked contrasting color is likely the result of the vessel's 300-year immersion in sea water. The vessel may have been used in Yard 5, or it may be that it was originally located upstairs—its thin walls and heavy glazed slip decoration suggest it would have been used at table—and was displaced to the yard by the earthquake/tidal wave.

Porringers are a common find in most 17th-century sites, performing most of the functions that would be taken up in the 18th century by deep plates or dishes made in white salt glaze, creamware, and pearlware (Yentsch 1990:40-41). Specific to Port Royal, Mayes (1972:89, 95, 97, Figure 27) recovered numerous porringer fragments from the 17th-century levels of the excavated naval dockyard. Brown (1996:209) records a minimum of 19 such vessels from the New Street tavern. An examination of Jamaican household probate inventories from the late 17th century similarly revealed to the author the popularity of these vessels (see, e.g., JPA 1686, 2:191; 1692, 3:424-427, 397).

Saucers

In the 17th century, saucers were used for serving sauces (hence sauce-r) or other condiments (e.g., oil, dry mustard, salt) at the table. It may well be that they were also used as drip dishes, placed under jugs or cups, i.e., in the modern sense of a saucer (Brown 1996:206). As would be expected, saucers are relatively small. Following Beaudry et al. (1983:34), the author has identified saucers as flatware forms with rim diameters of less than 15 cm.

The Building 4/5 analyzed assemblage contains at least five saucers, four of which are decorated. The decorated earthenware saucers include two tin-glazed forms and two slipware forms. One undecorated Chinese export porcelain saucer was also found.

One of the tin-glazed saucers (found in Room 2) is represented by a blue-bordered rim and is probably English (Figure 6.13*a*). The second form (see Figure 6.13*b*), recovered from Yard 4A, is a little more difficult to source. The small fragment shows two cobalt blue narrow lines surrounding a spray of several dark blue long leaf blades/flower stamens, which bend towards the left in an arc. To the right of this spray, a thin blue stem

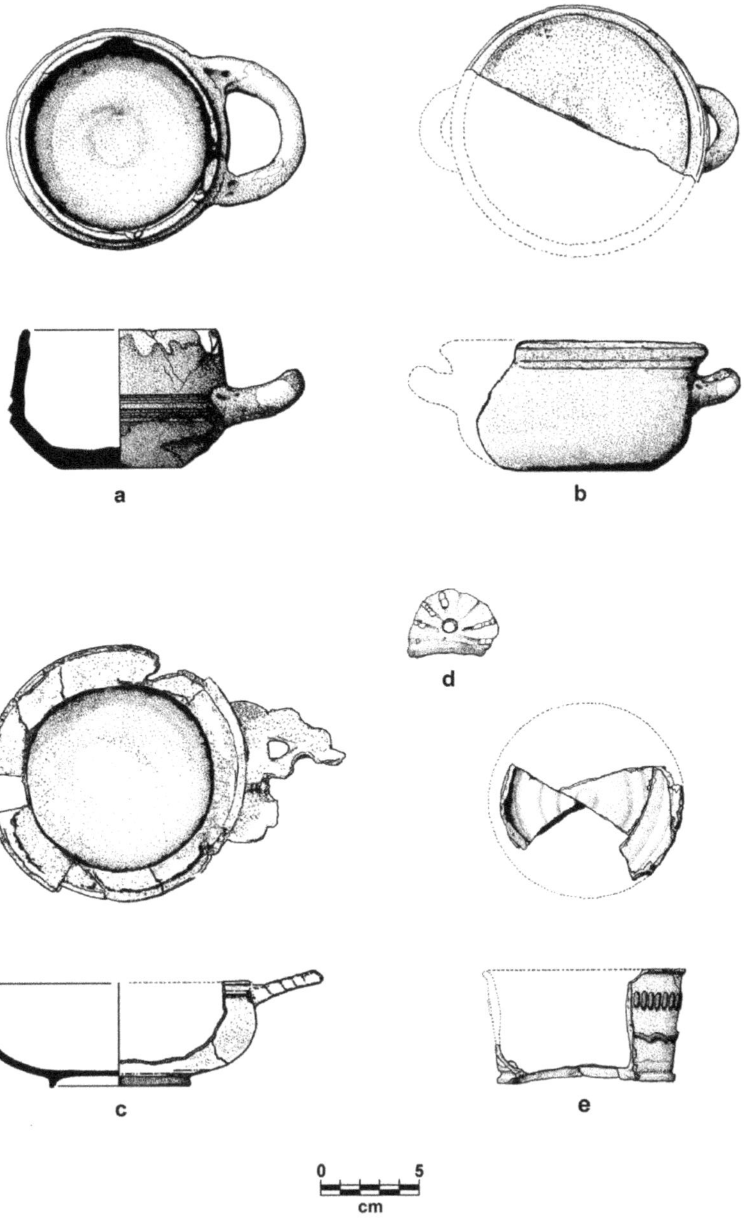

FIGURE 6.12. Examples of porringers recovered from the 17th-century context at Building 4/5. *a*, intact Red Border ware form found in Yard 4A; *b*, remains of a White Border ware form found in the alley; *c*, reconstructed undecorated English tin-glazed earthenware form found in Yard 5; *d*, one-holed tin-glazed earthenware 'shell' handle found in Yard 5; *e*, reconstructed Staffordshire slipware form found in Yard 5. (Illustrations *b-c* by Madeleine J. Donachie; illustrations *a*, *d*, and *e* by Helen Dewolf.)

supports a dark blue dot, possibly representing a stylized bud. The tin glaze is worn to matte finish but is still evident on both surfaces. This sherd is very similar to two illustrated in Lister and Lister (1987:90, Figure 102, cited in Wilcoxen 1992a:27), which are labeled as being of 16th-century date and attributed to Seville. However, the light cream/buff fabric argues more for an Italian provenance, as does the presence of a scar on the reverse, evidence for the use of saggar head pins to support and separate the vessel in the kiln (see Lister and Lister 1976a:37, 1978:12, 1982:59; Deagan 1987:61). It is suggested by the author that this saucer may be Sevilla ware (Italianate-Spanish, i.e., made in Spain by Italian potters). It may have arrived in Port Royal via England, since numerous Italianate-Spanish majolica are found on other Anglo-American sites. Alternatively (probably unlikely), it may have been shipped via Spain despite the strict Navigation Laws that stated that foreign goods were required to arrive at English colonies in English ships. Another possibility is that the saucer was simply brought to the port by a settler, merchant, or traveler.

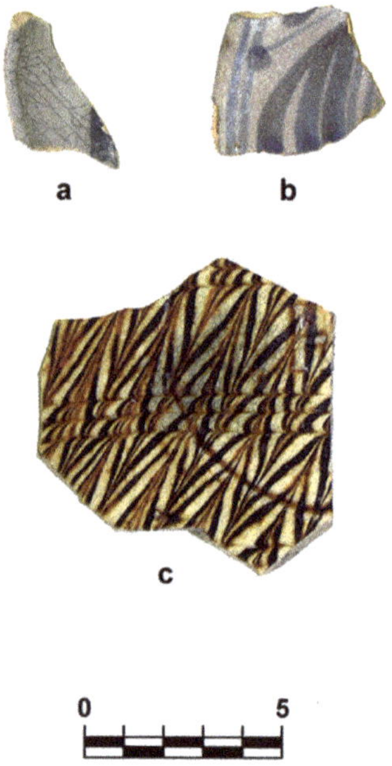

FIGURE 6.13. Examples of saucer fragments recovered from the 17th-century context at Building 4/5. *a*, decorated English tin-glazed earthenware rim found in Room 2; *b*, decorated Italiante-Spanish (Sevilla ware?) majolica rim found in Yard 4A; *c*, Staffordshire press molded slipware rim found in Room 1

One of the slipware saucers, represented by a small body sherd found in the alley to the east of Building 4, is possibly from North Holland, based upon its red (2.5YR 4/ -5/6) to reddish brown (5YR 5/4) sandy fabric and remnants of a white slip design. The second example in slipware, represented by a basal sherd from Room 1, is probably of Staffordshire provenance (see Figure 6.13*c*). Decorated on the obverse with dark reddish brown (5YR 2.5/2) and cream slips combed in a wavy pattern, the sherd shows the remains of the letter 'H' impressed into the clay fabric. The letter could be part of a motto or verse, common decorations on English slipware vessels of this period. Or, it could be part of the potter's name (last or first letter) or owner (unlikely). Its impression on the buff body rather than after the slip was applied indicates that the piece was press-molded, which suggests mass production, which, in turn, points to a slightly later date than the 17th century. Thus it may be intrusive from upper layers.

The remains of a saucer in undecorated white Chinese export porcelain were recovered from Yard 4B. It may be that the saucer was originally housed in Building 4 and was displaced during the disaster. Alternatively (less likely), it may be that it had been dumped in this area as trash, having broken during the life of the building.

Given the softness of earthenware and the costliness of porcelain, it seems reasonable to assume that the saucers in Building 4/5 were confined to serving table-top sauces, as noted above.

Dish

The remains of a fine, large (35 cm rim diameter) decorated dish, covered in a greenish blue glaze, were recovered from Rooms 1 and 2, with most of the fragments found in the former location (Figure 6.14). The decoration in the well is of cobalt blue painted panels, each of which is filled, in alternating patterns, with arabesques/floral motifs and a Chinese male figure seated in a landscape; the center shows a vestige of a pavilion canopy, a common central decoration on tin-glazed decorative dishes. Traces of widely spaced blue vertical lines are evident on the reverse. Wilcoxen (1992a:13) notes that similar lines are seen on 17th-century Portuguese majolica plates and dishes and are "crude simulations of the sometimes quite beautiful decorations on the under rims of fine kraak porselein of the late 16th century." Indeed, Wilcoxen (1992a:13) states, unequivocally, that this piece came out of Holland in the latter part of the 17th century and notes further that it represents "a very interesting type and pattern in the Dutch Delft genre, the prototype of which is a Chinese pattern of the Transitional Period that followed [the collapse of] the Ming Dynasty" (Wilcoxen 1992b:6). In its paneled arrangement, the well is reminiscent of the Ming bird-on-a-rock porcelain dishes. But unlike these, which feature Buddhist motifs, the design incorporates tulips, carnations, and other non-Chinese flowers. Indeed, this pattern, "borrowed by the Chinese from the Middle East, was some years ago designated as the 'Tulip Pattern' by the late Soame Jenyns [1955:25], a well-known English authority on Oriental porcelains" (cited in Wilcoxen 1992b:6).

The theme of the Chinaman in a rocky landscape was used from the late 17th century without much change over at least a quarter of a century (Britton 1982:168). In

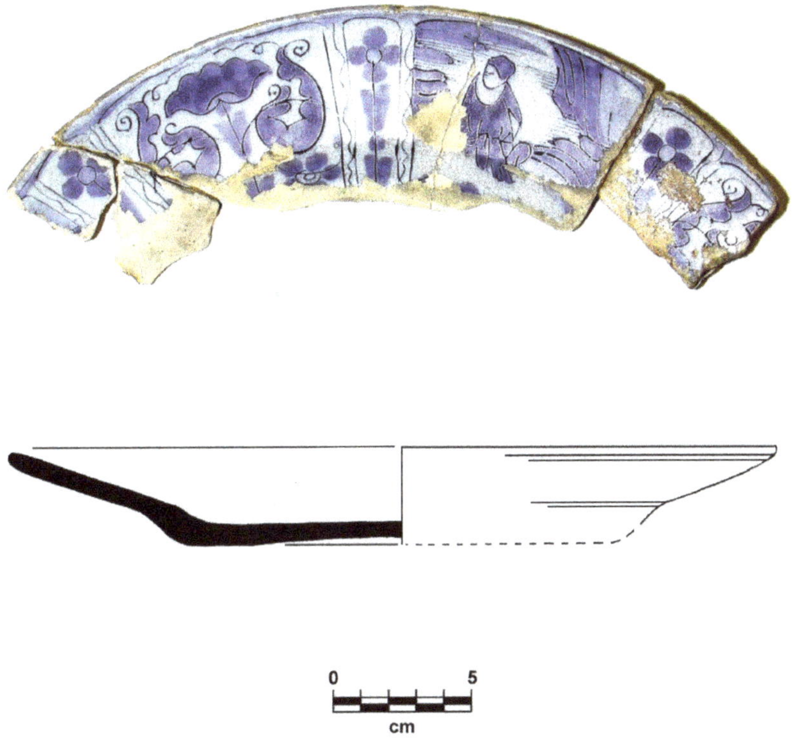

FIGURE 6.14. Reconstructed decorated Dutch tin-glazed earthenware serving dish recovered in fragments from Rooms 1 and 2 in the 17th-century context at Building 4/5

the beginning, a greenish blue glaze and trek outlining were characteristic, as were rather crude attempts at decorating the reverse of the rim. Gradually, during the 18th century, the tulip lost its distinctive form and virtually disappeared. Clear cobalt painting deteriorated to a dull, gray-blue wash on a white glaze. Decoration on the reverse discontinued, and other motifs came to be carelessly executed (Wilcoxen 1992b:7). Examples of English tin-glazed dishes with a similar pattern to that on the Building 4/5 piece are illustrated in Britton (1982:170), Lipski and Archer (1984:67, Figures 236, 237), and Archer (1997:Figures B.88, B.89, 204, Plates 84-85, 89). (At least four more dishes with similar designs were found by Hamilton's team in the 17th-century context at Port Royal.)

The dish from Building 4/5, with its elaborately painted scenes, may have been purely decorative. However, it cannot be assumed that it was not used at table as a serving vessel, like the platters and large dishes described by Randle Holme (1905 [1688]:14:4), which were "both for necessary use . . . to serve up to tables; as also to adorn their country houses; and court cupboards . . . set round about a Hall, Parlour and Kitchen."

Salt Stand

The final form to be discussed in the food consumption category is an almost completely reconstructed English tin-glazed salt stand, which was recovered in fragments from the intersection of the corners of Rooms 1, 3, and Yard 5 (Figure 6.15). The form is a common one of the period: a wide, circular base (13.5 cm diameter) that rises 8 cm to a round, rimmed platform (12.5 cm diameter) that, in turn, has a shallow recess at the center in which a small amount of salt would have been placed. Three curled upstanding knobs placed around the rim of the salt stand would have provided the support for a dish or a napkin thrown over the whole piece to protect the salt from dirt and to keep it from absorbing moisture (Britton 1990:87). In the Building 4/5 example, only two of the knobs were recovered. The salt stand may have been decorated, but so little remains of the glaze that this cannot be ascertained.

Salts such as these are referred to in potters' inventories of the period as 'curle salts' after the upstanding knobs (see, for example, Britton 1990:70). Many examples of curled-rim salts, as well as the 'plain rim' varieties and those with lugged foot supports, can be found in museum catalogues of 17th- and early 18th-century tin-glazed wares (see Britton 1987:115; Rackham 1987:2:119). They were also made in pewter and (particularly) silver, which, according to numerous probate inventories, were certainly popular in Port Royal. Arthur Turner, for example, owned a pewter 'Salt Seller' (JPA 1689, 3:307), while Thomas Stichbury, a Port Royal 'chyrurgeon' had a 'Silver Salt Seller' (JPA 1689, 3:254). John Campbell (JPA 1693, 3:503), Adam Weenan, a cooper (JPA 1690, 3:380), and tavern keepers Charles Cresso (JPA 1688, 3:217) and Charles Barre (1689, 3:255) also owned salt sellers made in silver. Ceramic forms, although not mentioned in the inventories, also must have been common, since Brown (1996:211) documents a total of 18 salt stands recovered from an excavated tavern in New Street.

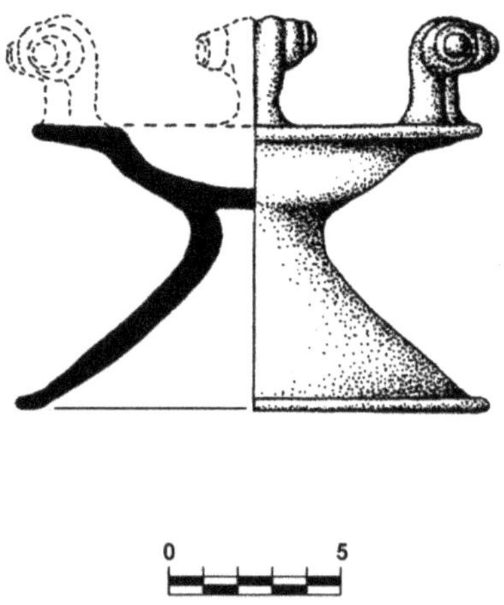

FIGURE 6.15. Reconstructed undecorated English tin-glazed earthenware salt stand recovered in fragments from the corners of Rooms 1 and 3 and Yard 5 in the 17th-century context at Building 4/5. (Illustration by Helen Dewolf.)

In the late medieval and early modern period, salt was a valuable commodity, prized as both a preservative and a seasoning, and although salt deposits are well distributed around the world, it has been suggested that salt became an important and expensive trade item due to the difficulties of transporting it in bulk (Multhauf 1978:10; Schaefer 1998:64). Certainly, salt played a central role in the cuisine at this time (Multhauf 1978; Caton 1999:10), and this is reflected in the elaborate forms in which these vessels were fashioned.

The salt stand also held a social symbolism—placed to the right of the host or in the center of the table, it created a hierarchy of seating, with more important diners above and less prominent ones 'below the salt' (Schaefer 1998:64; Caton 1999:75). In addition, to upend the salt was, and is still, considered a bad omen.

Although salt's prominence declined during the 17th century, as spices in general became more common, losing their value and exotic appeal, the salt stand remained an important piece of tableware. While only one ceramic salt stand was recovered from the building, the remains of a pewter specimen, as well as one of wood, were also found (from the brick-paved sidewalk and Room 1, respectively.) It is possible that there were more in pewter, and, perhaps, some in silver and even gold, which were were melted down and resold. Other wooden specimens may have simply decayed.

BEVERAGE CONSUMPTION VESSEL FORMS

A total of 214 sherds, or 18.5 percent of the analyzed sherds, were catalogued by the author as belonging to forms used for the drinking and serving of beverages. These sherds represent a minimum of 32 vessels, or 22.2 percent of the total number of ceramic vessels associated with the building's pre-1692 context. The vessel forms represented are cups (n=15), drinking pots (n=5), mugs (n=5), costrels (n=4), a ewer, a punch bowl, and a tea bowl. Most of the cups and the tea bowl presumably were used to drink the new and fashionable beverages of chocolate, coffee, and tea (although this may not have been always the case, see below). Drinking pots, mugs, and punch bowls are traditionally associated with the consumption of alcoholic beverages. Costrels and ewers were usually used to drink and serve water and/or wine.

The relative proportions of ceramic beverage consumption forms and their respective sherd counts are shown in Table 6.1. Each form and its distribution is now discussed.

Cups

A cup is distinguished here as "a small, [usually] handled drinking vessel of less than a pint in capacity" (Beaudry et al. 1983:29). Like their modern counterparts, cups in the 17th century came in a wide variety of shapes, and it is not possible to be categoric about how these vessels were used. Judging from contemporary pottery inventories, cups were generally for non-alcoholic liquids (chocolate, coffee, tea) (see Britton 1990). But there are also references to wine, syllabub, and caudle cups (Vaisey and Celoria 1974:27, 31, 33-34; Britton 1990:69-70). Indeed, there is ample evidence, as might be expected, that most shapes were, in fact, used for a variety of contents, as has been shown to be the case for many of the vessels already noted (e.g., bowls).

An array of cups is represented in the analyzed ceramic assemblage at Building 4/5. In fact, with at least 15 identifiable vessels, cups are the most prevalent form in the beverage consumption category. Further, they are the third most common form associated with food and drink in the entire analyzed assemblage (after storage pots and bowls). The majority of the cups are made in fine, decorated earthenware (tin-glazed and slipped). Two handleless Chinese export porcelain (*blanc de chine*) cups were also found. Three of the larger (plain) tin-glazed forms were possibly used for caudle; most of the others are likely chocolate and coffee (and possibly tea) cups. (Tea at this time was often drunk from small handleless bowls in the Chinese manner. The tea bowl recovered is thus included in a separate category and is discussed below, on page 95.)

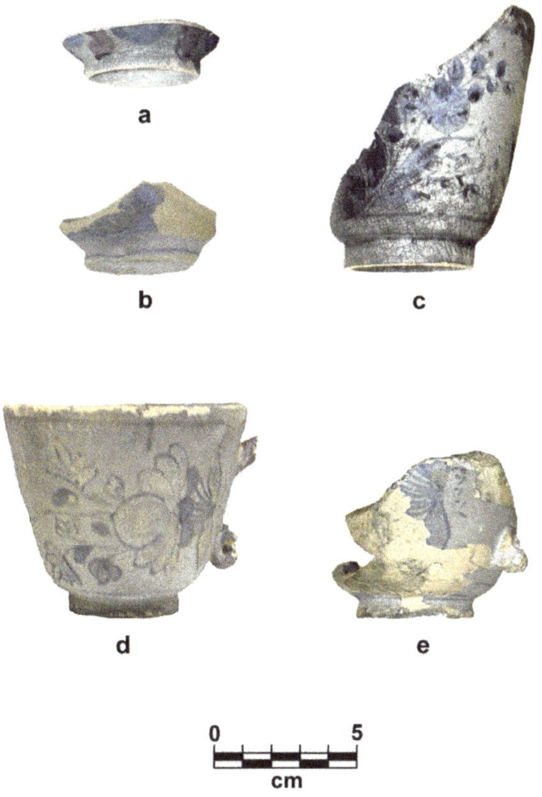

FIGURE 6.16. Remains of tin-glazed earthenware chocolate/coffee cups recovered from the 17th-century context at Building 4/5. *a*, English (?) basal sherd found in Room 1; *b*, English (?) basal sherd found in Room 3; *c*, near-intact English (?) form found in Room 4B; *d-e*, near-intact Dutch (?) matched set found in Room 2

Chocolate/Coffee Cups (n=10)
Five of the chocolate/coffee cups in the assemblage are fashioned in tin-glazed earthenware, and all appear to be of similar form (Figure 6.16*a-c*). Three are probably English (recovered from Rooms 1, 3, and 4B). Two examples (found in Room 2, see Figure 6.16*d*, *e*) are possibly part of a matched set and might be Dutch, since the design on the exteriors—an Oriental-inspired flower and insect motif painted in blue, purple, and yellow—resembles that on a Dutch cup illustrated in Britton (1987:132, Figure 17). The two (possible) Dutch cups show vestiges of a handle attachment. Such a handle would likely have been either a small plain loop or a more elaborate scroll, such as the one found in the southeast corner of Room 4B.

Cups of comparable size to the tin-glazed forms above are shown in an interior by J. van Aken of ca. 1720 (Figure 6.17), which shows a clear difference between tea bowls (a number of which are piled in a large slop basin on the table) and the handled cups used for drinking chocolate or coffee. A cup of similar shape to those from Building 4/5 is also shown on well-known trade card, dateable to about 1696-1700, in which James Morley, maker of salt-glazed stoneware at Nottingham, describes it as 'capuchine' (Oswald et al. 1982:Plate 57). (Indeed, the form was copied from Rhenish stoneware.

Capuchines were also made in Holland, and very similar silver cups, often with fine, chased chinoiserie decoration, were made in England in the late 17th century (Archer 1997:349, Plate 216). 'Eard coffees' are referred to in the inventory of London's Pickleherring Potteries (see Britton 1990:85, 87, 88, Figure 8). 'Coffee chainy cupps' and 'coffee dishes' are mentioned in English potter George Ecton's (1696) stock (Vaisey and Celoria 1974:29-30). The Bateman List of 1696 refers to chocolate cups (Wills 1967:443).

A cursory examination of 17th-century Jamaican household probate inventories also revealed to the author that many people in the colony owned these and other accroutements. The merchant Charles Booker, for example, had several 'Coffe dishes,' 'Coffe plates,' and 'coffe potts' (JPA 1688, 3:112), while Richard Roberts, a planter, owned '12

Chocolatto Cupps' (JPA 1690, 3:313-314). Isaac de

Fonseca Valle owned a 'cocoa cup tipt with silver' (JPA 1690, 3:361). 'Chocolatt pots' and 'Chocolatta stones' also appear to have been popular (see, e.g., JPA 1685, 2:93; 1690, 3:353; 1692, 3:402, 398; 1693, 3:446).

Three Staffordshire slipware cups (one of which is almost entirely reconstructed) are similar in size to those described above but have a more bulbous lower body (see Figure 4.8*a-b*). Most of the sherds of the reconstructed slipware cup were found in Room 4A. Fragments from the other two were found in the alley and Yard 5. All have the typical buff to pale yellow fabric characteristic of Staffordshire clays and are decorated on the exterior with the 'hallmark' combed/feathered dark reddish brown (5YR 2.5/2) slip. The clear lead glaze applied to these cups imparts to them a glossy yellow-brown (10YR 5/8) hue. Patches of olive green staining are also evident on the interior and exterior of most of the sherds. This could be a result of reduction in oxygen in the kiln during firing (Thompson et al. 1984:35) but is more likely sulfide staining caused by the underwater environment.

The Chinese export porcelain cups from Building 4/5 are of the plain *blanc de chine* tradition. (A total of 27 cups of similar shape was recovered from Building 8 [see Dewolf 1998:130, 135-139].) One was recovered intact from Yard 5 (Figure 6.18); the second vessel was found in fragments on the brick-paved sidewalk in front of Building 5. The intact piece is similar in form and size (but handless) to those made in tin glaze, but Dewolf (1998:108) notes that such cups were originally made for the consumption of wine, a popular beverage in China from at least the first century B.C. Seventeenth-century Dutch painter Osias Beert shows a similar-sized porcelain cup containing berries in a still life scene (Rijksmuseum, Amsterdam, illustrated in Van de Pijl-Ketel 1982:33).

That the intact porcelain cup was found in Yard 5 is interesting. Someone may have been drinking from it in the yard when the disaster struck. Alternatively, but perhaps less likely, given that it *is* intact, it may have been originally located in an upstairs room of Building 5 and was displaced by the earthquake/tidal wave. As for the fragmented pieces: they may have been parts of a cup originally housed in Building 8, which is located across the street, north of Building 4/5.

Caudle Cups (n=3)
Caudle (from the Old Norman French, *caudeler*, 'to heat gently') was made with gruel and spiced ale or wine and sweetened with sugar or honey. Taken warm, the drink was given chiefly to invalids (Markham 1986[1615]:26; Britton 1990:85; Hess 1995:127). It was also considered an appropriate beverage for festive occasions, such as weddings, baptisms, or celebrations on the birth of a child (Belden 1983:138; Archer 1997:261; Thirsk 1999:23).

The caudle or 'cawdle' cup was popular in the second half of the 17th century. It was more like a large mug than a cup and was characteristically of squat, globular shape with either one or two handles and often with a lid. Such cups were made in various sizes and were likely used for a number of purposes, even as containers for creams or baked custards, as illustrated in Belden (1983:Figure 4.8). Other examples of caudle cups are illustrated in Noël Hume (1977:Figures 6, 7), Britton (1987:122-125), and Rackham (1987:2:99). They are a frequent vessel form found in 17th-century sites.

FIGURE 6.17. Detail from 'An English Family at Tea,' by J. van Aken, ca. 1720. Tate Gallery, London. Chocolate/coffee cups similar to those found in Building 4/5 are shown on the table.

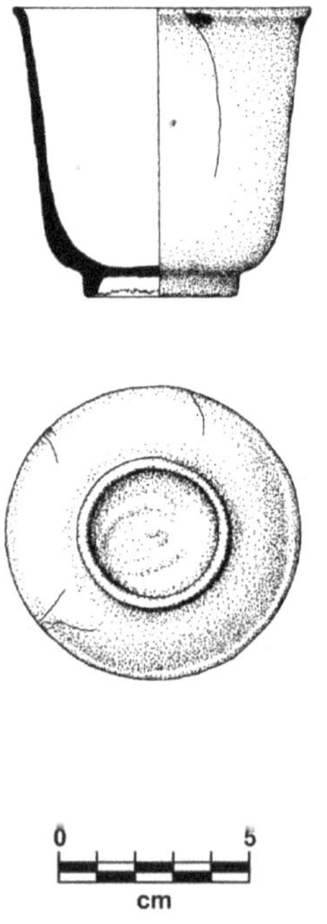

FIGURE 6.18. Intact Chinese export porcelain (undecorated *blanc de chine*) cup recovered from Yard 5 in the 17th-century context at Building 4/5. (Illustration by Helen Dewolf.)

At least three tin-glazed undecorated white caudle cups are represented in Building 4/5's analyzed assemblage; one, recovered in fragments from Rooms 2 and 3 (the majority found in the former location), has been almost completely reconstructed (see Figure 4.11*g*). It is a typical example of a (11 cm rim diameter) caudle cup with its globular lower body, which narrows towards the rim. It has one handle attached, and the plain undecorated glaze is in relatively good condition. Other caudle cup fragments (representing at least two vessels) were recovered from Yard 4A, directly behind Room 4A.

Cups - Use Unknown (n=2)
Two small, somewhat unusual cups were also recovered from Building 4/5's Layer 3. One, a Staffordshire slipware product, decorated on the exterior in the typical late 17th-century style (a dark reddish brown [5YR 2.5/2] combed and feathered slip) was recovered intact from the northeast corner of Yard 5 (Figure 6.19).

FIGURE 6.19. Intact Staffordshire slipware miniature cup recovered from Yard 5 in the 17th-century context at Building 4/5. (Illustration by Helen Dewolf.)

It measures only about 5 cm in diameter (excluding the handle) and is relatively short, at approximately 9 cm. The handle itself is very small, allowing room for only a forefinger, and the vessel's restricted orifice suggests a use other than beverage consumption. It may be that this 'cup' was used as a condiment/spice (sugar?) serving bowl. Or, perhaps, it held pharmaceutical substances, since a similar tin-glazed form (albeit with an inverted rim), illustrated in Caiger-Smith (1973:Figure 131), is recorded as an ointment pot. Alternatively (unlikely), the vessel may have been purely ornamental, brought to the port by an English settler as a sort of keepsake. That it was recovered intact from Yard 5, like the intact porcelain cup described above, suggests that someone was drinking from it in this area when the disaster occurred.

The second cup, which is, perhaps, the rarest piece in the analyzed assemblage, is a small, relatively shallow (5 cm deep) lobed quatrefoil vesse recovered from Room 3 (Figure 6.20). Charlotte Wilcoxen has made the following comments:

> It is shaped in a form that derives from China's Sung Period (A.D. 960-1279). Flowers have a prominent presence in Chinese legend and art, and the quatrefoil represents the begonia, a flower said to have sprung from earth watered by the tears of a woman disappointed in love. The exterior of [the cup] has no decoration, while a growing tulip in polychrome decorates the interior. The white glaze is stained and badly crazed, yet it adheres closely to the body (Wilcoxen 1992a:11).

FIGURE 6.20. Reconstructed decorated English (?) tin-glazed earthenware shallow, lobed quatrefoil cup recovered from Room 3 in the 17th-century context at Building 4/5

As noted in the discussion of the tin-glazed dish, the tulip motif was popular with Dutch potters in the mid 17th century. English potters, however, did not adopt the design enthusiastically until the second half of the century. As Wilcoxen (1992a:11) continues: "The possibility that the quatrefoil cup is Dutch was explored, but a reliable Dutch authority assured [Wilcoxen] that he has never encountered a similar cup of Dutch manufacture. The place of its recovery, the excessive crazing of its glaze, and the ruling out of a Dutch origin support an English source."

The stub of a handle is evident on one side of this piece; however, the opposing side is missing, and it may have had two handles. Its short, squat form indicates a use other than beverage consumption. It may have been used as a sugar or sweetmeat bowl.

Drinking Pots

Beaudry et al. (1983:30) define a drinking pot as a "one or multi-handled vessel, usually bulbous, but sometimes cylindrical in form, ranging in capacity from 1 pt to 2 qts or more." A tubular drinking/pouring spout was sometimes attached to the body, and often the vessels were made with a matching cover (see Archer and Morgan 1977:Figure 15; Britton 1982:69-74, 1987:Figure 125; Archer 1997:262-265 for examples). Drinking pots are common finds on sites of the 17th and early 18th centuries and were made in pewter and silver, as well as in clay. Poole (1995:32) notes that from the last quarter of the 17th century, some were even made in glass. Drinking pots were generally communal vessels, passed from hand to hand as each drinker took his fill, and are often illustrated in tavern scenes of the period, such as Adrian Brouwer's 'Fight over Cards' of ca. 1631-35 (Alte Pinakothek, Munich). In contemporary pottery inventories, they are often denoted by their contents, known variously as posset–, syllabub–, or wassail pots (see, e.g., Vaisey and Celoria 1974:33; Britton 1990, 1993:63).

Posset, basically milk curdled with spiced wine or other liquor, was related to caudle and was taken either warm or actually hot. From the outset, posset, like caudle, appears to have had both a medicinal and social function, depending on the particular mixture produced (Hess 1995:133; Poole 1995:32; Archer 1997:261). Thicker possets, which often included eggs and barley, oatmeal, or bread, were often eaten with a spoon. Those made with tubular spouts allowed the mix to be poured, or more commonly sucked, directly from the pot (Britton 1982:68; Belden 1983:137). Syllabub, at its simplest, was milk injected with some force into sweetened wine, cider, or beer/ale in order to create a cold frothy drink, which was drunk from an individual bowl or cup (Belden 1983:137; Hess 1995:138). Wassail differed significantly from posset and syllabub. It was a sweetened and heavily spiced ale into which the dissolved pulp of apples had been mixed and was served only on Twelfth Night in a ceremony known as Wassail (Archer 1997:261).

The five drinking pot examples from Building 4/5 are all in tin-glazed earthenware: one is almost completely reconstructed and is described further below; two are represented by lids; and at least two more can be identified by numerous diagnostic sherds, such as rims, elaborately scrolled handles, and broken spouts.[3]

The reconstructed piece (15 cm rim diameter x 12.5 cm tall) (Figure 6.21), a generously proportioned two-handled vessel that possibly originally had a lid, is covered in a blue glaze and was at first thought to be an example of the *bleu de Nevers* manner of decoration, which developed in France, at Nevers, in about 1630, and was occasionally used in England (Caiger-Smith 1973:113). On closer inspection, however, the pot shows no evidence of the white tin pigment, which in true *bleu de Nevers* style, is painted over the glaze. Further, the blue glaze is slightly lighter in tone than that used by the French. The fragments to the vessel were recovered from Rooms 3 and 4 and from the northwest corner of Yard 4A. It is suggested that these rooms were used as kitchen/food preparation areas. (Of the other drinking pot fragments: one lid was found in Room 1, while the other was found in Room 4A. The remaining sherds were found in Room 4 and Yards 4A and 4B.)

Mugs

At the end of the 17th century, mugs—single-handled, straight-sided drinking vessels (Beaudry et al. 1983:30)—were beginning to be favored over large, communal drinking pots, as the importance of individuality blossomed in the 18th century. As Yentsch (1991:42) notes: "If the phrase 'to each man his own plate' is seen by some archaeologists as exemplifying ceramic assemblages of the later 18th century, then 'to each man his own mug' was its forerunner." By about 1730, most sites in the New World show an increase in the numbers of mugs, as reflected in sherd counts. However, as Yentsch (1990:41, 1991:43) cautions, this increase may merely reflect a higher breakage rate. Mugs perhaps break more easily because they were subjected to more or rougher handling than other vessels (see David 1972).

While no more mugs than drinking pots are represented at Building 4/5, their very presence indicates that change "within the rituals of social bonding associated with drinking" (Yentsch 1991:43) was already become apparent prior to 1692. Significantly, all of the mugs identified are imported English and German stoneware. This is not surprising, given the ware's waterproof properties. (Salt glaze, when applied, is purely

[3] An almost intact (15 cm diameter x 12.5 cm tall) Staffordshire slipware drinking pot was recovered from Building 8. Its dark reddish brown (5YR 2.5/2) slip, applied in a feathered, vertical pattern, covers the vessel's lower exterior body. The plain straight-collar neck is inscribed with the potter's name: [RICHARD] MEIR, followed by the date: [1] . . . Dots of white slip (now brown/green) adorn the potter's name. The resulting 'jeweled' effect is common on many of Staffordshire's ornamental slipware vessels, but as Draper (1984:17) notes, this effect is also found on simpler vessels such as the Port Royal example.

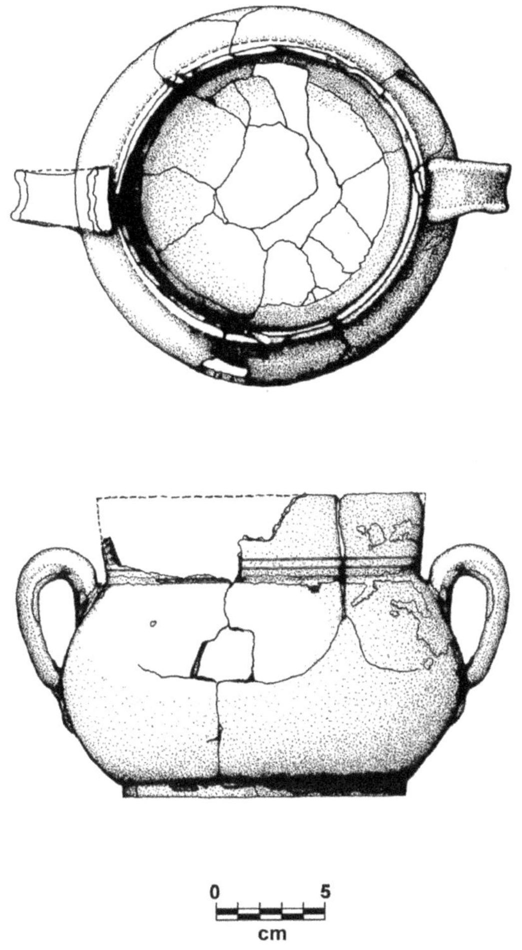

FIGURE 6.21. Reconstructed decorated English tin-glazed earthenware drinking pot recovered in fragments from Rooms 2 and 3 and Yard 4A in the 17th-century context at Building 4/5 (see Figure 4.11*c* for color photograph). (Illustration by Helen Dewolf.)

decorative, as are the appliquéd and impressed designs so often seen on this ware.) Further, as noted in the section on storage ware above, since stoneware has no interior lead glaze, prolonged contact with acid wine was not a health hazard as with lead-glazed earthenware.

Three of Building 4/5's five stoneware mugs were made in the Westerwald region of Germany (see Figure 4.16). As noted in the discussion of the German stoneware bottles, these vessels were possibly brought to Port Royal either via the English market or by Dutch traders, who were the most active distributors of German goods in the 16th and 17th centuries (Noël Hume 1970:276-285; Wilcoxen 1987:73; Gaimster 1997:98). Two of the mugs are represented by uniquely decorated body sherds. The third is a partially reconstructed rim and body section decorated in blue bands and mold-applied gray rosette motifs. All three vessels were recovered from Yard 4A. They may have been dumped in this area as trash, having broken during the life of the building. Alternatively, they were still in use in 1692 and were displaced from Building 4 by the earthquake/tidal wave.

Westerwald stoneware was more expensive than undecorated stoneware (which may account for its relative paucity at the site). When fitted with hinged pewter lids, for which the handle had a special hole, the value of these vessels was equal to those of pewter; some pieces even received silver fittings, often more valuable than the vessel itself (Schaefer 1998:68). Painted representations of Westerwald stoneware associate it both with wine and beer.

Two English slip-dipped stoneware mugs are represented in the analyzed assemblage and are illustrated in Chapter IV (see Figures 4.16 and 4.17). As noted in Chapter IV, it is possible that they are intrusive from upper layers, since both are similar to Staffordshire types illustrated in Mountford (1971:Plates 15, 52), which are dated to about 1710. London potter John Dwight, however, was making white-slipped wares, usually tavern mugs of this sort, in the late 17th century (Draper 1984:36). Thus it is equally possible that these vessels were part of the building's inventory in 1692. Both examples are represented by rim sherds, and both are covered with a white slip. One rim (represented by two sherds recovered from Room 4B and the alley) is edged with a thin band of dark reddish brown (5YR 2.5/2) slip; the other, found in Yard 6, is covered with an exterior freckled brown wash. This wash likely only covered part of the complete vessel (cf. Mountford 1971:Plate 15).

Costrels

A costrel is a bottle or flask with a bulbous body, narrow neck, and two ears or strap handles, each rising from opposing sides of the vessel at the shoulder (Beaudry et al. 1983:32). A cork or parchment/bladder cover stoppered the mouth, and a cord or leather thong was threaded through the handles so that the vessel could be carried or tied to a belt or saddle. Costrels were "most used among the shepherds and harvest people of the countrey," according to one early 17th-century chronicler (Chappell 1859:515). An alternate scenario that would be more common in an urban mileau such as Port Royal would be it use by travelers or as a household water container.

Four ceramic costrels were recovered from Layer 3 at Building 4/5: two in earthenware and two in stoneware. The earthen forms are represented by a red coarse ware basal sherd, possibly of Iberian provenance, and an almost completely reconstructed piece (15 cm diameter x 32 cm tall) in North Italian marbled slipware. The basal sherd was recovered from the brick-paved sidewalk in front of Building 5, and so it is possible that it originated in either the upper story(s) of Building 5 or in Building 8, located across the street. The slip-decorated costrel was found in Room 1, an area which may have served either as a place to entertain guests and/or as an commercial eating establishment. Its white/cream marbling design covers a soft light red (10R 6/8) fabric, and the whole vessel is lead glazed on the exterior. A loop handle is evident on either side of the shoulder; each handle is molded in the form of animal/human mask. This vessel likely had a pedestal foot, and its neck would have been topped with a rounded mouth. The costrel form with this decoration has been found on many colonial New World sites and seems to have been produced from around 1610 to 1660 (Noël Hume 1970:77; Lister and Lister 1976a:33-34). An example was also recovered from the 17th-century levels at a site in Aldgate, London (Thompson et al. 1984:Figure 23, Number 88). Specific to Port Royal, a similarly decorated bottle was recovered by Mayes (1972:82, Figure 22, Number 20) in the 17th-century context of the old naval dockyard site.

Both stoneware costrels were found in the yard areas, specifically in the northwest corner of Yard 4A and next to the hearth in Yard 7. Both are fairly thin and relatively lightweight, and like the earthen forms, probably were used to carry water. One stoneware form, represented by several body sherds, has a light yellowish brown (10YR 6/4) salt-glazed fabric and a short, ridged high-arched handle of a size and orientation that suggests that it would have held a cord so that the vessel could be carried. The second form, also represented by body sherds, has a light brown to mottled pink unglazed exterior. It has a similar handle with three depressed ovals at one end.

The stoneware costrels are possibly Dutch, since vessels with similar handles were recovered from the Dutch East Indiamen *Zeewijk*, which wrecked in 1727, and the *Witte Leeuw*, which wrecked in 1613 (Ingelman-Sundberg 1978:74, 146, 197; Van der Pijl-Ketel 1982:223-242). A similar handle illustrated in Green (1977:1:Figure GT 913) was found among the remains of the Dutch ship, the *Vergulde Draeck*, which wrecked off the coast of Australia in 1656. Alternatively, they may be of South East Indian provenance, since they are similar to storage jars from that region known as *martavans* (Van der Pijl-Ketel 1982:220-221; Wilcoxen 1992a:38).

Ewer

Several cross-mended sherds belonging to a ewer, which is possibly of Spanish provenance, were recovered from Rooms 1 and 2. The ewer is made in white unglazed earthenware and shows reeded decoration on its neck, and two incised circular lines are evident along the shoulder (see Figure 4.3*b*). The two-handled, globular ewer form is common in the Middle East and Mediterranean regions, and while no other vessel of this type has been found in Port Royal, similar vessels have been found on other contemporary American sites that were in contact with Spanish shipping (Lister and Lister 1987:13, 19, 27, Figures 10, 14, 21; Wilcoxen 1992a:43). A similar form in glass is illustrated in Deagan (1987:130, Figure 6.1).

Punch Bowl

As was noted earlier in this chapter, prescribing the function of a bowl from a handful of sherds is notoriously difficult, and it is with caution that the author suggests at least one of these forms, recovered in fragments from Room 2 (tin-glazed basal section in blue and white and possibly red floral decoration), may have been used to serve punch.

That a punch bowl would be found in the building may be expected, since the form became common in the late 17th century, and according to Beaudry et al. (1983:33), ranged in capacity from a half pint to several gallons. The larger forms were used as serving containers from which cups or wine glasses were filled with the drink using a ladle. The smaller forms were often used as drinking containers (Beaudry et al. 1983:33). It has been suggested that the word 'punch,' which according to the Oxford English Dictionary was first used in 1632, derives from the Hindu word *panch*, meaning 'five,' the number of ingredients used to make the drink: generally rum or brandy and red wine to which were added sugar, spices, and lime or lemon juice, diluting the mixture with hot or

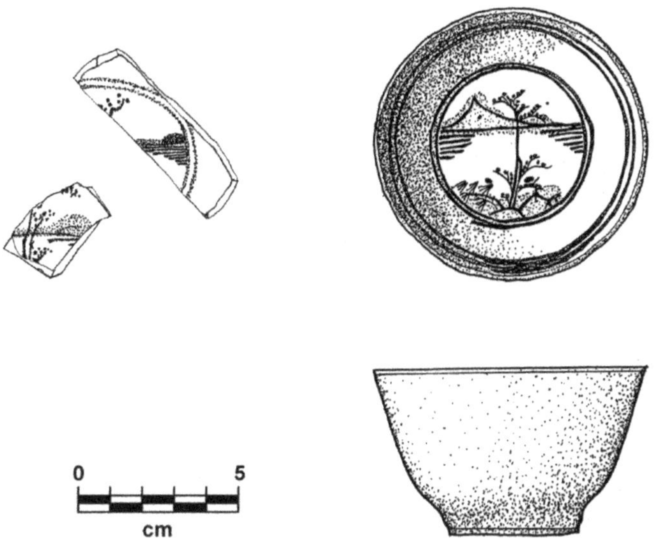

FIGURE 6.22. Basal fragments from a Chinese export porcelain (Batavia ware) tea bowl (and hypothectical reconstruction) recovered from the brick-paved sidewalk in front of Building 4/5. (Illustration by Helen Dewolf.)

cold water "according to the season and the capacity of the drinkers" (Belden 1983:237). The word 'puncheon' has also been suggested as the origin for the name of the drink, since punch in early references is frequently a sailor's drink of lime juice, sugar, and water mixed with a ration of rum doled out from a puncheon (Belden 1983:237).

Punch certainly seemed to be popular among 17th-century Port Royalists. A cursory examination of household probate inventories for the period, for example, revealed that many of the colonists owned punch bowls (e.g., JPA 1689, 3:317; 1690, 3:333, 355; 1693/4, 3:606). Brown (1996:181-182) records 12 of these forms from an excavated 17th-century tavern located in the city's New Street.

Tea Bowl

The only identifiable tea bowl in the analyzed assemblage is in Chinese export porcelain (Batavia ware) (Figure 6.22). It is represented by two basal sherds recovered from the sidewalk in front of Building 5. It may be that the vessel was part of Building 8's ceramic inventory, since it included at least six similarly decorated tea bowls in its 17th-century context (see Dewolf 1998).

Just as there is some doubt as to the varying shapes of cups used for the different beverages, tea bowls may not have been used for drinking tea. Indeed, evidence exists to show that their use in other roles was probably just as common. Dutch genre paintings, for example, show them being used to serve berries and nuts (see, e.g., a 17th-century still-life painting by Dutch artist Floris van Dijck [Rijksmuseum, Amsterdam, illustrated in Van der Pijl-Ketel 1982:35]). Schaefer (1998:75) points to their use for wine, or even caudle, consumption.

It is entirely possible, too, that some of the unidentified tin-glazed fragments in the assemblage may also be parts of tea bowls, since a cheaper version of the form was also produced in this ware at the end of the 17th century (see examples in Archer 1997:Figures H.1, H.3, Plate 219). Certainly the thinness and whiteness of porcelain could not be copied, but tin-glazed examples usually imitate the gracefully flowing everted rims. In fact, as Yentsch (1990:43) notes, "delft [i.e., tin-glazed earthenware] was an acceptable substitute [for the porcelain form] until ca. 1720-1740. No one suffered significant loss of prestige by serving tea in delft vessels because the ability to purchase and serve tea was sufficient in and of itself to denote status."

Tea was as expensive in Port Royal as it was in Europe at the end of the 17th century. In Charles Booker's inventory, for example, 'four pound of tea' was valued at £4. Compare this to some of his other inventoried items, such as 'fourteen leather chaires' for £4, 'six dozen pint bottles of Canary' for £3.10s., or 'fifty pound of pewter,' valued at £2.10s. (JPA 1688, 3:112). The inventory of Joseph Bedow, a Port Royal merchant, also shows the value given to tea: in 1689, two pounds of the brew was appraised at £1.10s, equal in value to Bedow's '4 caine chairs and 2 tables' and slightly more than his '14 bottles of oyle' (valued at £1.8s.) and even his '5-6 oz. wrought gold' (valued at £1.4s) (JPA 1689, 3:347).

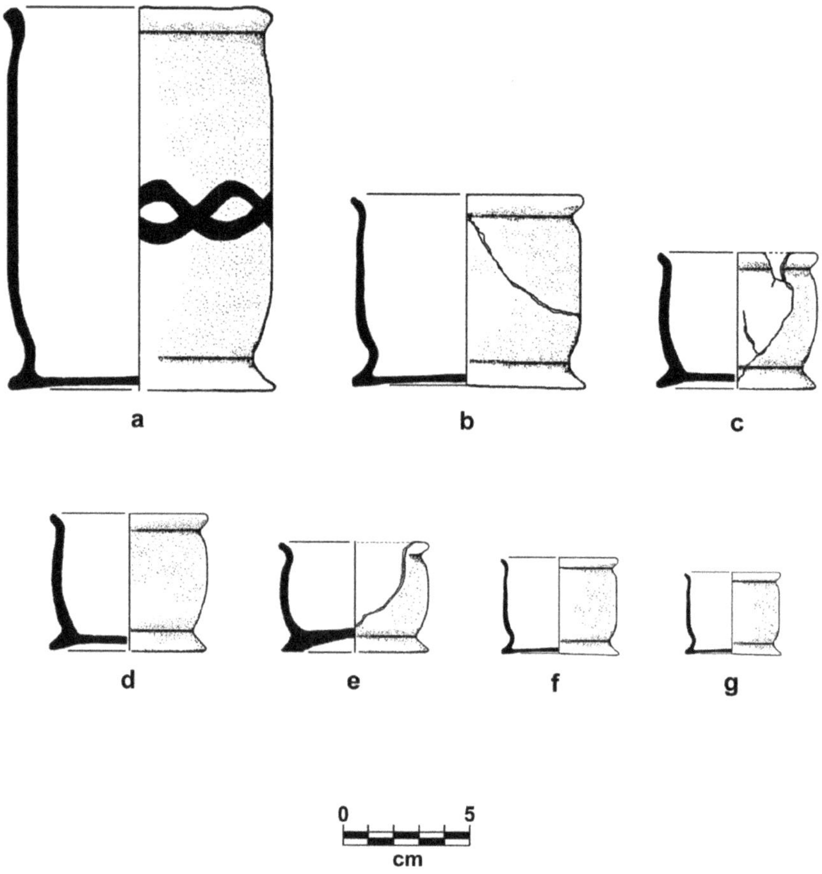

FIGURE 6.23. Examples of English tin-glazed earthenware apothecary pots recovered from the 17th-century context at Building 4/5. *a*, intact decorated form found in Yard 6; *b-g*, intact, near-intact, and reconstructed forms found in and around Room 3 and Yard 5. (Illustration by Madeleine J. Donachie.)

VESSEL FORMS RELATED TO HEALTH AND PERSONAL HYGIENE

Ceramic vessel forms related to health and personal hygiene were also identified in the Building 4/5 assemblage. A total of 103 sherds, or 8.9 percent of the analyzed sherds, were catalogued by the author as belonging to forms used for such purposes. These sherds represent an estimated minimum of 24 vessels, or 16.7 percent of the total number of ceramic vessels taken to be associated with the building's pre-1692 context. The vessel forms represented are ointment/drug storage jars, here combined and referred to as apothecary pots (n=16); chamber pots (n=6); and washbasins (n=2). While the majority of the forms are made in tin-glazed earthenware, a few are in coarse ware.

The relative proportions of ceramic forms related to health and personal hygiene and their respective sherd counts are shown in Table 6.1. Each form and its distribution is now discussed.

Apothecary Pots

Apothecary pots, often known as gallipots, supposedly after the Venetian rowed *gallées* that first brought them to northern Europe, are cylindrical, usually decorated, vessels that were used for storing drugs and ointments and sometimes also condiments (Noël Hume 1970:203-204; Beaudry et al. 1983:37; Archer 1997:377-380). They were made in an array of sizes by European tin glaze potters from the late 16th through the early 18th centuries, and demand for them was enormous. As delftware scholar Michael Archer comments:

> When Jasper Andries and Jacob Jansen petitioned Queen Elizabeth in 1570 for a patent to make tin-glazed earthenware in London, they specifically mentioned that they made 'Galley pavinge tyles and Vessels for potycaries.' In his application for the patent he was in awarded in 1628 Christian Wilhelm included 'Apothecaries . . . pott[es]' in his list of the types of delftware he was making. By the end of the century apothecaries wares dominated the

> output of the Pickleherring Pottery, representing no less than 28 per cent of the total production. The Gravel Lane inventory of 1727 includes 'Apothecary's Pots,' and it is clear from all this evidence that such jars were the single most important product in some delftware factories and among the staple wares made in all of them (Archer 1997:377).

Given their popularity—there is ample evidence that apothecary pots were frequently treated as general storage containers by society as a whole—it is not surprising that at least 16 were recovered from Building 4/5.[4] In fact, together with bowls, they are the second most prevalent form in the analyzed ceramic assemblage (after coarse and stoneware storage pots). All of the excavated vessels have the characteristic tin glaze (it was not until about 1800 that tin-glazed pharmaceutical containers were replaced by equivalents in lead-glazed earthenware or glass [Archer 1997:380]), and various shapes and sizes are represented. Some were recovered intact, some are near-intact, and some were identified by the author by their respective rim and basal sherds.

While most of the apothecary pots have lost their original decoration, a tall, intact form (recovered from Yard 6, near the cistern) shows the very faint remains of a cobalt blue figure-of-eight chain (a common contemporary motif) around its lower body (Figure 6.23*a*, see also Figure 4.11*e*). (Apothecary pots of the period were often decorated over their white tin glazes in blue and manganese purple, and also often yellow geometric patterns.) Blue dots are faintly evident towards the vessel's base. A vessel of similar shape and size (ca. 10 cm diameter x 14 cm tall) is shown in Britton (1982:Figure 5.5) and is attributed to Bristol and dated to ca. 1700. In the Pickleherring Potteries inventory, this type of vessel is referred to as a 'stall jar,' after the Dutch *stallen* ('to put away') (Britton 1990:91). Indeed, Frank Britton (1987:Figure 9a) shows a selection of variously sized apothecary pots recovered from sites in London that are similar to the collection from Building 4/5 (see also Archer 1997:381-386 for other examples).

A second painted (possible) apothecary pot (also found in Yard 6, associated with the cistern) is represented by a large rim and body sherd (see Figure 4.11*b*). (The original vessel may have been a ginger jar or vase.) The colors are more pristine than the above example, and the design, in a light, soft blue on a bluish white ground, is a typical European adaptation of a common Chinese theme: a Chinese figure standing among rocks and clouds in an oriental landscape. Like all of the apothecary pots, this vessel probably had a cover, a parchment or a piece of bladder, which could be tied tightly down over the rims to keep the contents fresh (Archer 1997:377; Schaefer 1998:89). Its shape is similarly common for the time (see examples in Britton 1982:Figure 5.4, 1987:Figure 118; Archer 1997:391-395).

The distribution of apothecary pots throughout the building complex appears to be somewhat arbitrary. However, a concentration is apparent in and around Room 3 and Yard 5. It might be that the larger forms were used in roles associated with food preparation, since many of these activities occurred at the rear of dwellings and in the back yard areas. Seventeenth-century recipes certainly point to the use of apothecary pots as convenient containers for "conserves of prunes and damsons," "sweet meates," "sweet sucket," quinces, currants, mustard, prunes, "pippins" (apples) and "Musk Sugar" (cited in Vaisey and Celoria 1974:27-28; see also Hess 1995:161, 238). Britton (1987:114) suggests that some small apothecary pots may have been used on the table to hold salt, in the fashion of modern salt cellars. Of course, they could also have been used as containers for the drugs and ointments for which they were originally made. Cleansers, perfumes, cosmetics, and medicinal waters and oils are some other possible contenders for these pots' original contents (Crellin 1970:197; Archer 1997:380; Schaefer 1998:89).

It is possible that some (n=4) of the small apothecary pots recovered from Building 4/5 are intrusive from upper layers, since their slightly pedestal-footed bases are suggestive of an 18th-century date (Archer 1997:385-386). If so, it may be that they were dumped in Kingston Harbor as trash by the hospital staff at the naval base, which lies close to the area excavated by Hamilton and which was in operation in Port Royal from 1743 to 1905.

Chamber Pots

Six intact and near-intact chamber pots made in a variety of wares and types were also recovered from the building complex. They are all of similar form—one-handled vessels with convex sides and a sturdy, flat-topped rim or brim—and of similar size. Two intact examples are in Border ware. Two (intact and near-intact) are made in lead-glazed, undecorated English Redware, and two tin-glazed examples have been reconstructed from cross-mended sherds.

Both Border ware chamber pots are in particularly good condition. The red variety (Figure 6.24*a*) (16 cm rim diameter x 13 cm tall), a squat-bodied flat-rimmed vessel typical of the late 17th century (rims on earlier forms are often everted and may also have an internal ledge, which allowed fro the fitting of a lid) is glazed green (2.5Y 4/4) on the interior, with runs and dribbles of glaze also apparent on the exterior surfaces. (The latter characteristic suggests the pot was stacked in an upright position in the kiln for firing [see Pearce 1992:85].) The thick vertical strap handle has a pinched lower terminal with a central thumb-print impression. The White Border

[4] Of the 77 large and small apothecary pots recovered from a 17th-century tavern on New Street, Port Royal, 60 of the larger forms are thought to have been used as general storage containers (see Brown 1996:216-222).

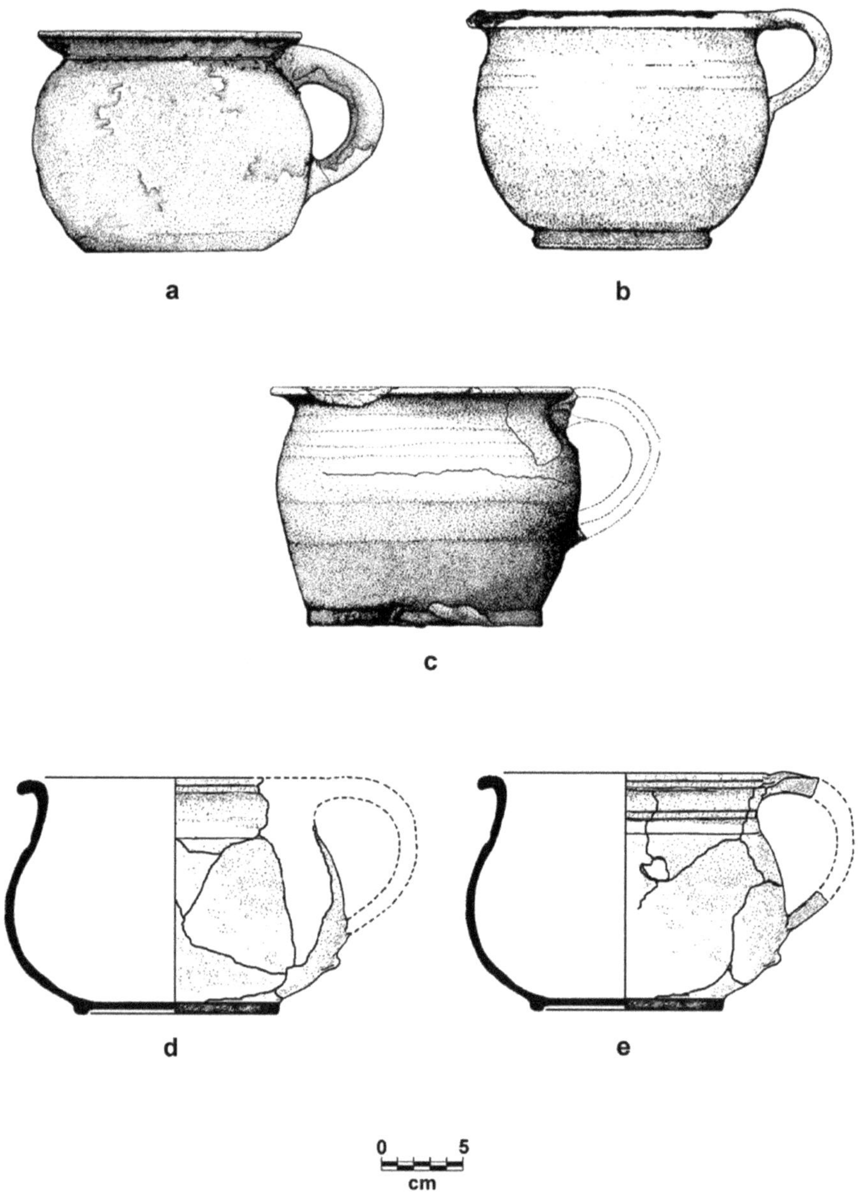

FIGURE 6.24. Examples of chamber pots recovered from the 17th-century context at Building 4/5. *a*, intact Red Border ware form found in Yard 4A; *b*, intact White Border ware form found in Yard 5; *c*, intact English Redware form found in the alley; *d*, reconstructed undecorated English tin-glazed earthenware form found in Room 1; e, reconstructed undecorated English tin-glazed earthenware form found in Yard 5. (Illustrations *a-b*, *d*, and *e* by Madeleine J. Donachie; illustration *c* by Helen Dewolf.)

ware form (see Figure 6.24*b*) (18 cm rim diameter x 14 cm tall) is also covered in a dark green- to olive- (2.5GY 6/8 - 7.5Y 8/4) colored glaze but only on its interior surfaces. It is slightly larger than the red form, and although it has flat, broad rim, it has the slightly ovoid shape of earlier vessels. It too has a broad strap handle, which is attached at the rim and mid body. A thumb-print impression is also evident on the handle's lower terminal. Various kiln scars are present on the White Border ware chamber pot's base and interior surfaces. Those on the base are associated with small patches of green glaze, likely from another vessel next to it in the kiln. The scars on the interior provide evidence that smaller vessels were sometimes placed within larger ones during firing in an effort to increase kiln productivity (Pearce 1992:85).

Both English Redware chamber pots are typical of the period. The intact example is illustrated in Figure 6.24*c*. The second example is represented by several rim and body sherds.

The undecorated tin-glazed chamber pots are not in as good condition as the heavier Border ware types. Much of the tin glaze is crazed and discolored and has also crawled significantly. The example illustrated in Figure

6.24*d* has a wider, slightly more flared rim than that shown in Figure 6.22*e* (20 cm vs. 16 cm), but both vessels are approximately the same height (15 cm), and the handle of each has a pinched lower terminal with a central thumb impression. Such handles are typical of English tin-glazed chamber pots of the late 17th century. The 'better' pots usually have handles with a little more character—an ornamental terminal, perhaps, or a single or double reeding (Noël Hume 1970:146). Chamber pots at this time were also often decorated, as noted in contemporary pottery inventories (see Wills 1967:443; Britton 1990:68, 70). Noël Hume (1970:147) notes that some were elaborately painted with motifs of cobalt blue in the Chinese manner.

Four of the six chamber pots were found in and around Building 5: the White Border ware example and the near-intact Redware form were recovered from the northwest corner of Yard 5, near the remains of a two-hole privy located behind Room 1. One of the tin-glazed specimens (Figure 6.24*e*) was also found in the yard but in its northeast corner. Most of the sherds from the second tin-glazed form were recovered from Room 1. Associated with Building 4 was the Red Border ware chamber pot (found in Yard 4A) and the intact Redware form (recovered from the alley).

Building 4/5's privy, a small triangular structure, was built on a brick foundation above the level of Yard 5 and contained the remains of what appears to be a wooden seat with two circular openings. A chamber pot, or even a bucket, would have been placed under each opening, periodically being emptied in an appropriate place. Indeed, it may be that all Port Royal privies were constructed in this way, since comparable privies have been excavated in other parts of the 17th-century city (see Brown 1996:155-156). As Pawson and Buisseret (1975:94) note: "water lies everywhere just a few feet (or even sometimes inches) below the surface," so the more common earth pits dug below the ground could not be constructed. Certainly, it appears that the chamber pot was a common item in 17th-century Port Royal dwellings. Brown (1996:222), for example, notes that 10 such vessels were recovered from an excavated tavern on New Street. Entries for chamber pots or 'close stools' in Jamaican household probate inventories from the years just before the earthquake substantiate the archaeological finds (see, e.g., JPA 1687, 3:26, 54; 1689, 3:64, 254, 259, 309, 317; 1690, 3:323, 380; 1692, 3:427).

Washbasins

What appears to be the remains of two tin-glazed earthenware washbasins is an interesting find. One washbasin has been almost fully reconstructed. The other is represented by a large, undecorated rim sherd. The reconstructed piece (28 cm rim diameter x 12.5 cm deep), illustrated in Figure 6.25, is glazed a greenish blue and is decorated on the interior with a carelessly applied rim-to-rim oriental scene that shows a Chinese figure seated in a landscape among rocks and spiky, sparse decoration. As already noted, this pattern was popular with both English and Dutch tin glaze potters in the late 17th century. Interestingly, however, the small anomalous wooden structure shown in the central foreground is a slight divergence from the formula (Wilcoxen 1992a:17-18). The exterior is devoid of decoration, except for one diagonal slash of cobalt near the rim.

The reconstructed washbasin is probably of English provenance and dates to ca. 1650-1700, based upon the crude, sketchy drawing (which suggests a Liverpool or Bristol attribution, see Archer 1997:Figure B.220, Plate 144). The painter may have copied the design from Chinese porcelain or from Japanese wares, which in the later 17th century were also popular in Europe. Japanese wares copied Chinese Transitional Period originals, except they were sketchier and simplified, with a more schematic treatment (Archer 1997:30).

Recovered in parts from Room 1 and Yard 5, this washbasin was probably used in Room 1 or one of the upstairs rooms. (The rim from the second basin was found in Room 2). That it can be identified as a washbasin rather than, for example, a punch bowl or a large mixing bowl, is due to the vessel's shape: it narrows quite significantly towards its base, suggesting that it may have fitted into a circular aperture cut into a three-legged wooden washstand that was made to hold it. Its rolled-over rim suggests the same. In the 18th century, washbasins were often made *en suite* with tall, long-necked water bottles (sometimes known as gugglets) with bulbous lower sections, which stood on a small shelf below (see Britton 1982:Figures 165a-b, 1987:93-95; Archer 1997:Figure 35). Perhaps this was also true in the 17th century. Washbasins were certainly being made in the 17th century, as shown by an entry for a 'waish bowell' in the 1696 inventory of English potter George Ecton (Vaisey and Celoria 1974:35).

OTHER VESSEL FORMS

Remains of the final three identifiable vessels recovered from the 17th-century context at Building 4/5 are considered here in a distinct 'other' category. One, a tin-glazed earthenware flower vase, has no practical function beyond the decorative. The second, a cross-mended tin-glazed earthenware lid, is from an unknown vessel form. The third vessel (also of unknown form) is represented by a single North Devon Plain ware (i.e., untempered) sherd.

Flower Vase

The reconstructed vase (ca. 10.5 cm rim diameter x 15 cm tall) has a hollow-stemmed trumpet foot and globular upper section with a delicate fluted rim (Figure 6.26). Three sockets for flowers alternate on the shoulder with two twisted horn-like projections (originally there were three). The glaze has almost disappeared from the action of water and abrasives, but traces retained in a crease around the exterior of the base indicate that the piece originally had a greenish blue glaze and was possibly decorated in either a Chinese or floral design in blue and,

Figure 6.25. Reconstructed decorated English (?) tin-glazed earthenware washbasin recovered in fragments from Room 1 and Yard 5 in the 17th-century context at Building 4/5

perhaps, yellow and purple (see examples in Rackham 1987:1:205, Number 1643; Archer 1997:Figure I.2). An initial examination of the vase by Charlotte Wilcoxen resulted in the following comments:

> This piece is one of the most important items in the collection and represents a significant form in English 17th-century tin-glazed earthenware. It is stylistically related to the far more ornate nozzled tulip vases made at Delft in the Netherlands and imported into England for use in the royal palaces by the monarchs William and Mary in the final decade of the 17th century. Simple, plain white vases in this nozzled form were made in England as early as 1643 (Archer and Morgan 1977:Numbers 12, 41), and by the second half of the century, they had become more sophisticated both in form and decoration (Wilcoxen 1992a:20).

Although these vessels are usually labeled flower vases, it is possible that it was filled with earth, and a growing plant placed inside. The water poured into the bowl section would drain through the hollow stem to exit by way of the hole in the base (the hole would be stopped with a cork or hardened candle wax if the vessel was used to hold cut flowers). Noël Hume (1977:30, 70), for example, labels such vessels as "posy holders" or "flower potts." It is equally possible, as Schaefer (1998:86) suggests, that the vase, as today, was considered an *objet d'art* and displayed without flowers. Despite the existence of an entire genre of scenes of flower paintings, very few vases with flowers appear in 17th-century Dutch domestic interiors. That only one (recognized) vase should be found in the building complex (in Room 4A) is not surprising, since any jug, mug, or bottle could be used for the same purpose. (Indeed only two other vases were recovered by Hamilton's team from the Port Royal site as a whole. The first, found in Building 2, is of a narrow, tubular form and has a blue floral decoration painted on a white ground. The second, found in Building 8, may be of Italian provenance, according to Wilcoxen [1992a:27]. Its upper body is molded in a lobed sexfoil, and it has vestiges of ornate double-scrolled handles. Blue-and white floral motifs and scenes of a house and ship appear on opposing sides of its lower body.)

FIGURE 6.26. Reconstructed decorated English tin-glazed earthenware flower vase recovered in fragments from Room 4A in the 17th-century context at Building 4/5. (Illustration by Helen Dewolf.)

Lid (Unknown Vessel)

A tin-glazed earthenware lid (ca. 9 cm diameter), cross-mended from two sherds, was recovered from Room 2 (Figure 6.27). It is painted with the typical oriental-inspired figure-in-a-landscape scene: a male figure is seated among rocks by water and is dressed in a yellow cloak and blue hat and holds above his head a parasol, with its strange, high-crowned shape typical of parasols found on Chinese Transitional Period porcelain, as well as on tin-glazed copies of it produced in the second half of the 17th century (Wilcoxen 1992a:19). It is possible that this piece is of Dutch provenance. The combination of colors (blue, yellow, and purple) is considered "exceedingly rare" on English tin-glazed earthenware of this early date (Archer and Morgan 1977:37). The fine quality of decoration and of potting also vies against English manufacture. The vessel to which this lid belonged was probably very expensive and possibly was acquired as ornament.

Body Sherd (Unknown Vessel)

A North Devon Plain coarse ware body sherd is too small to allow identification of its form. That it is glazed on both surfaces (olive [5Y 5/3] on the interior, grayish brown on the exterior) suggests that it may be from a storage rather than a cooking pot. Its context of recovery, in Room 1, also hints at a use other than food processing.

FIGURE 6.27. Remains of a decorated Dutch (?) tin-glazed earthenware lid recovered from Room 2 in the 17th-century context at Building 4/5

SUMMARY

Table 6.2 and Figure 6.28 summarize the functional classifications for the analyzed ceramic sherds and the vessels they represent. A total of 23 individual vessel forms was identified in the assemblage. A total of 924 sherds, or 80 percent of the analyzed sherds, was identified as belonging to separate and distinct vessels. The remainder (n=231) were too fragmentary to ascertain the type of vessels of which they were originally a part and probably belong to vessels already noted. The estimated minimum ceramic vessel count for Layer 3 at Building 4/5 is calculated to be 144 vessels. The functional classifications used in this study then yield the following 'bottom lines':

A. Food processing forms are represented by:
   (a) 8 cooking pots, 1 colander, and 1 pipkin (all in coarse ware)
   (b) 1 pudding/pastry pan (in slipware)

B. Food and beverage storage forms are represented by:
   (a) 25 storage pots (23 in coarse ware and 2 in stoneware)
   (b) 8 bottles (7 in stoneware and 1 in tin-glazed earthenware)

C. Food consumption forms are represented by:
   (a) 16 bowls (10 in coarse ware, 2 each in slipware, tin-glazed earthenware, porcelain)
   (b) 13 plates (all in tin-glazed earthenware)

(c) 5 porringers (2 each in coarse ware and tin-glazed earthenware and 1 in slipware)
(d) 5 saucers (2 each in slipware and tin-glazed earthenware and 1 in porcelain)
(e) 1 dish (in tin-glazed earthenware)
(f) 1 salt stand (in tin-glazed earthenware)

D. Beverage consumption forms are represented by:
(a) 15 cups (including 3 caudle cups) (9 in tin-glazed earthenware, 4 in slipware, and 2 in porcelain)
(b) 5 drinking pots (all in tin-glazed earthenware)
(c) 5 mugs (all in stoneware)
(d) 4 costrels (2 in stoneware and 1 each in coarse ware and slipware)
(e) 1 ewer (in coarse ware)
(f) 1 punch bowl (in tin-glazed earthenware)
(g) 1 tea bowl (in porcelain)

E. Health and personal hygiene forms are represented by:
(a) 16 apothecary pots (all in tin-glazed earthenware)
(b) 6 chamber pots (4 in coarse ware and 2 in tin-glazed earthenware)
(c) 2 washbasins (both in tin-glazed earthenware)

F. Other forms are represented by:
(a) 1 flower vase (in tin-glazed earthenware)
(b) 1 lid (in tin-glazed earthenware)
(c) 1 body sherd (in coarse ware)

The distribution of the analyzed ceramics within and around the complex shows certain patterns. For example, most of the heavy-duty, easily replaced earthenware, such as the utilitarian coarse wares, appears to be concentrated in the yards and other back areas of the two structures. Many of the forms identified are cooking and storage vessels; as was shown in Chapter III, the kitchen/food preparation areas were located at the rear of the complex. It is not unreasonable to suggest that these areas were also used for storage. (Indeed, remains of large wooden barrels were also found in these locations.)

The finer wares, in contrast, i.è., those ceramics used at table or in the chamber, appear to be concentrated at the front of the complex, again suggesting that they were found where they may have been used or stored. Much of the slipware, for example, was found in and around Room 1, which possibly served as a parlor or as a place that provided the public with food and drink. Room 2 showed another concentration of slipware, tin-glazed earthenware, and porcelain. This room may have been the main entrance to the residential portion of Building 5 (its excavation revealed it had a brick floor laid in a decorative fashion) and also possibly in part a storage space. Tin-glazed fragments recovered from Rooms 3 may have either been used in this area or were displaced there from Building 5's upper story(s) by the earthquake/tidal wave. (Few ceramic remains [only two identifiable vessels] were found in Room 4.)

The functional versatility of stoneware, which allows it to serve both utilitarian and social roles, has been noted already. That stoneware sherds at Building 4/5 were found throughout the structure supports this interpretation. Those fragments found in the rooms are presumably parts of vessels that were still being used in 1692.

That finer ceramic sherds were recovered from the yards and the alley presented the author with ambiguities. Is their presence in these locations due to post-depositional disturbance? Were they displaced to these areas by the earthquake/tidal wave? Or, were they simply thrown in these areas as trash during the life of the building? In the case of Building 4's yards (Yards 4A and 4B), both of which have sand floors (and certainly in the case of the alley), these concentrations of broken vessels may be the remains of accumulated surface middens. (Dumping trash in back areas and public streets was common practice in England and the American colonies in the 17th and 18th centuries, see South 1977:47-80.) Alternatively, these fragments may be parts of vessels that were originally housed in the front rooms of Building 4 (Rooms 4A and 4B), having been blown out on to the yards as the ship rammed through the structure. In the case of Yard 5, which was brick-paved and probably regularly swept of refuse, and which appears to have being used, at least in part, as an extended work, and perhaps social, activity area, some of the fine ware fragments may be parts of vessels that were, indeed, being used in this location. It is equally possible (and probable), however, that some are from vessels that were originally housed in Building 5's upper story(s).

Another ambiguity was the presence of some fine wares (particularly porcelain) on the brick-paved sidewalk in front of Building 5. While garbage was also commonly deposited at the entrances of early modern dwellings, this seems not to have been the case in this instance (comparatively few ceramic remains and other artifacts were recovered from this area, indicating that the sidewalk, like Yard 5, was probably regularly swept). Instead, the author has suggested that the fragments on the sidewalk may be parts of vessels that were originally located either in Building 5's upper story(s) or in Building 8, which lies directly to the north of the complex. Building 8, which may have functioned partly as a shop dealing in fine-quality tableware, contained a high proportion of Chinese porcelain items.

TABLE 6.2. Vessel Forms, as Represented by the Analyzed Sherds (n), Estimated Minimum Number of Vessels (MNV), and Relative Percentages (%), from the 17th-Century Context at Building 4/5 (in Order of Frequency)

Vessel Form	n	% Total n	MNV	% Total MNV
Storage Pot	202	17.5	25	17.4
Bowl	107	9.3	16	11.1
Apothecary Pot	39	3.4	16	11.1
Cup	115	10.0	15	10.4
Plate	53	4.6	13	9.0
Bottle	30	2.6	8	5.6
Cooking Pot	49	4.2	8	5.6
Chamber Pot	51	4.4	6	4.2
Drinking Pot	40	3.5	5	3.5
Mug	15	1.3	5	3.5
Porringer	33	2.8	5	3.5
Saucer	9	0.8	5	3.5
Costrel	30	2.6	4	2.8
Washbasin	13	1.1	2	1.4
Flower Vase	75	6.5	1	0.7
Dish	17	1.5	1	0.7
Salt Stand	13	1.1	1	0.7
Pudding/Pastry Pan	8	0.7	1	0.7
Ewer	7	0.6	1	0.7
Pipkin	7	0.6	1	0.7
Punch Bowl	5	0.4	1	0.7
Colander	1	0.1	1	0.7
Tea Bowl	2	0.2	1	0.7
Subtotal	**921**	--	**142**	--
Unknown Vessels	3	0.2	2	1.4
Unknown Sherds[a]	231	20.0	--	--
Total	**1155**		**144**	

[a] most of these probably belong to vessels already identified

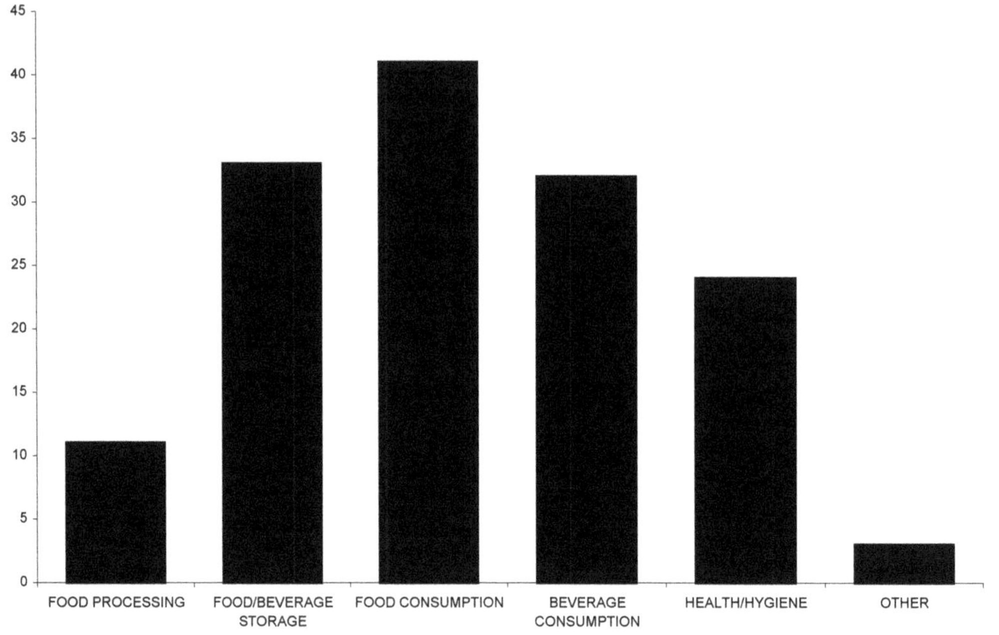

FIGURE 6.28. Estimated minimum number of vessels by functional classification, as represented by the analyzed sherds from the 17th-century context at Building 4/5

CHAPTER VII

INTRA-SITE COMPARISONS

The previous chapter records the location of the ceramic finds from Layer 3 at Building 4/5. Their distribution helps us to a broad picture of the building complex as it was in 1692, and the reader is referred to the author's brief remarks on this at the end of the chapter. For the sake of completeness, the author now looks for a moment at the relationships *inter se* of the various ceramics inventories, inwards at Building 4/5 and outwards—digressing a little—at the larger excavations conducted by Dr. Hamilton.

BUILDING 4 TO BUILDING 5

It has been suggested throughout this study that the individual structures that comprise Building 4/5 may have served different functions: Building 5 being part moderately affluent residence, with at least one room perhaps serving as a place in which food and/or drink was served to the public, with Building 4 being a lesser property, which may have also been a residence(s), housing either tenants or, perhaps, the domestic help who worked in Building 5. There is always the possibility that Building 4 had a commercial aspect to it as well.

Certainly there is nothing in the distributions of the ceramic wares and forms within each of the structures to contradict these suggestions. (A full list of the vessel types and forms recovered from each room and yard in the building complex is presented in Appendix A.) In Building 5, for example, Room 1 and Yard 5 contain about half (n=45) of the total number of vessels recovered from that structure and its associated areas (n=88). Nearly half of Room 1's inventory consists of tin-glazed earthenwares, many of them being the decorated and fancier forms in the entire complex's assemblage (e.g., the decorated serving dish, the decorated bottle, the elaborate salt stand). This seems to coincide, then, with the suggested use(s) of this room, as either a best room, or parlor, or, and perhaps more likely given the building's location in the commercial center of town, as housing some sort of retail establishment connected with eating and/or drinking. As to the vessels from Yard 5 (n=24): many of them are storage pots, which brings to mind the work by historical archaeologist Anne Yentsch (1991:45), where she notes that "the greater complexity of the elite cuisine and the larger stocks of food it required . . . created a demand for more vessels, *especially storage vessels* [emphasis added]." Yard 5 also contained several fine decorated forms, such as the remains of a Mexican majolica plate, an intact Chinese export porcelain cup, and an intact slipware cup. These forms (particularly the two cups which were found intact) may very well have been used in this area or were displaced there from Building 5's upper story(s) during the disaster. Regardless, their presence in the structure points to an establishment whose occupants had the desire, as well as the means, to acquire quality tableware. This is confirmed by the presence of much tin-glazed earthenware in various vessel forms throughout the rest of the building.

As was noted in Chapter III, much of Building 4 was destroyed by the earthquake, and thus its ceramic assemblage (n=56) is a veritable mix of wares and forms. Most noticeably, perhaps, is that nearly half (n=27) of Building 4's ceramic inventory was recovered from Yard 4A. Many of these are coarse ware cooking and storage pots, some of which must have been in active use at the time of the earthquake. The author believes that the finer wares found in this location were probably displaced there from Rooms 4A and 4B, which are located at the front of the structure.

It may also be useful to compare, by ceramic ware and functional classification, the total number of vessels identified in each of the structures that make up the complex. As shown in Table 7.1, tin-glazed earthenware makes up almost half of Building 5's assemblage but only one-third of that at Building 4. There is also slightly more porcelain associated with Building 5. On the other hand, coarse ware is the most frequent ware at Building 4, while at Building 5 it accounts for only 34.1 percent. Furthermore, and what is not shown on the table, a significant proportion of the coarse ware at Building 4 is the presumably more easily acquired and cheaper African slave-made Yabba.

With regard to vessel forms, it can be seen in Table 7.2 that food processing, or cooking, forms, while not frequent in the overall assemblage, are more common at Building 4, while Building 5, presumably because it was bigger and also possibly because it may have been used commercially, has more storage ware. Building 5 also contained more vessels associated with health and personal hygiene, such as the two tin-glazed washbasins. This suggests that its residents were, perhaps, more concerned with personal appearance than those in Building 4, since the simple act of washing was not generally considered part of one's toilet in the 17th century. As to the drinking forms: while similar amounts of these vessels were found in both buildings, many more of Building 5's are fine earthenware cups of the shape used for the then new, fashionable and popular beverages of chocolate and coffee. Building 4, in contrast, contains proportionately more drinking vessels associated with more traditional beverages, such as beer and ale. Indeed, the only thing that seems to contradict the suggestion that Building 5 was not only a little more 'comfortable' than Building 4 but that it possibly contained in part of it some kind of eatery is in the food consumption category, whereby more eating and serving vessels were recovered from Building 4. This discrepancy, though, can be partly explained when it is taken into account that several of these forms are actually undecorated coarse ware bowls rather than decorated quality tableware. These figures thus seem to bear out the suggestions that have been made in this study as to how the two structures may have functioned.

TABLE 7.1. Ceramic Wares by Estimated Minimum Number of Vessels (MNV) and Relative Percentages (%) from the 17th-Century Context at Building 4/5 (including Yards 4A and 4B and the Alley) and Building 4/5 (including Yards 5, 6, and 7 and the Sidewalk)

Ceramic Ware	Building 4		Building 5	
	MNV	%	MNV	%
Coarse Ware	22	39.3	30	34.1
Slipware	5	8.9	6	6.8
Tin-Glazed Earthenware	19	33.9	40	45.4
Stoneware	8	14.3	8	9.1
Porcelain	2	3.6	4	4.5
Total	**56**		**88**	

TABLE 7.2. Ceramic Functional Classifications by Estimated Minimum Number of Vessels (MNV) and Relative Percentages (%) from the 17th-Century Context at Building 4/5 (including Yards 4A and 4B and the Alley) and Building 4/5 (including Yards 5, 6, and 7 and the Sidewalk)

Functional Classification	Building 4		Building 5	
	MNV	%	MNV	%
Food Processing	6	10.7	5	5.7
Food/Beverage Storage	9	16.1	24	27.3
Food Consumption	22	39.3	19	21.6
Beverage Consumption	12	21.4	20	22.7
Health/Personal Hygiene	6	10.7	18	20.4
Other	1[a]	1.8	2[b]	2.3
Total	**56**		**88**	

[a] n=1 flower vase
[b] n=2 unknown vessels

BUILDING 4/5 TO THE REST OF THE PORT ROYAL SITE

It may be of interest, too, to consider the wider context of rest of the Port Royal site excavated by Hamilton. It must be noted, however, that there is a difficulty with the larger ceramic count, since no breakdown into vessel numbers for the sherds from the other buildings has been undertaken. (Table 4.3 shows how the sherd count from Layer 3 of Building 4/5 compares with the sherd count from Layer 3 of all of the other excavated buildings combined [taken from Hamilton's general data base].) One could hazard a guess at the larger count, based upon the sherd/vessel ratio for Building 4/5. As shown in Table 7.3, this would give vessel counts across the board for Layer 3 of 190 for coarse ware, 24 for slipware, 110 for tin-glazed earthenware, 104 for stoneware, and 88 for porcelain. But since the source figures for the calculation are so few, this can hardly be satisfactory. Consequently, the functional classifications that are fundamental to the present study cannot be attempted.

SUMMARY

Deductions the author has made regarding Building 4/5's ceramics are as follows: (1) the vessel forms and where they were found coincide broadly with the site's physical remains and other finds; (2) two activities can be distinguished: residential and something of a public or semi-public nature connected with eating and/or drinking; a possible third activity may be some form of domestic support system; and (3) the householders, at least in Building 5, seem to have lived a reasonably affluent and sophisticated way of life. In short, the author found nothing to contradict other research as to suggested building use or the standard of the people using it for residential and commercial purposes. It should be remembered, however, that we are dealing with a total of only 144 pottery vessels, as far as the author is able to judge. Obviously, it is not possible to make a leap from this number of vessels to a whole lifestyle with any certainty. All that can be looked for is a scenario that (a) seems to fit the facts and (b) is consistent.

TABLE 7.3. Ceramic Wares by Estimated Minimum Number of Vessels (MNV) and Relative Percentages (%) from the 17th-Century Context at Port Royal, Jamaica (Hamilton's Excavations, 1981-1990)

Ceramic Ware	Building 4/5		Other Excavated Buildings[a]		Total	
	MNV	% Total	MNV	% Total	MNV	% Total
Coarse Ware	52	36.1	138	72.6	190	36.8
Slipware	11	7.6	13	54.2	24	4.7
Tin-Glazed Earthenware	59	41.0	51	46.4	110	21.3
Stoneware	16	11.1	88	84.6	104	20.2
Porcelain	6	4.2	82	93.2	88	17.0
Total	**144**		**372**		**516**	

[a] the MNV of the other excavated buildings are based on sherd/vessel ratio for Building 4/5

CHAPTER VIII

INTER-SITE COMPARISONS

This chapter compares the analyzed inventory of Building 4/5 with ceramic assemblages from two sites from the same period: a governor's residence in James City County, Virginia, and an artifact-refuse pit from an inn in Surrey, England. The primary aim of this comparison is to investigate ceramic use by comparing the range of vessel forms present at each site. A second goal is to consider ceramic sources by looking at the incidence of pottery types from different parts of the world as disclosed at these sites.

All three sites produced assemblages dated to the late 17th and very early 18th centuries. The Drummond site at Governor's Land in Virginia is associated with William Drummond, who was governor of North Carolina in the third quarter of the 17th century. Drummond was a wealthy planter and active politician, who became a supporter of and participant in Bacon's Rebellion of 1676 and who was hanged for treason in 1677. The Tun Inn in Guildford, Surrey, was an important post-meeting rendezvous for the town's movers and shakers. It was torn down and rebuilt in 1817 and thereafter served as the town's police station throughout much of the 19th century.

The reader should be aware that since the author did not have physical access to the comparison collections, she has used the best available information on minimum vessel counts from the selected sites. As has been often noted (see Beaudry et al. 1983; Yentsch 1990, 1991; Pittman 1993), historical archaeologists have not yet agreed upon form designations for their excavated ceramic vessels. Accordingly, the differences within each of the functional categories of vessel forms as presented in Tables 8.1 and 8.2 reflect to some extent the variation in reporting methods. For example, the 'tea wares' in the Drummond assemblage is understood to be tea bowls, teapots, and (possibly) saucers. In the Building 4/5 assemblage, saucers are listed separately, since they were probably used as condiment dishes (no teapots were recovered from the building complex). Drummond's 'butter pots' are equivalent in terms of form to 'storage pots' and thus have been listed as such. The English inn report was particularly troublesome in its vessel nomenclature, but, fortunately, the report includes illustrations of most of the finds. For comparison purposes, therefore, the author examined all of the drawn forms in that report and renamed some to reflect more closely the descriptive categories used in POTS. These changes are included as footnotes in Table 8.2.

The Virginian ceramic assemblage and that from the English site are each described and compared separately with the Building 4/5 ceramics before a final discussion brings all of the evidence together. The descriptions are supported by tables, which present the numbers, percentages, and forms of vessels of each ware, grouping each of the forms into the functional categories allocated to the Building 4/5 assemblage. The exact location of the comparison sites is not relevant to this discussion, and location maps and excavation plans are not presented.

THE DRUMMOND II SITE, VIRGINIA

The Drummond site is situated within James City County, in the lower James River Basin of tidewater Virginia. Excavations at the site were conducted by the Virginia Research Center for Archaeology from 1977 to 1980 (Outlaw 1989). Deposits dating from the early 17th through the mid 18th century were explored, producing a large, well-stratified ceramic assemblage. The ceramic assemblage considered here is from the 1680-1710 strata, which was identified as the location of Governor Drummond's residence and is known as Drummond II. After the governor's death, the site continued to be occupied by his widow and family (Alain Outlaw 2001, pers. comm.).

The Ceramic Assemblage

Ceramic Ware

Table 8.1 shows the nature of the Drummond II assemblage.[1] It quantifies the wares and forms recognized at the site and shows the minimum number of vessels identified for each ware. A total of at least 268 vessels was recovered. Coarse and tin-glazed wares comprise the majority of the assemblage and are represented almost equally (n=92 and 106, respectively). Other wares include stoneware (n=31), porcelain (n=24), and slipware (n=15).

Ceramic Function

As already noted in this study, coarse ware was mostly used for utilitarian kitchenware, such as cooking and storage pots. This is reflected at Drummond II, where almost all (96.7 percent) of the coarse ware vessels are found in the food processing and food/beverage storage groups.

As earlier noted too, tin-glazed earthenware was primarily used for dining forms and for those vessels whose function was ornamental. It was also commonly used for vessels associated with health and personal hygiene. At Drummond II, plates, dishes, cups, and tea wares represent a significant part of the tin glaze assemblage (34.9 percent).

[1] There is no formal report on the excavations at Drummond II. Table 8.1 was thus taken from a secondary source, written by Anne Yentsch (1991:58, Table 2). Dr. Yentsch extracted the ceramic data from the Drummond site artifact catalogue, an enormous six-volume tome housed in the Virginia Department of Historic Resources (Alain Outlaw 2001, pers. comm.). The author has it on good authority—Dr. Alain Outlaw, the principal investigator at the site—that the data from the Drummond II phase are correct.

TABLE 8.1. Ceramic Wares, Forms, and Functional Classifications by Estimated Minimum Number of Vessels from the 1680-1710 Strata at the Drummond Site (Drummond II), James City County, Virginia (after Yentsch 1991:Table 2)

Vessel Form and Functional Classification	Coarse Ware	Slipware	Tin-Glazed Earthenware	Stoneware	Porcelain	Total	%
Food Processing							
Milk Pan	33	--	--	--	--	33	
Cooking Pot	10	--	--	--	--	10	
Bowl	7	--	--	--	--	7	
Pipkin	2	--	--	--	--	2	
Chafing Dish	1	--	--	--	--	1	
Colander	1	--	--	--	--	1	
Pinch Pot	1	--	--	--	--	1	
Subtotal	**55**	--	--	--	--	**55**	**20.5**
Food/Beverage Storage							
Storage Pot[a]	34	--	--	--	--	34	
Bottle	--	--	--	4	--	4	
Subtotal	**34**	--	--	**4**	--	**38**	**14.2**
Food Consumption							
Dish	--	6	10[b]	--	--	16	
Porringer	3	--	11	--	--	14	
Plate	--	--	12	--	--	12	
Bowl	--	3	--	--	--	3	
Subtotal	**3**	**9**	**33**	--	--	**45**	**16.8**
Beverage Consumption							
Mug	--	1	10	14	--	25	
Tea Wares[c]	--	--	11	--	11	22	
Jug/Pitcher	--	--	1	12	2	15	
Cup	--	2	4	--	7	13	
Punch Bowl	--	--	3	--	4	7	
Drinking Pot	--	2	--	--	--	2	
Ewer	--	--	--	1	--	1	
Subtotal	--	**5**	**29**	**27**	**24**	**85**	**31.7**
Health and Hygiene							
Apothecary Pot	--	--	34	--	--	34	
Basin[d]	--	1	10	--	--	11	
Chamber Pot	--	--	--	--	--	--	
Subtotal	--	**1**	**44**	--	--	**45**	**16.8**
Total	**92**	**15**	**106**	**31**	**24**	**268**	
%	**34.3**	**5.6**	**39.5**	**11.6**	**9.0**		

[a] n=14 butter pots
[b] n=4 are identified in the report as chargers/platters, which are simply large dishes (no 'chargers/platters' category in POTS)
[c] presumably includes tea bowls, teapots, and saucers; specific information was not available to the author
[d] may include barbers' basins and/or washbasins; specific information was not available to the author

Numerous apothecary pots and basins were also found.

Stoneware at Drummond II is, as could be expected, confined to those vessels used for the storage and consumption of beverages. The mug is the most common form in this ware at the site.

As to the slipware: like the tin-glazed earthenware, it is principally represented by tableware, with bowls, dishes, cups, communal drinking pots, and a single mug identified. One slipware vessel, a basin, which may have been used for washing or for grooming, was also found.

All of the porcelain forms, like those in stoneware, are associated with beverages (of different sorts), with tea wares the most common. Porcelain chocolate/coffee (or wine) cups and punch bowls were also found.

At the Drummond II site, the 'storage pot' and 'apothecary pot,' which includes large drug jars and small ointment pots, are the most common forms, each with a total of 34 vessels identified. Milk pans traditionally used for dairying practices follow a close second at 33 vessels, but since dairying was a peripheral activity to tobacco growing on many Chesapeake plantations (see Main 1982; Bowen 1994), this may not have been how all of them were used. Mugs (n=25) and tea wares (n=22) are also prominent.

Ceramic Sources

With regard to Drummond II's coarse ware, while some of the bowls were identified as Colono ware (open-hearth fired vessels made by Indians and/or African slaves using local clays) and two of the storage pots (olive jars) are Iberian, information on the source of most of the coarse ware vessels was not available to the author. Some may have been made by Virginian potters, since coarse pottery manufacture in the North American colonies had begun by the mid 17th century (see Watkins 1950; Noël Hume 1963; Kelso and Campbell 1974; Turnbaugh 1983, 1985). English coarse ware pots were almost certainly used. As historian James Horn (1994:307-308) notes: "Cargoes from London, Bristol, and lesser ports remind us how dependent Virginia and Maryland settlers were on English manufactures throughout the colonial period and their close association with the produce and commerce of the hinterlands of English ports."

With regard to the site's tin-glazed earthenware, it is probably a mixture of vessels of English and Dutch manufacture. As was noted in Chapter IV, it is extremely difficult to distinguish between tin-glazed vessels from these sources, since clay was often imported for mixing with local clays.

All of the Drummond II slipware is English and could be from either Staffordshire or Bristol, the two main 17th-century slipware-producing centers in England. Other possible English source areas are North Devon or Kent.

Slightly over half (n=16) of the site's stoneware was identified as of German provenance. The bottles and pitchers likely were made in England, but this information was not available. American stoneware production began only in the second quarter of the 18th century (Noël Hume 1970:100).

Drummond II's porcelain is, as expected, all Chinese export (i.e., made specifically for the Western market).

Discussion

At a time when material goods were beginning to be seen as an index of social position and status within a society, the character of the Drummond II assemblage is indicative of an extremely affluent colonial planter-merchant. First, the collection is large, especially when compared to other sites in the region (see, e.g., Kelso 1984:Table 5; Yentsch 1990:Tables 2 and 3, 1991:Table 1). With the increasing social importance given to goods, a high rate of ceramic consumption in a household is significant: "Being rich meant having more," according to Horn (1994:326). Second, most of the wares (as expected) are European imports. This indicates that the Drummond residence had overseas trade connections. Third, much of the assemblage consists of fine wares (particularly Chinese export porcelain) designed for dining and/or display, providing yet further insight on the economic and social standing of the household. The coarse ware is relegated to those 'hidden' vessels that would be used in the kitchen, for food preparation and cooking, and in the pantry/back rooms, as storage ware.

The relatively large number of tea wares is particularly telling of the social status of the assemblage's owner(s), these vessels being one of the best indicators of wealth at that time: "In the early 18th century, the rarity and expense of tea and sugar, the costly equipment to serve it, and leisure considered necessary for entertainment all served to limit tea drinking to the very upper class" (Smart Martin 1994:172). The shape of some of the cups indicates that they may have been used for chocolate or coffee, possibly further evidence that the household had the means, as well as the connections, to purchase these exotic new caffeine drinks.

The Drummond II assemblage also includes punch bowls, mugs, and communal drinking pots, suggesting that ceremonial toasting and possibly entertaining was part of the social scene at the governor's house. The large tin-glazed dishes may have been displayed in the parlor, on the walls or on the dresser. Alternatively, both these and the plates may have been used at table, indicating that the household was aware and responding to the new style of dining, which was to dominate the 18th century. This new etiquette, which emphasized individuality, in specialized dining spaces with individual place settings and a 'one person one dish' relationship, is thought to reflect a new formal, segmenting behavior departing from the medieval tradition of shared dishes and communal eating areas. This has been seen as mirroring the wider ideological transformations that were occurring in every aspect of life, as the dominant elites came under the rationalization of the Enlightenment:

> Not only must one now own a proper set of accoutrements for smart living, but know a complex set of rules on how to use them. For it was in this same era [in the early 1700s] that notions of gentility began to spread, as more of the English gentry invested time and money in proper education, manners, and leisure. Many Chesapeake elites, anxious to be considered up to date, modeled their behavior and their life-styles on their peers in England. Genteel and social behavior became the hallmark of the well

> bred or wealthy, including accomplished dancing, games of skill, tea drinking, and elaborate dining (Smart Martin 1994:171).

The numerous washing/barbers' basins further suggests that the household was responding to these newly defined rules, since this growth of 'manners' or 'civilized' behavior also introduced frequent bathing and meticulous grooming (McCracken 1988:16-21; Shackel 1993:161). (That no chamber pots were found may simply reflect personal/social choices. The Drummond household may have chosen to use metal ones.)

Finally, when it is considered that the household also probably owned numerous pewter vessels (indeed, before 1760, three-fourths of the plates stocked in Chesapeake stores were pewter [Smart Martin 1994:174]), as well as wooden bowls and plates, metal cooking utensils, glassware, and also probably silver, the quantity of moveable goods in the dwelling was likely extremely large.

THE TUN INN SITE, SURREY, ENGLAND

The site of the Tun Inn is in the center of the town of Guildford, which lies south of London, in the county of Surrey, England. Specifically, an artifact-filled refuse pit was discovered in 1991 during the rebuilding of offices which presently occupy the site. It was excavated by the Guildford Museum Volunteer Excavation Unit, who carried out a watching brief at the location and adjoining properties between 1991 and 1993. The pit yielded a discrete group of ceramics dating from ca. 1650-1714, and deposited ca. 1702-1714. It is believed that they, along with the other artifacts recovered, came from the inn or from a nearby satellite property. The Tun Inn as such appeared in local records in 1535 (although a building occupied the site in earlier times) and was demolished in 1817. The excavation of the pit and a discussion of all the finds are documented in Fryer and Selley (1997).

The Ceramic Assemblage

The reader should note that since most of the ceramic assemblage is composed of intact or near-intact pots, the investigators concluded that the deposit had been discarded as trash on a single occasion or over a very short period of time, after which the pit was quickly sealed. Fryer and Selley (1997) suggest that the pit represents a 'tavern clearance group,' i.e., the pottery appears to have been intentionally thrown out to make way for more fashionable wares (to compete with its rival inns). Alternatively, the vessels were broken in an 'accident,' such as the riot that occurred during the county election held in the Tun Inn on 11 October 1710 (see Turner 1971:131-151; Corke 1997:150). It is reiterated that some of the vessel forms have been renamed by the author to more closely reflect the terminology used in POTS.

Ceramic Ware

Table 8.2 shows the nature of the Tun Inn assemblage. It quantifies the wares and forms recognized at the site and shows the minimum number of vessels identified for each ware. A total of at least 211 vessels was recovered.[2] Coarse ware, at 156 vessels, comprises the largest proportion (over 70 percent) of the assemblage. Tin-glazed earthenware is the second most common type (n=37). Stoneware is represented by 16 vessels, while only one each in slipware and porcelain is present.

Ceramic Function

Unlike at Drummond II, where all of the coarse ware vessels are in the food processing and food/beverage storage groups, the coarse ware from the pit appears in a wide range of forms, both utilitarian and dining. The remains of coarse ware chamber pots and two vessels provisionally identified as close stool pans were also found. Numerous 'unknown' vessels are also in the coarse ware inventory.

The tin-glazed earthenware forms include decorated plates and a dish, as well as porringers, tea bowls, and two cups. Chamber pots and apothecary wares (small ointment pots and larger drug jars) are also represented.

While far fewer stoneware vessels were recovered from the pit than were noted at Drummond II, all were similarly used for the storing and serving/drinking of beverages (jugs, bottles, mugs). A stoneware teapot was also found.

The slipware cup recovered from the site is straight-sided, rather short and squat (ca. 5 cm tall x 5 cm diameter), and is similar to a mug. While its shape is not typical of 17th-century chocolate/coffee cups, which tend to be tall and narrow or have a bulbous body and flaring mouth, it would be misguided to make any definitive statements about its use. Perhaps, given its small size and the context in which it functioned (i.e., a tavern), it may have been used for a stronger drink. As Archer (1997:240) notes: "The only tenable assumption about the shapes of containers and the liquids consumed from them is that stronger drinks such as wine and the more powerful ales would have been served in smaller vessels."

The single porcelain vessel, identified as a 'dish' by Fryer and Selley (1997:161, 185), more closely resembles the shape of a large, shallow bowl. It has an underglaze blue floral decoration similar to that on a bowl recovered from the wreck of the Portuguese frigate, the *Santo António de Tanná*, which foundered in 1697, in front of Fort Jesus, Mombasa, Kenya (Sassoon 1981:Figure 4).

[2] The original report specifies the minimum ceramic count at 193 vessels. On closer inspection, however, the author found that this number reflects only those vessels which were more than 5 percent complete. According to the report's finds inventory, a further 18 vessels, mostly represented by rim and base sherds, were also recovered.

TABLE 8.2. Ceramic Wares, Forms, and Functional Classifications by Estimated Minimum Number of Vessels from the 1702-1714 Deposit at the Tun Inn Site, Guildford, Surrey, England

Vessel Form and Functional Classification	Coarse Ware	Slipware	Tin-Glazed Earthenware	Stoneware	Porcelain	Total	%
Food Processing							
Pudding/Pastry Pan[a]	9	--	--	--	--	9	
Pipkin	4	--	--	--	--	4	
Chafing Dish	1	--	--	--	--	1	
Colander	1	--	--	--	--	1	
Subtotal	**15**	--	--	--	--	**15**	**7.1**
Food/Beverage Storage							
Storage Pot[b]	11	--	--	--	--	11	
Bottle	--	--	--	9	--	9	
Subtotal	**11**	--	--	**9**	--	**11**	**9.5**
Food Consumption							
Bowl[c]	22	--	--	--	1[d]	23	
Porringer	17	--	4	--	--	21	
Dish	16	--	1	--	--	17	
Plate[e]	--	--	11	--	--	11	
Salt/Sweetmeat Stand	1	--	--	--	--	1	
Subtotal	**56**	--	**16**	--	**1**	**73**	**34.6**
Beverage Consumption							
Cup	7[f]	1	2	-	--	10	
Jug	3	--	3	3	--	9	
Mug	7	--	--	2	--	9	
Drinking Vessel[g]	6	--	--	-	--	6	
Tea Wares	--	--	3	1	--	4	
Subtotal	**23**	**1**	**8**	**6**	--	**38**	**18.0**
Health and Hygiene							
Chamber Pot	14	--	2	--	--	16	
Apothecary Pot	--	--	10	--	--	10	
Stool Pan (possibly)	2	--	--	--	--	2	
Subtotal	**16**	--	**12**	--	--	**28**	**13.3**
Other							
Unknown Vessel Forms	30	--	--	1	--	31	
Bowl (unknown size)	5	--	1	--	--	6	
Subtotal	**35**	**1**	**1**	**1**	--	**37**	**17.5**
Total	**156**	**1**	**37**	**16**	**1**	**211**	
%	**73.9**	**0.5**	**17.5**	**7.6**	**0.5**		

[a] identified in original report as deep dish, but form is that of a pudding/pastry pan in POTS
[b] n=2 froms identified in report as deep bowls, but their form suggests a storage function
[c] n=20 forms are relatively large and thus may have been used in preparatory function (identified in original report as deep bowls, handled bowls, and wide bowls)
[d] identified in original report as dish, but form is that of a bowl in POTS
[e] n=4 identified in original report as dishes, but their diameters are smaller than those for dishes as outlined in POTS
[f] n=6 caudle cups, n=1 pedestal cup
[g] identified as such in original report, with no illustration given

The bowl (n=29) is the most common vessel form in the Tun Inn refuse pit. Porringers, totaling 21 vessels, are the second most frequent. Chamber pots (n=16) are also prominent.

Ceramic Sources

All of the site's coarse ware is Border ware, with the red and white varieties represented in almost equal amounts (n=80 and 76 vessels, respectively). As Fryer and Selley (1997:160) note: Red Border ware was "more commonly used for heavy-duty items such as larger dishes and storage jars," while White Border ware was "used for more delicate items such as wide bowls and porringers."

The site's tin-glazed earthenware is a mixture of English and Dutch, with Fryer and Selley (1997:170-171) suggesting that the majority (nearly 80 percent) is English and was probably made in London. The Dutch type appears to be confined to a few dishes and plates and a single jug. However, as has been noted before in this study, making distinctions between Anglo and Dutch tin-glazed wares is difficult.

As to the stoneware: fortunately, its provenance is more easily identified, since, as we have seen, the German industry monopolized the ware throughout the 17th century. The stoneware bottles recovered were probably made in the factories at Frechen or Raeren. The Westerwald type is represented by jugs and a mug, the latter of which shows a portrait of William III, who reigned from 1689-1702. English stoneware, although present, is minimal at the site and is dated to the early 18th century. Specifically, the teapot and a jug are attributed to John Dwight's Fulham factory and date to about 1700-1710. A salt-glazed mug has a detailed 'AR and crown,' a well-known English ale measure mark used throughout Queen Anne's reign (1702-1714). The provenance of the 'unknown' stoneware form is not identified in the report.

The slipware cup is English, from either Staffordshire or Bristol. It is very difficult to distinguish between the wares produced by these potteries in the late 17th century.

The porcelain bowl is Chinese export.

Discussion

The Tun Inn refuse pit ceramic assemblage is, as might be expected, markedly different in character from that recovered from Drummond II. The range of imported pottery is small (10.4 percent by minimum number of vessels), and the context includes a high proportion of coarse ware. That this coarse ware is all Border ware is not surprising, considering the inn's location, close to the production source. The Border ware industry flourished throughout the 1600s and was well noted in and around London for producing good-quality and attractive pottery. (Contemporaneous English deposits similarly carry large amounts of Border ware, see, e.g., Orton and Pearce 1984; Pearce 1992, 1998.)

That the Border ware was used for all types of vessel forms, both utilitarian and dining, is also not surprising, since in a busy inn, sturdy earthenware vessels would be more serviceable than finer wares such as tin-glazed pottery. Indeed, as Fryer and Selley (1997:161) point out, the tin-glazed plates and dish "are in remarkably good condition and show no knife-marks or other signs of heavy use and were almost certainly for display." Eating and serving, it seems, was from coarse ware bowls and porringers. This suggests that the inn's fare was simple, likely stews and broths, rather than the finer dining that, perhaps, would have been seen at Drummond II. Drinking, too, appears to have been of the type traditionally associated with inns (beer, ale, wine), judging by the types and quantities of beverage forms recovered (n=24 mugs, caudle cups, bottles, and jugs).

That tea wares were found at the site, regardless of how few, sheds some light on the social rank of the inn, or certainly of the assemblage's owner(s), since, as has been noted above and in Chapter VI, tea was an expensive luxury item until the 18th century: "The price of tea c. 1706 was between 12*s*. and 36*s*. a pound. At this time [in London] 12*s*. or 14*s*., the price of a fairly standard tea, was a week's wages for many a master craftsman" (Emmerson 1992:4). In this, the archaeological evidence corroborates the documentary record for inns in Guildford generally: "Guildford, the half-way house between London and Portsmouth, was famous for its good inns. John Aubrey (in the 1670s) said they were 'the best perhaps in England'. . . . Daniel Defoe, traveling through England fifty years later, noted that Guildford was a town 'very well furnished with inns for accommodation of travelers'" (cited in Corke 1997:152). That decorative tin-glazed plates and a dish and a delicately painted Chinese export porcelain bowl were also found further suggests that the assemblage's owner(s) "must have been relatively wealthy, knew about the tastes and fashions of the day, and had access to them" (Fryer and Selley 1997:156).

GENERAL DISCUSSION: BUILDING 4/5, DRUMMOND II, TUN INN

Few archaeological reports provide the information necessary to make a legitimate comparison between ceramic assemblages in terms of vessel counts, forms, and functional categories. Differences in reporting methods, as noted above, further complicate the issue. But the decision to compare Building 4/5's ceramic assemblage with those from Drummond II and the Tun Inn was not arbitrary. The author was interested to see how the ceramic inventory of an urban dwelling in a large colonial port fares when compared with the dwelling of a rural colonial elite. The comparison with the English site attempts to show the differing requirements of a large, popular inn located close to London and those of a relatively large building in Port Royal that may have functioned partly in a similar manner. (It is worth mentioning here that the ceramic inventory of a 17th-century Port Royal tavern has also been examined [Brown 1996], but the author wished to contrast the Old and New Worlds as part of the second issue in this study.)

Drummond II and Building 4/5

The Drummond II assemblage is similar in many ways to that at Building 4/5. Both are heavily weighted towards imported wares, particularly those from England. Further, as shown in Table 8.3, the same types of wares are represented at both locales in more or less similar relative quantities, with coarse and tin-glazed wares predominating. The coarse ware at both sites is a mix of locally produced vessels and imports from Europe; the slipware is from England. At both sites, too, the tin-glazed vessels are probably a mix of those made in England and Holland (although Building 4/5 also includes some Mediterranean and Mexican majolica), while most of the stoneware is German. The porcelain is,

TABLE 8.3. Ceramic Wares by Estimated Minimum Number of Vessels (MNV) and Relative Percentages (%) at Building 4/5, Drummond II, and the Tun Inn Deposit

Ceramic Ware	Building 4/5		Drummond II		Tun Inn Deposit	
	MNV	%	MNV	%	MNV	%
Coarse Ware	52	36.1	92	34.3	156	73.9
Slipware	11	7.6	15	5.6	1	0.5
Tin-Glazed Earthenware	59	41.0	106	39.5	37	17.5
Stoneware	16	11.1	31	11.6	16	7.6
Porcelain	6	4.2	24	9.0	1	0.5
Total	**144**		**268**		**211**	

of course, Chinese export. European porcelain manufacture was not accomplished until later, in Germany, in the 18th century.

Indeed, the range of wares at both Drummond II and Building 4/5 indicates a variety of ceramic choices were available to New World settlers at the end of the 17th century. Drummond (or his widow or staff) probably obtained the wares directly from overseas traders, since large tobacco planter-merchant households in the Chesapeake did most of their shopping in England via their London agents until the mid 18th century (Walsh 1983:116; Kelso 1984:14; Carr and Walsh 1988:139; Smart Martin 1994:173). It is possible that the occupants of Building 4/5, living as they did in a thriving merchant port at the center of colonial trading networks, purchased their ceramics at the local marketplace or obtained it as barter from incoming shipmasters.

Of particular interest is that both sites contain an array of fine, decorative earthenware and, particularly at Drummond II, Chinese export porcelain. This indicates that both Governor Drummond (or his widow and family) and those who occupied Building 4/5 were in a position to buy non-essential goods for a high style of living. This suggests a concern with status, and what has been said about Chesapeake households appears to be reflected at Port Royal: "Through social display or conspicuous consumption [i.e., leisure spending], Chesapeake households conveyed a family's social, economic, or political power and its position in the social hierarchy" (Yentsch 1991:27). Or, put another way, the consumption of material goods was seen by relatively wealthy colonists "as one means of separating themselves from their compatriots who ranked below them on the social ladder" (Yentsch 1991:30; see also Carr and Walsh 1988; Shackel 1993, 1994).

Social rank was also reflected in the vessel forms used in dwellings of the period, and certainly, in their range of forms, both sites point to upwardly mobile households. As shown in Table 8.4, 'fine dining' ware comprises a significant proportion of each site's ceramic inventory. In the case of Drummond II (see also Table 8.1), where the fine/social dining forms (i.e., tin-glazed dishes [chargers/platters] and plates; and tin-glazed and porcelain cups, jugs, punch bowls, and tea wares) comprise half (n=65) of the food and beverage consumption forms and nearly one-fourth of the total assemblage, a certain amount of household sociability and social display seems to have occurred. Such activities would certainly befit a wealthy and influential planter, who also worked in the public domain dealing with affairs of state. Similarly, at Building 4/5 (see also Table 6.1), where such forms amount to 34.2 percent (n=25) of the food and beverage consumption forms (17.4 percent of the total assemblage), entertaining and/or hospitality would seem to have been a fairly conspicuous activity.

Table 8.4 also shows that a considerable proportion of the ceramics at both sites consists of apothecary pots. Indeed, these vessels—which were just as often used as general storage containers—are the most common form at the Drummond II site. They are also found in the Building 4/5 assemblage. In fact, if drug jars are considered primarily as storage ware, this functional category of vessel forms is second only to Building 4/5's food consumption group and second only to Drummond's beverage consumption group. A large amount of storage ware in a domestic assemblage is significant. As was noted in the previous chapter, "the greater complexity of the elite cuisine and the larger stocks of food it required . . . created a demand for more vessels, especially storage vessels" (Yentsch 1991:45).

Differences between the two sites with regard to their ceramic inventories are also to be noted. Most apparent is, of course, the significantly greater quantity of vessels at Drummond II (n=268) compared to that at Building 4/5 (n=144). This can be explained, in part, by the fact that Drummond II is an accumulated deposit, whereas the ceramics recovered from Building 4/5 are those that were (for the most part) currently in use. The difference between the two sites may also reflect personal/social choices. The occupants of Building 4/5, for example, may have used more pewter or wooden vessels, some of which have, indeed, made it into the archaeological record (see, e.g., Gotelipe-Miller 1990:48-49). It must be remembered, too, that Building 4/5, located in a tightly

TABLE 8.4. Estimated Minimum Number of Ceramic Vessels by Form and Functional Classification at Building 4/5, Drummond II, and the Tun Inn Deposit

Functional Classification	Vessel Form	Building 4/5		Drummond II		Tun Inn Deposit	
		MNV	%	MNV	%	MNV	%
Food Processing	Bowl	--		7		--	
	Chafing Dish	--		1		1	
	Colander	1		1		1	
	Cooking Pot	8		10		--	
	Milk Pan	--		33		--	
	Pinch Pot	--		1		--	
	Pipkin	1		2		4	
	Pudding/Pastry Pan	1		--		9	
	Subtotal	**11**	**7.6**	**55**	**20.5**	**15**	**7.1**
Food/Beverage Storage	Bottle	8		4		9	
	Storage Pot	25		34		11	
	Subtotal	**33**	**22.9**	**38**	**14.2**	**20**	**9.5**
Food Consumption	Bowl	16[a]		3		23[a]	
	Dish	1		16		17	
	Plate	13		12		11	
	Porringer	5		14		21	
	Salt/Sweetmeat Stand	1		--		1	
	Saucer	5		--		--	
	Subtotal	**41**	**28.5**	**45**	**16.8**	**73**	**34.6**
Beverage Consumption	Cup[b]	15		13		10	
	Costrel	4		--		--	
	Drinking Pot	5		2		--	
	Drinking Vessel	--		--		6	
	Ewer	1		1		--	
	Jug/Pitcher	--		15		9	
	Mug	5		25		9	
	Punch Bowl	1		7		--	
	Tea Wares	1		22		4	
	Subtotal	**32**	**22.2**	**85**	**31.7**	**38**	**18.0**
Health/Personal Hygiene	Apothecary Pot	16		34		10	
	Basin	2		11		--	
	Chamber Pot	6		--		16	
	Stool Pan	--		--		2	
	Subtotal	**24**	**16.7**	**45**	**16.8**	**28**	**13.3**
Other	Bowl (unknown size)	--		--		6	
	Flower Vase	1		--		--	
	Unknown Vessels	2		--		31	
	Subtotal	**3**	**2.1**	--	-	**37**	**17.5**
	Total	**144**		**268**		**211**	

[a] some of these bowls may have served a prepatory function
[b] some of these are caugle cups (n=3 at Building 4/5, n=6 in the Tun Inn deposit); n=1 pedestal cup at the Tun Inn

packed port town, had much less available domestic space than would be found in a Great House on a plantation. Another notable difference is the preponderance of food processing forms at Drummond II.

Economic prosperity meant that households in the late 17th century were able to enjoy items that were rare and expensive, and although, as noted above, dining vessels appear in relatively large amounts at both sites, proportionately more of these forms at Drummond II are (possibly) associated with the consumption of the new, hot beverages, such as chocolate, coffee, and, especially, tea. These popular and fashionable drinks, which prompted the use of a range of specialized utensils in distinctive forms, were introduced into Europe and then the American colonies in the late 17th century (Yentsch 1990:44, 1991:50-51; Archer 1997:346; Thirsk 1999:23). In particular, the adoption of tea was accompanied by an influx of exotic oriental porcelain, which is more evident at the governor's house (n=18 cups and tea wares) than it is at Building 4/5 (a single tea bowl and 2 cups). Indeed, the higher proportions at Drummond II of porcelain, and of tea accoutrements in both this ware and tin-glazed earthenware, speak to the growing significance of the tea ceremony at the opening of the 18th century. It also reveals that the Drummond household, unsurprisingly perhaps, was responding to the dictates of fashion.

Finally, we look at the counts of those ceramics connected with cleanliness/personal hygiene, in particular, chamber pots. While there is a significant presence of these forms (n=6) at Building 4/5, Governor Drummond and his family apparently did not own any, at least none in ceramic. His household, in fact, is not too different from many 17th- and early 18th-century Virginian plantations, where there is a distinct lack of these vessels (see Kelso 1984:214). Chamber pots, of course, were made in other materials, most noticeably pewter and copper, which were often melted down and which possibly explains their absence in the assemblages at Drummond II and these other sites. Alternatively, and perhaps more likely, privy pits were the most common form of waste disposal in the Chesapeake region. As noted earlier in this study, this form of waste management was not as easily implemented at Port Royal, since the water table was very high.

Tun Inn and Building 4/5

As shown in Table 8.3, all of the ceramic wares found in Building 4/5 were also recovered from the Tun Inn deposit, with coarse and tin-glazed wares the most frequently occurring. At both sites, the coarse ware and slipware is English, while most of the stoneware is German. English tin-glazed vessels, as well as some made in Holland, are likewise represented (although as noted above, Building 4/5 includes some Mediterranean and Mexican tin-glazed wares). The porcelain in both assemblages is Chinese export.

The English site's domestic vessel forms also represent the same functional groups in broadly similar proportions as in Building 4/5, food consumption forms being the most common category in both (although, it must be noted that the inn site contains a far greater number of those forms), and vessels used for food processing (i.e., cooking and preparation) the least (see Table 8.4). In both assemblages, too, beverage consumption forms are generally comparable, both in quantities and types of vessels (i.e., many of the drinking forms are those used for traditional drinks, such as ale, beer, and wine) (although, again, it must be noted that more of these are present in the inn). Tea wares, although present, are relatively few in each collection. Health- and hygiene-related wares are also broadly comparable.

Despite such general similarities, however, it is the differences between the two assemblages that is, perhaps, most striking. The English inn site includes significantly more ceramic vessels than does Building 4/5 (n=211 vs.144). As was noted in the previous comparative discussion, this may simply reflect choice of material—a large number of ceramic vessels is expected from a tavern assemblage (see Bragdon 1988; Brown 1996). Another obvious difference is that not only does coarse ware comprise by far the majority of the English assemblage, but it is all of the same type (Border ware). In contrast, at Building 4/5, Border ware vessels represent only 4.9 percent of the total analyzed ceramics. The proportion of tin-glazed earthenware is also rather different at both sites (17.5 percent of the English assemblage vs. 41.0 percent at Building 4/5). Slipware vessels, too, are less prevalent in the pit (n=1) than at Building 4/5 (n=11). Similarly, Chinese export porcelain: only 1 vessel at the English site vs. 6 at Building 4/5, although as noted earlier in this study (see Chapters IV and VI), some of the latter's porcelain may have been part of another building's inventory.

The presence of so much Border ware in the inn deposit has already been explained. It made economic sense for the inn holder to use a local product rather than an import (although determining what pots actually cost the consumer at this time is difficult, since factors of mass production and marketing could competitively lower the price of superior or imported products). As noted earlier too, hardy coarse ware vessels would obviously be more serviceable in a busy, popular inn than would fine ceramics. In contrast, the richer, more diverse assemblage at Building 4/5 points to the wider social and commercial contacts that was afforded by its location in a large, international port.[3] It may also strengthen the case for a multi-functional role.

It should also be remembered that the English site's assemblage was intentionally discarded as trash at a single point in time, while much of the wares at Building 4/5 were clearly still being used when they were buried in the 1692 earthquake. The Drummond II assemblage was deposited throughout the years the site was occupied (1680-1710). From this perspective, the Tun Inn site is, perhaps, more telling as to the proliferation of consumer goods and the upsurge in consumer spending habits, which appears to have had its roots at the beginning of the 18th century (see Weatherill 1988; Shammas 1990). That a large quantity of ceramics could be simply thrown away shows that household objects, at least in Europe, were increasingly seen not just as purely functional items but as items that fulfilled new social and cultural needs, such as chocolate, coffee, and tea drinking, and changing dining habits in general (see McCracken 1988:16-21;

[3] That Port Royal was, indeed, at the forefront of 'fashion' is similarly reflected in the ceramic assemblage from an excavated 17th-century tavern on New Street, Port Royal, whereby 95 percent (n=244) of the total number of vessels (n=266) identified are tin-glazed earthenware (Brown 1996:165, Table 4).

Shackel 1993:159-169).

Patterns of Use: Building 4/5, Drummond II, and the Tun Inn

Food Processing

As shown in Table 8.4, it is apparent that vessels used for food processing are the least common group at both Building 4/5 and the Tun Inn deposit. In contrast, they are substantially more frequent in the Drummond II assemblage (n=55 vessels, or 20.5 percent). More cookware should, perhaps, be expected from a rural, self-sufficient site like Drummond. It may be that the governor, too, who was of the rural elite, had his staff prepare a variety of elaborate meals. That relatively little ceramic cookware was found at Building 4/5 may either be because of the use of metal pots and pans (which were also found) and/or because meals at the dwelling required less preparation. The use of metal ware/simple cooking methods may also account for the relative lack of these forms in pottery at the English site. It should be remembered, too, that the English site may represent a clearance of mostly tableware.

Beverage Consumption

In proportions of beverage forms, Building 4/5 and the inn are also generally comparable (22.2 vs. 18.0 percent, respectively). At the governor's house, these vessel forms are more than twice as frequent and, in fact, at 31.7 percent, comprise the largest functional group at the site. What is also apparent from Table 8.4 is that the drinking vessel types are quite different at all three sites. The Tun Inn site (unsurprisingly) contains a significantly higher number of vessels associated with traditional alcoholic beverages. (Chocolate/coffee and tea drinking, which was also becoming popular in taverns and inns of the period [Emmerson 1992:5], does not seem to have been an important element in the day-to-day working life of this particular establishment, although, of course, it may be that the tea and chocolate/coffee wares were not part of the discard.) Building 4/5's most common drinking form is the chocolate/coffee cup (in slipware and tin-glazed earthenware and also [possibly] the porcelain examples), while tea wares are especially prominent at the Drummond II site.

Food and Beverage Storage

Building 4/5 and Drummond II are more akin in their respective amounts of storage wares. Storage vessels at the Tun Inn site are comparatively less frequent (only 9.5 percent of the total assemblage). This reflects the location of the two colonial sites, dependent as they were on imported goods from Europe. It should be remembered, too, that the English site may represent a clearance of mostly tableware.

Food Consumption

In terms of numbers, food consumption forms at Building 4/5 and Drummond II are also broadly comparable (n=41 vs. 45, respectively). This category at the Tun Inn is much greater, including nearly twice as many vessels as either of the other two sites. It must be noted, however, that some of the plates and dishes in all three assemblages may well have served only as ornament. This certainly seems to have been the case for the decorated tin-glazed and porcelain forms recovered from the inn, since the excavators of the site note that none showed any wear use (Fryer and Selley 1997:161). (Pewter plates and porringers and pewter and wooden bowls were more common eating vessels in the early modern period; certainly, numerous pewter tableware items were, indeed, recovered from Building 4/5.)

Health and Personal Hygiene

Health- and hygiene-related wares are broadly comparable in terms of proportions in all three assemblages, averaging about 16.0 percent. It will be noted, however, that far more of these wares were recovered from the Drummond II site than were found at either Building 4/5 or the inn.

Ceramic Quantities

The Drummond II site includes the largest collection of ceramic vessels in this comparative study. Indeed, when compared to the average number of vessels per site in the Chesapeake region—76 in the 17th century and 125 in the early 18th century (Yentsch 1991:49)—Drummond II is atypical. Notwithstanding that the assemblage is an accumulated deposit, this large quantity is significant, since being rich at this time 'meant having more.' That most of Drummond's ceramics are tin-glazed earthenware vessels compares favorably with his role and position in society. That he or his surviving family should also own Chinese export porcelain in significant amounts and have several tea ware sets further illuminates his elite social standing.

The deposit from the English inn also includes a large ceramic collection, predictable for an inn assemblage. It is also to be expected that a tavern/inn's assemblage would include a large number of drinking vessels, particularly those associated with alcohol (Bragdon 1988:90). When the glass ware from the deposit is included in the vessel count (n=61 tumblers, wine glasses, and bottles), the proportion of its drinking forms dramatically increases. It would seem that the Tun Inn also served meals. That many coarse ware bowls and porringers were found suggests that the inn's fare was simple rather than what would be considered 'fine dining.' This is not to say that the inn was of lowly status. That it served wine in glass stemware and had many decorative plates and dishes suggests that it catered to a well-heeled middle-class clientele.

Building 4/5 contains the fewest ceramic vessels in this comparative study. Unlike the other two sites, its assemblage is more 'balanced' in its proportions of functional categories, indicating that it is, in fact, representative of vessels in current or very recent use, as opposed to accumulated refuse. In particular, those vessels used for food and beverage consumption are more closely correlated with each other than is shown at either the Drummond II site or in the inn deposit. Indeed, not only does the Tun Inn deposit include a far greater *quantity* of eating forms than does Building 4/5, but proportionately more of them are, as already noted, associated with the traditional beverages of beer and ale. This, perhaps, indicates that if Building 4/5 did have in part of it (namely Building 5) some sort of retail establishment dealing with food and/or drink, it does not seem to have been on the same scale as a 'full-blown' tavern or inn. As to eating/drinking forms at Building 4/5 vs. the Drummond II site: many of them, as at Drummond, are in tin-glazed earthenware; a few pieces of Westerwald stoneware and Chinese export porcelain are also present. Unlike at Drummond, however, tea wares and punch bowls are not prevalent, suggesting that the building complex served a different class of occupant. It would seem the assemblage belonged to members of the 'middle ranks' rather than to more affluent members of society. It should be added that the 'middling' parts of society at this time encompassed a wide range of groups, who "were neither at the bottom (servants, labourers, and wage-earners) nor at the top (country gentry and aristocracy) of the social hierarchy. . . . They therefore include the lesser gentry, professions, merchants, shopkeepers, farmers, yeoman, husbandmen, and craftsmen" (Weatherill 1988:13). It is important to realize also that a great gulf existed between the wealthy merchants at the top and the small shopkeepers at the bottom, who might still have lived relatively well.

When comparing ceramic assemblages, it must be borne in mind that we are dealing with *estimated* vessel counts, based upon cross-mended sherds, rim and basal sherds, and other diagnostic ceramic fragments. While no information is available for the Drummond II site, the author reminds the reader that some of Building 4/5's analyzed sherds (n=231, or 20.0 percent of the assemblage) could not be identified to specific vessel form. (The English inn site's ceramic count appears to be fairly accurate, considering that most of the assemblage was made up of intact or near-intact vessels.) It must also be remembered that the differences in the ceramic quantities at all three sites may simply reflect choices regarding their household containers: "Consumers based their choices on a number of variables: what their neighbors or superiors used, what one ware cost in relation to another, what they thought was attractive or practical, or what was considered 'suitable' for a certain function" (Smart Martin 1994:183). Product availability, as well as marketing strategies, are just some of the other many factors that should be considered. The Drummond II site was occupied through the early 18th century. The Tun Inn deposit also straddles the turn of the century. The early years of the 1700s is well noted as a watershed for rapid changes and advances in pottery technology and manufacture and use of ceramic vessels in the household (see Weatherill 1986).

Patterns of Supply: Building 4/5, Drummond II, and the Tun Inn Refuse Pit

The ceramic wares from each of the three sites have been combined and are presented in Table 8.5. They have also been discussed above in detail. There are contrasts between the three assemblages, but there are also telling similarities. The occupants of all three sites appear to have owned significant amounts of English wares. In the Tun Inn site, English products represent 89.6 percent of the assemblage by minimum vessel count; at Building 4/5, they amount to at least 65 percent; at Drummond II, they are not specifically known, but one may assume that a fair amount is English.

The proportion would be expected at the Tun Inn site, since it makes economic sense to use local products. That so many English ceramics were found at both Drummond II and Building 4/5 is also not surprising, given the Anglo ethnicity of both sites and the fact that ceramics were not manufactured *en masse* in the American colonies until the 18th century. It also shows us how dependent New World Anglo settlers were on English goods, a result, in part, of the Navigation Acts (enacted 1651, 1660, 1663, 1696), which stipulated that imports into English ports and colonies could be carried only in English ships.[4] Indeed, as one colonist noted in the early 18th century: "gentlemen of Virginia received goods from 'London, Bristol, etc. with less trouble and cost, than to one living five miles in the country in England'" (cited in Horn 1994:308). Historian James Horn (1994:418) continues: "[Virginian and Maryland] settlers down to the final decades of the century were overwhelmingly English . . . by birth and brought with them a miscellany of English traditions, customs, and attitudes that were enormously influential in shaping their responses to conditions encountered in the Chesapeake. Cultural continuities—their national characteristics and assumptions—were more important than environment in shaping the societies that emerged." According to historian Jack Greene (1988:164), settlers in Jamaica exhibited a similar pattern of social development throughout most of the 18th century, since "they imported large quantities of British finished goods, built elaborate public buildings and private houses, created expensive social infrastructures and showy cultural institutions, and otherwise sought to live in the British manner."

It is not surprising, then, that all three sites contained large quantities of English wares. What is particularly striking, however, and what emerges from Table 8.5, is that the English inn has only 22 vessels (10.4 percent of the total assemblage) that are identifiably non-English. Apart from the practical reason for this, the author suggests in Chapter IX that this demonstrates differences between the Old World and the New that may have had far-reaching consequences.

[4] It must be remembered, however, that these laws were often ignored, as evidenced by Anglo-Jamaican merchants' papers and letters from the period 1688-189 (Zahedieh 1986b:210).

TABLE 8.5. Estimated Minimum Number of Ceramic Vessels by Wares and Manufacturing Provenance at Building 4/5, Drummond II, and the Tun Inn Deposit

Ceramic Ware	Estimated Minimum Number of Vessels		
	Building 4/5	Drummond II	Tun Inn Deposit
Coarse Ware:			
English	32	--	156
Iberian	6	2	--
Other	14[a]	90[b]	--
Subtotal	**52**	**92**	**156**
Slipware:			
English	9	15	1
Continental European	2	--	--
Subtotal	**11**	**15**	**1**
Tin-Glazed Earthenware:			
English/Dutch	51	106	37[c]
Mediterranean/Mexican	8	--	--
Subtotal	**59**	**106**	**37**
Stoneware:			
English	4	15	3
German	9	16	13
Other	3[d]	--	--
Subtotal	**16**	**31**	**16**
Porcelain:			
Chinese Export	6	24	1
Subtotal	**6**	**24**	**1**
Total	**144**	**268**	**211**

CHAPTER IX
SUMMARY AND GENERAL DISCUSSION

Pottery is usually identified with a settled, domestic existence, and its value to humans often operates at two levels—the utilitarian and the aesthetic. It is, therefore, a material of great interest to the archaeologist, since it can provide information on humans' needs at a practical/physical level and on their aspirations or needs at a psychological level. Think of Caesar's legions on their great campaigns, for example, carrying with them most of their basic utensils, while their generals brought their personal dining ware, often extremely fine and often used for political purposes. When we think of the New World colonies and the tiny merchant vessels which brought them shipments of wares, we remember that throughout the voyage, the shipmasters often dined off their own monogrammed dishes. Theirs, after all, was a domestic existence, although one lived on the high seas.

Pottery is, of course, not the only artifact in the domestic environment that functions in this way—metal is another—but it is certainly one of the most valuable to the archaeologist. As historian Jean Mudge (1986:12) notes, "by separating the elements of date, quality, form, decoration, and use, a [ceramic] piece may be understood in context, its language may be read, and the message of the objects enjoyed beyond physical appeal alone."

SUMMARY

The author's goal in this study has been to obtain information not so much about the pots in Building 4/5 as about the day-to-day habits of the people who used them. It is for this reason that the function of a vessel, so far as can be determined, has a priority. However, in looking at the source areas of the vessels, the author has also detected human tendencies from which she has drawn some tentative conclusions of a more general cultural nature.

The significance of Building 4/5 is that it constitutes a relatively uncontaminated deposit, a deposit moreover that was located in Port Royal, Jamaica, the 'best emporium and mart' of England's New World colonies. The advantages of the site are obvious. But there is a downside. Inevitably, valuable evidence has dissolved or has been washed away. Human remains are conspicuous by their absence—only the remains of three children were found associated with the structure. And while organic material, including wood, rope, textiles, hair, feathers, grains, and nuts, has been preserved due to the site's underwater location, much must inevitably be missing. Any material on paper located in the building complex (e.g., notebooks, letters, etc.) has degraded. All of these would have contributed greatly to our understanding of the day-to-day life of the structure's occupants. Nevertheless, the author believes that the ceramic remains that are available yield important information.

That the original location of the individual artifacts is not always beyond doubt is to some extent irrelevant for this study. What we are seeking to reconstruct, after all, is the lifestyle of Port Royal, or at least of its central business area. It is, of course, very useful to be able to examine closely the microcosm of Building 4/5, but any contemporary material that may have been washed into it during the earthquake improves our understanding of its wider context.

The study begins with the data base of 1617 sherds recovered from the three layers at the Building 4/5 site and the approximately 70 percent (n=1155) thereof which was taken to be associated with the living building (including accumulated refuse in the yards and associated areas). This 70 percent, therefore, formed the core of the study. Having analyzed these sherds by reference to diagnostic rim, base, and body fragments and to intact and cross-mended forms, the author concluded that at least 144 ceramic vessels were identifiable. The ware types and forms of these vessels were then categorized and the primary function of each postulated. The following paragraphs summarize the findings.

Ceramic Ware

Five ceramic wares were recognized (see Chapter IV). Tin-glazed earthenware is the most common, comprising 40.9 percent of the vessels. Coarse ware, almost as prevalent, comprises 36.1 percent. Stoneware is far less frequent, amounting to 11.1 percent. Slipware vessels are even fewer, at 7.6 percent. Porcelain, at 4.2 percent, is the least frequent ware in the assemblage.

The manufacturing sources of most of the vessels were also recorded. Of those in tin-glazed earthenware, the vast majority was probably made in England, although some may be Dutch. Only a few are Mediterranean and Mexican.

Of the coarse earthenware vessels, almost 60.0 percent is English, with undecorated Redwares the most frequent type. Mediterranean examples were also identified, Iberian being the most prominent. Hand-coiled Yabba vessels represent about one-fourth of the coarse ware assemblage. Yabba was made by African slaves using local clays.

German stoneware vessels outnumber those from England by almost three to one. Two stoneware vessels have been tentatively identified as either Dutch or South East Asian. One is possibly Iberian.

Almost all of the slipware, like the tin-glazed earthenware, is English. Only two vessels are from the Continent, one from northern Italy, the other (possibly) from North Holland.

All of the porcelain is Chinese export (i.e., made specifically for the Western market). This is to be expected in a 17th-century site, since true porcelain was not made in Europe until 1710.

Vessel Form and Function

Various vessel forms were assigned to the sherds. These forms and their functional descriptions are presented in Chapter III (Table 3.1) and are based upon a vessel form typology established by Beaudry et al. (1983), known as the Potomac Typological System, or more commonly, POTS. Table 6.1 (Chapter VI) quantifies the analyzed sherds and the vessels they represent. The relative proportions of vessels by functional classification are also summarized in Chapter VI (Figure 6.27). The detailed analysis of the ceramic inventory shows the following (in order of frequency):

Food Consumption Forms

These forms—dining and service tableware—comprise the majority of the vessels (28.5 percent of the analyzed assemblage), with bowls the most common form, followed by plates. Porringer and saucer counts are comparatively small. One serving dish (which may have been used for display) and one salt stand have also been identified.

With the exception of bowls, the majority of which are coarse ware, together with two Border ware porringers, all of the food consumption forms are made in fine, decorated ware. Most of these are tin glazed, likely of English provenance, and decorated in blue and white, the most common palette in use by the late 17th century. Iberian and Mexican majolica are represented by fragments of plates, a saucer, and a bowl. Dutch tin-glazed earthenware is represented by the large, decorated serving dish noted above. Staffordshire and North Holland slipware and Chinese export porcelain are present in limited amounts.

Food and Beverage Storage Forms

Storage wares are the second most common forms, comprising 22.9 percent of the analyzed assemblage. Within this category, variously sized storage pots are by far the most prevalent. A comparatively small number of bottles are also represented.

Of the identified storage pots, all but two are made in coarse ware, the remainder being stoneware. Over half of the coarse ware vessels are English (mostly undecorated Redware). Iberian and African-Jamaican Yabba are also present. One of the stoneware forms is English, the other is (possibly) Iberian.

All but one of the identified bottles are made in stoneware and are mostly of German provenance. The exception is a tin-glazed (possibly Dutch) wine bottle. That relatively few ceramic bottles were found suggests that the household used glass, which, from about 1680, became more affordable and began to be used for short-term storage and serving of wine and other liquids. Indeed, this suggestion is supported by the recovery of 10 glass bottles from Room 2.

Beverage Consumption Forms

These are nearly as common as the storage wares (22.2 percent of the analyzed assemblage). Within this category, cups are by far the most frequent, followed by mugs and communal drinking pots, which are represented in equal, but much fewer, numbers. Four costrels, a ewer, a punch bowl, and a tea bowl were also identified.

As with the food consumption category, very few drinking vessels are made in coarse ware. Like the eating/serving forms, too, most of the vessels in this category are tin-glazed and (likely) of English provenance. Plain and painted (blue-and-white and polychrome) tin-glazed forms are represented in equal amounts. Slip-decorated vessels (Staffordshire cups and a North Italian marbled costrel) and those in Chinese export porcelain are present in limited amounts. A distinct difference between the food and beverage consumption categories at Building 4/5 is the presence in the latter of stoneware vessels, which are second to the tin-glazed forms in terms of frequency.

Forms Related to Health and Personal Hygiene

These comprise 24 vessels (16.7 percent of the analyzed assemblage). Within this category, apothecary pots are by far the most prevalent, followed a distant second by chamber pots and an even more distant third by washbasins.

With the exception of four chamber pots (in English Redware and Border ware), all of the vessels are tin-glazed and are all probably English. While most of the vessels appear to be plain white, which would have been produced in thousands and was generally reserved for these more 'ordinary' forms, this cannot be ascertained, since many have lost a significant proportion of their glaze. The exceptions are two apothecary pots and one washbasin, which have retained their original decoration in blue and white.

Food Processing Forms (Preparation and Cooking)

Food processing vessels are the least common in the analyzed assemblage, with a total of 11 vessels identified (7.6 percent of the analyzed assemblage). All but one (a Mottled slipware pudding/pastry pan, which may have been [also] used at table) are made in coarse ware (English Redware and Border ware). All except four Yabba cooking pots are of English provenance.

That relatively few cooking/kitchen wares were found is not unexpected, since by the 17th century, most households had the much more utilitarian option of metal pots for cooking. Ceramics are subject to breakage as a result of too rapid heating or cooling. Removing a hot earthenware pot from the fire and placing it on a cold surface, or leaving it on the fire until its contents have dried up and burnt, also causes cracking that will ruin the vessel for cooking and almost any other form of adaptive reuse.

Ceramic Distribution
Isolating the vessel forms and examining their distribution in and around Building 4/5 reveals patterns that seem to confirm the functional layout of the complex, as suggested by architectural details and other finds. Most of the coarse ware vessels were found in the rooms and yards at the back of the structure, where they appear to have been used for food preparation, cooking, and food and beverage storage. That relatively few of these vessel forms were found at the front of the structure suggests that little or none of these activities took place in these areas.

Tin-glazed earthenware and slipware vessels—predominantly dining and drinking forms—were recovered in quantity from in and around Rooms 1, 2, and 3, strongly suggesting these areas were used, at least in part, for food and beverage consumption and also possibly for storing tableware. As was noted in Chapter VI, some of these decorative wares were also recovered from the site's back yards. In the case of Building 5's yard (Yard 5), it may be that some of these vessels were also used in this area, since other food-related artifacts found in this location and the fact that it is brick-paved suggests that it served more than just a place in which to dump refuse. Such throw-away activity, resulting in surface middens, seems to have been more common in Building 4's yards (Yards 4A and 4B) (e.g., several hundred discarded used clay tobacco pipes were recovered from these locations, see Fox [1998:76, Table 3]). It is, of course, entirely possible that some of the fine wares recovered from all three yards may have been displaced during the disaster. At least some of those in Yards 4A and 4B surely came from the front two rooms of that structure (Rooms 4A and 4B), having being blown out on to the yards as the ship rammed through the building. At least some in Yard 5 were probably originally housed in that building's upper story(s).

Stoneware at the site (mostly bottles and mugs) was concentrated around Building 4, specifically in the yards and in the alley located to the east of the structure. It may be that some (or all) of these vessels, like those in fine earthenware, originated in Building 4, or indeed in the upper floor(s) of Building 5, and were displaced during the disaster. Again, the broken vessels (two stoneware bottles were found intact) may be from surface middens that had accumulated in these yards.

Porcelain at the site is conspicuous by its near absence. Some of the vessels identified (a bowl, a cup, and a saucer) were recovered from the yards. Another bowl came from Room 2, while a tea bowl was found on the brick-paved sidewalk in front of Building 5. The author believes the pieces from the sidewalk were originally housed either in Building 5's upper story(s) or in Building 8, which is located across the street, north of the complex.

With regard to particular forms, many of the plates (both plain and decorated) were recovered from Room 3 and Yard 5. Those found in Room 3 were undoubtedly used in this location (they were recovered from under a fallen interior wall which lay across the entire floor). Those in Yard 5 may have also been used in this area. It is again entirely possible that some (if not all) of those from the yard were originally housed in Building 5's upper story(s), since as noted in Chapter V, 17th-century diners often set their tables in the upstairs bedrooms, "and kept their eating and drinking vessels in upstairs or downstairs cupboards, wherever convenience and the season suggested" (Belden 1983:5-6).

The washbasins may have also come from the upper story(s) in Building 5, as suggested in Chapter VI. Their presence at the site at a time when cleanliness was not achieved "with soap and water, but rather 'created' with cosmetics and perfumes" (Schaefer 1998:90) suggests that the Building 4/5 household, or at least those who lived in Building 5, could afford to pay attention to personal etiquette, which in the 17th century demanded that the hands be washed before meals (Gunn 1973:80; Schaefer 1998:91). This interpretation is enhanced by the recovery of the ewer (in fragments) from Rooms 1 and 2.

GENERAL DISCUSSION

The Building 4/5 Complex: Site Function
The majority of the larger houses in Port Royal was "invariably owned or occupied by merchants, factors for overseas merchants, or prosperous artisans like sailmakers and coopers" (Pawson and Buisseret 1975:107). Building 4/5, however, may have not been in single ownership, based on documentary evidence in the form of land plats and archaeological evidence in the forms of various ownership marks on artifacts (see Chapter III and Appendix C). On the face of it then, it probably falls into a middle-income category.

It is, of course, not easy to deduce the role(s) of Building 4/5 from 144 ceramic vessels. What is impressive, however, is the sheer *variety* of their forms and sizes, which seems to confirm the multi-purpose function of the structure, especially when one remembers that other materials, namely metal, glass, and wood, were also in use, as revealed by excavation. It is entirely credible to say that part at least of the building complex (namely Building 5) may have served in part as some type of rudimentary restaurant (a 'victualing house') or small inn/public house (an 'ordinary'). The range and numbers of eating/drinking/serving vessels (in glass and pewter, as well as in ceramic) and the pewter spoons support the hypothesis. Porringers, bowls, and plates, important if meals are being served, are present in fair amounts. Those meals need not have been elaborate, hence the apparent lack of cookware. Quantities of bottles, caudle cups, and mugs are indicative of an association with ale or beer. Chocolate/coffee, and perhaps even tea, appears to have also been served. The recovery of stacks of pewter plates, several new tobacco pipes, uncorked glass bottles, and stemmed drinking glasses (see McClenaghan

1988, Fox 1998) also lends credence to the hypothesis of the complex containing in part of it something of a public or semi-public nature connected with eating and/or drinking and socializing.

There are indicators, too, that Building 4/5 contained a residential element, with washbasins featuring in the ceramic inventory. Apothecary pots, some of which may have contained medicines and/or perfumes, were also found. Indeed, remnants of an internal staircase reveal that Building 5 was at least two-storied, and records of the town indicate that mercantile/dwelling combinations were common, with the living quarters often located on an upper floor.

Indeed, the ceramic assemblage seems to be consistent with the physical lay-out suggested by the structural remains, and the author has not found anything to contradict the hypothesis that the Building 4/5 complex comprised a moderately affluent (probably multiple occupant), structurally sound residence and/or commercial establishment dealing in food and drink (namely Building 5), with an attached one-story building of poorer construction that may have housed tenants and/or the domestic help who worked in Building 5. (There may have been some commercial activity attached to Building 4 as well.) This view is strengthened by other evidence at the site: At Building 5, physical evidence on the ground points to a well-founded establishment (e.g., decorative brick floors, plastering of walls [in Rooms 1 and 2] and also of the floor in Room 1; lead caming on the windows with the remnants of panes; and the presence in front of a brick-paved sidewalk). At Building 4, two cast-iron cooking pots and a trivet, a brass mortar, and pewter plates and spoons were recovered from in around and the site. Faunal remains were also recorded, indicating food preparation and butchering. These artifacts and residues are certainly explainable by a hypothesis of residential occupation.

The Building 4/5 Complex*: *Occupants' Lifestyle

The small size of the ceramic assemblage makes interpretation of the lifestyle elusive also, but both the everyday and fashionable aspects of life at Port Royal can be detected. The households of most people of middle rank in the 17th century were sustained by servants (or slaves), and their ghostly presence certainly hovers over the coarse earthen cooking pots and the Yabba bowls. The predominance of storage vessels tells us that the provisioning of the household relied heavily on goods that were not immediately accessible and that the economic use of resources was a significant feature of life.

There are indicators, too, that the occupants of Building 4/5 (certainly those in Building 5) formed settled households of reasonably comfortable, although not overly affluent, means: hardly any decoratively matched pieces were found, pointing to the acquisition over time of individual pieces of quality rather than sets, or partial sets, of wares. Also, it appears that some of the porcelain recovered may have come from Building 8, and that the Building 4/5 residents appear to have owned very little, although the use of Chinese porcelain is well documented as a 'middle-market' phenomenon (see Weatherill 1988). On the other hand, tin-glazed earthenware (a luxury item) appears in relative abundance, and Westerwald stoneware (a carefully executed product) is present, together with silver forks and spoons. The tin-glazed saucers may also point to the use of salt and other spices, such as saffron, cinnamon, cloves, nutmeg, mace, and pepper, all considered 'luxuries' in the 17th century.

In trying to assess the value of the Port Royal finds, it is necessary to bear in mind the location of the town as a port of discharge for the trans-oceanic trade. The amount of decorative European pottery at the site suggests a history of local trading in such products stretching back over a good number of years. The residents at Building 4/5 were well-placed to benefit from this trade and appear to have done so. For example, there is some English slipware, although trade in this ware in the New World was in its infancy in 1692. Continental European and Mexican wares are also represented, yet official mercantile policies meant that non-English ceramics could get through only in comparatively small numbers. Further, as was noted in Chapter IV, no significant Spanish shipping took place in the Caribbean after 1630.

Despite the particular circumstances of Building 4/5, the author thinks it is possible to deduce a society at a point of change. That the residents in at least part of it found it desirable to acquire such utensils and other items as were found implies that concepts of lifestyle were important. Certainly, we know from numerous sources that the end of the 17th/ beginning of the 18th centuries was a period of marked change in domestic habits. The tin-glazed ware assemblage, in particular, is diverse, appearing in table- and hygiene/storage forms, as well as decorative ornament. The numerous tin-glazed plates, quite possibly used at table and not simply as ornament, suggest that the fashionable practice of individualized dining had been adopted. The chocolate/coffee cups suggest that these new hot beverages were being consumed. The tin-glazed punch bowl was another new vessel form, and it, as well as a rather elaborate salt stand, confirms consciousness of social position. The reconstructed decorated tin-glazed flower vase indicates the ability (and desire) to acquire inessentials.

These new attitudes were not confined to 'public' display. It would appear that a range of cooking methods was being practiced (a silver spice grater and a large brass mortar and strainer were found), and that distinctions were being made in relation to the foods and liquids brought to table. Tin-glazed vessels, such as the washbasins and medicine/ointment/cosmetic storage pots, evidence a concern and value for appearance and health. (Other artifacts related to personal appearance include two bone combs, a brass buckle, and pewter buttons.)

What is equally apparent nevertheless is the evidence of 'on-going' habits (particularly in Building 4). The presence of communal drinking pots confirms this, as does the large (probably shared) coarse earthenware bowls.

It must be remembered, of course, that the relationship between vessel form, function, and use is complex. Many vessels probably served multiple functions, particularly if access to replacement vessels was restricted to the arrival of merchant fleets. In the case of Building 4/5, we are fortunate to have a 'snapshot' of the living conditions in 1692, which provides a most valuable context for the recovered ceramics. Jonathan Swift's satire on the behavior of English servants provides insight into the misuses to which ceramic vessels might have been put:

> When you have broken all your earthen Drinking Vessels below Stairs [in the servants' rooms and workrooms] (which is usually done in a Week) the Copper Pot will do as well; it can boil Milk, heat Porridge, hold Small-Beer, or in Case of Necessity serve for a Jordan [a chamber pot]; therefore apply it indifferently to all of these Uses; but never wash it or scour it, for fear of taking off the Tin (Swift 1958[1726]:499-500).

The Comparison Assemblages at Virginia and Guildford, England

Table 8.4 in Chapter VIII, which compares the Building 4/5 assemblage with two others from contemporary sites, assists us in trying to determine what Building 4/5 was used for, although it is the *ratio* of vessel forms in the table, not the actual numbers, that must be borne in mind. The findings from the Virginian governor's site confirmed that while some of Building 4/5's residents were probably reasonably affluent, holding a certain position in society, they were definitely not playing on the same status field as a governor. The most obvious indicator for this was the far greater quantity in Drummond II of Chinese export porcelain. Furthermore, most of the Drummond II porcelain is in the form of tea wares, which are not prevalent in Building 4/5. This last point is particularly significant, since tea at the end of the 17th century/ early 18th century was a luxury and the drinking of it was a mark of the well bred or wealthy.

As for the English site (a refuse dump from a popular inn located in Guildford in Surrey), it was probably most helpful in confirming that Building 4/5 was, in large part, residential, and that if there was a commercial aspect to a portion of it in the dealing of food and/or drink, it was probably not a tavern or inn as such; much more likely is that it was a smaller restaurant-type establishment. The most obvious indicator for this was that a far greater proportion of the inn's drinking forms were those used for the traditional alcoholic beverages of ale and beer. Tavern type mugs and communal drinking pots are far less frequent in Building 4/5. Instead, nearly half of its drinking forms are cups of the shape used to drink chocolate or coffee.

A more illuminating aspect of the three-way comparison, however, relates to ceramic ware type. While the colonial deposits include relatively even proportions in all categories of wares, the English site is almost wholly made up of coarse ware. It is true, of course, as noted above, that the Tun Inn assemblage is from a refuse pit, but the discrepancy is not without significance. It strengthens the conjecture that Building 4/5 was a place of multiple use and housed, in part of it at least, residents with a good standard of living.

This study has focused on the practical use of ceramic vessels, but what has become apparent from the external comparisons is the emerging 'style consciousness' manifested in them. It is not especially surprising that a wealthy state official possessed elegant and decorative wares, but it is rather unexpected to find that a house in a bustling seaport possessed them also. Historian Lorna Weatherill (1988:25) notes that the period post-1690 was one of changing consumption, as exemplified by the great expansion in the ownership of china and utensils for hot drinks. If we look at the chocolate/coffee cups at Building 4/5 (12 to Drummond's 13 and the inn's 3), as one example, we can see that the Port Royal site was well in the vanguard of this change.

Deference

But what to the author is the most interesting and, perhaps, most significant finding in Chapter VIII (see Table 8.5) is the preponderance of locally produced ware in the English inn's assemblage. Apart from some German stoneware and one Chinese export piece, the vessels are all home-grown products (if one lays aside some doubt regarding some of the tin-glazed earthenware). This is in sharp contrast to the colonial assemblages, where only some of the coarse ware is locally made. By far, the largest proportion of their vessels consists of English imports. This is, of course, not unexpected. The tendency for colonial settlers to bring their culture with them is well recorded, and in the households of prosperous Port Royal, positioned as it was at a crossroads of trading routes, it would be strange if goods from the Old World, especially England, did not predominate.

It is only when we see that the ceramic assemblage of an English inn is dominated by goods produced locally that an important implication of the *variety* of ceramic wares and types found at Building 4/5 shows itself. The English inn-keeper was happy to patronize English potters. The residents at Building 4/5, on the other hand, went to market and chose their finer wares from foreign wares and appear to have been aware of what were the 'latest things' and the differences between them.

Indeed, the author believes it is possible to say that the increasing availability of ceramics from Europe, coupled with a deliberate policy by England to tie the colonies to the English trade in pottery, must have led to a significant loss of skill among colonial potters: England saw to it that colonial production was confined to cheap coarse ware for all practical purposes. (Cromwell was not alone in recognizing the value of the colonies to the English export market.) Therefore, as we have seen, for colonial families, the acquisition of European ceramics became a matter of quality and the owning of them a mark of social success.

How could these things not undermine the evolution of an indigenous colonial pottery industry? Why would the descendants of the settlers care about what their own craftsmen could produce when commercial and social pressures pointed them to Europe? How could Americans not feel diffident about the ability of their own society to produce fine decorated work when they looked for it always across the seas? The author believes that deference to Europe affected succeeding generations across many other art forms as well. Indeed, it can be argued, can it not, that the perception of Europe as the repository of 'culture' and of excellence in the arts generally did not begin to be challenged by Americans until the second half of the 18th century.

REFERENCES CITED

Allan, J. P., and J. Barber
1992 A 17th-Century Pottery Group from Kitto Institute, Plymouth. In *Everyday and Exotic Pottery from Europe*, edited by D. R. M Gaimster and M. Redknap, pp. 225-245. Oxbow Books, Oxford.

Amis, P.
1968 Some Domestic Vessels of Southern Britain: A Social and Technical Analysis. *Journal of Ceramic History* 2. George Street Press, Stafford, England.

Andersen, A. B.
1974 A Dutch Galliot which Struck a Rock in 1677. *The International Journal of Nautical Archaeology and Underwater Exploration* 3(1):91-100.

Archer, M.
1997 *Delftware: The Tin-Glazed Earthenware of the British Isles, A Catalogue of the Collection in the Victoria and Albert Museum.* The Stationery Office/Victoria and Albert Museum, London.

Archer, M., and B. Morgan
1977 *Fair as China Dishes: English Delftware from the Collection of Mrs. Marion Morgan and Brian Morgan.* International Exhibitions Foundation, London.

Armstrong, D. V.
1985 An Afro-Jamaican Slave Settlement: Archaeological Investigations at Drax Hall. *In The Archaeology of Slavery and Plantation Life*, edited by T. A. Singleton, pp.261-287. Academic Press, San Diego.

1990 *The Old Village and the Great House: An Archaeological and Historical Examination of Drax Hall Plantation, St. Ann's Bay, Jamaica.* University of Illinois Press, Urbana.

Arnold, B., and R. Weddle
1978 *The Nautical Archaeology of Padre Island: The Spanish Shipwrecks of 1554.* Academic Press, New York.

Ayers, J.
1980 *Far Eastern Ceramics in the Victoria and Albert Museum.* Sotheby Park Bernet/ Victoria and Albert Museum, Totowa, New Jersey.
1985 *Chinese Ceramics: The Koger Collection.* Sotheby, London.

Barber, V. C.
1977 The *Sapphire*, a British Frigate, Sunk in Action in Bay Bulls, New Foundland in 1696. *The International Journal of Nautical Archaeology and Underwater Exploration* 6(4):305-313.

Bardenheuer, F. H.
1960 Gedanken über Bartmannkrüge. *Keramos* 10:39-42.

Beaudry, M. C.
1988 Words for Things: Linguistic Analysis of Probate Inventories. In *Documentary Archaeology in the New World*, edited by M. C. Beaudry, pp. 43-50. Cambridge University Press, Cambridge.

Beaudry, M. C., J. Long, H. M. Miller, F. D. Neiman, and G. Wheeler Stone
1983 A Vessel Typology for Early Chesapeake Ceramics: The Potomac Typological System. *Historical Archaeology* 17(1):18-42.

Beckmann, B.
1974 The Main Types of the First Four Production Periods of Siegburg's Pottery. In *Medieval Pottery from Excavations*, edited by V. I. Evison, H. Hodges, and J. G. Hurst, pp. 183-220. St. Martin's Press, New York.

Beebe, L.
1980 Rhenish Stoneware of the Renaissance. *American Ceramic Circle Bulletin* 2:125-140.

Belden, L. C.
1983 *The Festive Tradition: Table Decoration and Desserts in America, 1650-1900.* W. W. Norton, New York/London.

Berry, F.
1933 *Wine Trade of Drinking Vessels, also Books, Documents, Etc.* (Exhibition 23 June-6 July.) Vintner's Hall, London.

Bimson, M.
1961a References to John Dwight in a Seventeenth-Century Manuscript. *Transactions of the English Ceramic Circle* 4(5):10-12.
1961b John Dwight. *Transactions of the English Ceramic Circle* 5(2):95-109.

Binford, L. R.
1962 Archaeology as Anthropology. *American Antiquity* 28(2):217-226.
1965 Archaeological Systematics and the Study of Culture Process. *American Antiquity* 31(2):203-210.

Black, C. V.
1965 *The Story of Jamaica: From Prehistory to the Present.* Second edition. Collins, London.

Blacker, J. F.
1922 *The ABC of English Salt Glaze Stoneware, from*

Dwight to Doulton. S. Paul, London.

Blome, R.
1678 *A Description of the Island of Jamaica*. Published by the author, London.

Bowen, J.
1994 A Comparative Analysis of the New England and Chesapeake Herding Systems. In *Historical Archaeology of the Chesapeake*, edited by P. A. Shackel and B. J. Little, pp.155-167. Smithsonian Institution Press, Washington D.C.

Bragdon, K. J.
1988 Occupational Differences Reflected in Material Culture. In *Documentary Archaeology in the New World*, edited by M. C. Beaudry, pp. 79-91. Cambridge University Press, Cambridge.

Bratton, J.
1992 Yabba Ware: The African Presence at Port Royal, Jamaica. Manuscript on file, Department of Anthropology, Texas A&M University, College Station.

Braudel, F.
1973 *Capitalism and Material Life, 1400-1800*. Harper and Rowe, New York.

Britton, F.
1982 *English Delftware in the Bristol Collection*. Sotheby, London/New Jersey.

1987 *London Delftware*. Jonathon Horne, London.

1990 The Pickleherring Potteries: An Inventory. *Post-Medieval Archaeology* 24:61-92.

1993 Delftware Inventories. *Transactions of the English Ceramic Circle* 15(1):59-64.

Bronitsky, G., and R. Hamer
1986 Experiments in Ceramic Technology: The Effects of Various Tempering Materials on Impact and Thermal-Shock Resistance. *American Antiquity* 51:89-101.

Brown, M.
1996 An Archaeological Study of Social Class as Reflected in a British Colonial Tavern Site in Port Royal, Jamaica. Unpublished M.A. thesis, Department of Anthropology, University of Texas at San Antonio.

Brown, M. R.
1973 Ceramics from Plymouth, 1621-1800: The Documentary Record. In *Ceramics in America*, edited by I. M. G. Quimby, pp. 41-74. University Press of Virginia, Charlottesville.

Bushell, S. W.
1910 *Description of Chinese Pottery and Porcelain*. Clarendon Press, New York.

Buys, S., and V. Oakley
1993 *The Conservation and Restoration of Ceramics*. Butterworth-Heinemann, Oxford.

Cackette, M., J. M. D'Auria, and B. E. Snow
1987 Examining Earthenware Vessel Function by Elemental Phosphorous Content. *Current Anthropology* 28(1):121-127.

Caiger-Smith, A.
1973 *Tin-Glaze Pottery in Europe and the Islamic World: The Tradition of 1000 Years in Maiolica, Faience, and Delftware*. Faber and Faber, London.

Calado, R. S., and J. Baart
1987 *Faiança Portuguesa, 1600-1660*. Ministerio dos Nogocios Estrangeiras Secretaria de Estado da Cultura, Lisbon, Portugal.

Carr, L.G., and L. S. Walsh
1988 The Standard of Living in the Colonial Chesapeake. *William and Mary Quarterly* 45(1):135-159.

Caton, M.
1999 *Fooles and Fricassees: Food in Shakespeare's England*. Folger Shakespeare Library, Washington, D.C.

Chaffers W.
1850 On Medieval Earthenware Vessels. *Journal of the British Archaeological Association* 5:22-29.

1946 *Marks and Monograms on European and Oriental* [1856] *Pottery and Porcelain*. Borden, Los Angeles.

Chappell, W.
1859 *Popular Music of the Olden Time: A Collection of Ancient Songs, Ballads, and Dance Tunes*. Cramer, Beale, and Chappell, London.

Church, A.
1911 *English Earthenware Made During the 17th and 18th Centuries*. Her Majesty's Stationery Office/Victoria and Albert Museum, London.

Clark, G. R.
1995 Swallowed up. *Earth Magazine*. April 1995:34-41.

Clifford, S. A.
1993 An Analysis of the Port Royal Shipwreck and its Role in the Maritime History of Seventeenth-Century Port Royal, Jamaica. Unpublished

M.A. thesis, Department of Anthropology, Texas A&M University, College Station.

Coles, J.
1973 *Archaeology by Experiment*. Charles Scribner's Sons, New York.

1979 *Experimental Archaeology*. Academic Press, New York.

Cooper, R. G.
1968 *English Slipware Dishes, 1650-1850*. Alec Tiranti, London.

Corke, S.
1997 The Documentary Evidence. In Excavation of a Pit at 16 Tunsgate, Guildford, Surrey, 1991, by K. Fryer and A. Selley, pp. 147-154. *Post-Medieval Archaeology* 31:139-230.

Courtney, P.
1986 Documentary Evidence for the Use of Ceramics and Vessel-Glass at Two Great Houses: Tredegar House, Gwent and Chirk Castle, Clwyd. *Medieval and Later Pottery in Wales* 9:7-15.

1989 Some References to Ceramics and Glass in the Wynn of Gwydir Papers. *Medieval and Later Pottery in Wales* 10:65-68.

Cox, O.
1992 Seventeenth-Century Port Royal: Its Urban Form and Architectural Character. Paper presented at the Annual Meeting of The Society for Historical Archaeology Conference on Historical and Underwater Archaeology, Kingston, Jamaica.

Cox, O., and J. Cox
1987 Port Royal, Jamaica: Upgrading and Renewing the Historic City, Report on the First Stage of Implementation. Report submitted to the Jamaica National Heritage Trust and the Urban Development Corporation, Kingston, Jamaica.

Crellin, J. K.
1970 Medical Ceramics: Their Scope and Significance. *Transactions of the English Ceramic Circle* 3(3):191-199.

Crocket, Captain
1692 *A True and Perfect Relation of the Most Sad and Terrible Earthquake at Port-Royal in Jamaica, Which Happened on Tuesday the 7th of June, 1692*. R. Smith, London.

Crossley, D.
1990 *Post-Medieval Archaeology in Britain*. Leicester University Press, Leicester/London.

Cunningham, C. M.
1985 The Pottery. In Post-Medieval Sites and their Pottery: Moulsham Street, Chelmsford, England. Chelmsford Archaeological Trust Report 5, edited by C. M. Cunningham and P. J. Drury. *Council for British Archaeological Report* 54:63-78.

Curtis J. B.
1988 Perceptions of an Artifact: Chinese Porcelain in Colonial Tidewater Virginia. In *Documentary Archaeology in the New World*, edited by M. C. Beaudry, pp. 20-31. Cambridge University Press, Cambridge.

Darrington, G.
1994 Analysis and Reconstruction of Impermanent Structures in the 17th and 18th Centuries. Unpublished M.A. thesis, Department of Anthropology, Texas A&M University, College Station.

David, N.
1972 On the Life Space of Pottery, Type Frequencies, and Archaeological Inference. *American Antiquity* 37:141-142.

Dawson, G.
1978 Two Delftware Kilns at Montague Close, Southwark. *The London Archaeologist* 1(10):228-231.

Deagan, K.
1983 *Spanish St. Augustine: The Archaeology of a Colonial Creole Community*. Academic Press, New York.

1987 *Artifacts of the Spanish Colonies of Florida and the Caribbean, 1500-1800*. Vol 1, *Ceramics, Glassware, and Beads*. Smithsonian Institution Press, Washington, D.C.

Deetz, J.
1973 Ceramics from Plymouth, 1620-1835: The Archaeological Evidence. In *Ceramics in America*, edited by I. M. G. Quimby, pp. 15-40. University Press of Virginia, Charlottesville.

1977 *In Small Things Forgotten: The Archaeology of Early American Life*. Anchor Press/ Doubleday, New York.

Dewolf, H.
1998 *Seventeenth-Century Chinese Porcelain at Port Royal, Jamaica*. Ph.D dissertation, Department of Anthropology, Texas A&M University, College Station. University Microfilms International, Ann Arbor, Michigan.

Di Zerega Wall, D.
1991 Sacred Dinners and Secular Teas: Constructing Domesticity in Mid-19th Century New York. *Historical Archaeology* 25(4):69-81.

Draper, J.
1984 *Post-Medieval Pottery, 1650-1800.* Shire, Buckinghamshire, England.

Du Boulay, A.
1984 *Christie's Pictorial History of Chinese Ceramics.* Prentice-Hall, Englewood Cliffs, New Jersey.

Dufour, P. S.
1685 *Traitez nouveaux & curieux du café, du thé, et du chocolate.* Jean Giron and B. Riviere, Lyons, France. Microfilm, Evans Library, Texas A&M University, College Station.

Eden, M., and V. Eden
1999 *Slipware: Contemporary Approaches.* A&C Black, London.

Edwards, R.
1974 London Potters circa 1570-1710. *Journal of Ceramic History* 6. George Street Press, Stafford, England.

Elliot, G. B.
1986 The Success of Frechener Flagons Exported as 'Cologne Ware' in Late Medieval Trade from the Rhine. *The International Journal of Nautical Archaeology and Underwater Exploration* 15(2):85-91.

Emmerson, R.
1992 *British Teapots and Tea Drinking, 1700-1850.* The Stationery Office, London.

Fairbanks, C. H.
1973 The Cultural Significance of Spanish Ceramics. In *Ceramics in America*, edited by I. M. G. Quimby, pp. 141-174. University Press of Virginia, Charlottesville.

Ferguson, L.
1992 *Uncommon Ground: Archaeology and Early African America, 1650-1800.* Smithsonian Institution Press, Washington, D.C.

Forster, W. A., and K. B. Higgs
1973 The *Kennermerland*, 1971: An Interim Report. *The International Journal of Nautical Archaeology and Underwater Exploration* 2(2):291-300.

Fox, G.
1998 *The Study and Analysis of the Kaolin Clay Tobacco Pipe Collection from the 17th-Century Archaeological Site of Port Royal, Jamaica.* Ph.D. dissertation, Department of Anthropology, Texas A&M University, College Station. University Microfilms International, Ann Arbor, Michigan.

Fox, R., and K. G. Barton
1986 Excavations at Oyster Street, Portsmouth, Hampshire, 1968-1971. *Post-Medieval Archaeology* 20:31-256.

Franklin, M.
1992 Wrought-Iron Hand Tools in Port Royal, Jamaica: A Study Based Upon a Collection of the Tools Recovered from Archaeological Excavations and Listed in the Probate Records of Colonial Port Royal, c. 1692. Unpublished M.A. thesis, Department of Anthropology, Texas A&M University, College Station.

Fryer, K., and A. Selley
1997 Excavation of a Pit at 16 Tunsgate, Guildford, Surrey, 1991. *Post-Medieval Archaeology* 31:139-230.

Gaimster, D. R. M.
1997 *German Stoneware, 1200-1900: Archaeology and Cultural History.* British Museum Press, London.

Gaimster, D. R. M., and D. R. Hook
1995 Post-Medieval Stoneware Manufacture and Trade in the Rhineland and Southern Britain: A Programme of Neutron Activation Analysis at the British Museum. In *Trade and Discovery: The Scientific Study of Artefacts from Post-Medieval Europe and Beyond*, edited by D. R. Hook and D. R. M. Gaimster, pp. 69-90. The British Museum, London.

Gardner, W. J.
1971 *A History of Jamaica: From its Discovery by Christopher Columbus to the Year 1872.* Frank Cass, London.

Garway, T.
1660 *An Exact Description of the Growth, Quality, and Vertues of the Leaf Tea.* Printed by the author, London.

Goggin, J. M.
1960 The Spanish Olive Jar: An Introductory Study. *Yale University Publications in Anthropology* 62. Department of Anthropology, Yale University, New Haven, Connecticut.

1968 Spanish Majolica in the New World: Types of the 16th-18th Centuries. *Yale University Publications in Anthropology* 72:115-135. Department of Anthropology, Yale University,

New Haven, Connecticut.

Gotelipe-Miller, S.
1990 Pewter and Pewterers from Port Royal, Jamaica: Flatware before 1692. Unpublished M.A. thesis, Department of Anthropology, Texas A&M University, College Station.

Grant, A.
1983 *North Devon Pottery: The Seventeenth Century.* University of Exeter Press, Exeter, England.

Green, J. N. (editor)
1977 The Loss of the Verenigde Oostindische Compagnie Jacht *Vergulde Draeck*, Western Australia 1656: An Historical Background and Excavation Report. Two volumes. *British Archaeological Reports Supplementary Series* 36. Oxford.

Greene, J. P
1988 *Pursuits of Happiness: The Social Development of Early Modern British Colonies and the Formation of American Culture.* University of North Carolina Press, Chapel Hill.

Griffiths, D.
1978 Use-Marks on Historic Ceramics: A Preliminary Study. *Historical Archaeology* 12:68-81.

Grigsby, L. B.
1990 *English Pottery: Stoneware and Earthenware, 1650-1800: The Henry H. Weldon Collection.* Philip Wilson for Sotheby, London.

Gunn, F.
1973 *The Artificial Face: A History of Cosmetics.* Newton Abbot, David and Charles, London.

Hailey, T. I.
1994 *The Analysis of 17th-, 18th-, and 19th-Century Ceramics from Port Royal, Jamaica for Lead Release: A Study in Archaeotoxicology.* Ph. D. dissertation, Department of Anthropology, Texas A&M University, College Station. University Microfilms International, Ann Arbor, Michigan.

Hally, D. J.
1983 Use Alteration of Pottery Vessel Surfaces: An Important Source of Evidence for the Identification of Vessel Function. *North American Archaeologist* 4(1):3-26.

Hamilton D. L.
1984 Preliminary Report on the Archaeological Investigations of the Submerged Remains of Port Royal, Jamaica, 1981-1982. *The International Journal of Nautical Archaeology and Underwater Exploration* 13(1):11-25.

1986a The City Under the Sea. *Science Year, 1986: World Book Science Annual*, pp. 92-109. World Book, Chicago.

1986b Port Royal Revisited. Underwater Archaeology: In Proceedings of the 14th Conference on Underwater Archaeology, edited by C. R. Cummings. *Special Publication* 7:73-81. Fathom Eight, San Marino, California.

1986c Underwater Archaeology Field School Manual: Port Royal, Jamaica, 1986 Season. Manuscript on file, Department of Anthropology, Texas A&M University, College Station.

1987a The Port Royal Project: A 17th-Century Learning Laboratory for Future Underwater Archaeologists. Port Royal 1981. *INA Newsletter* 14(1/2):4-6. Institute of Nautical Archaeology, Texas A&M University, College Station.

1987b Underwater Archaeology Field School Manual: Port Royal, Jamaica, 1987-1990 Seasons. Manuscript on file, Department of Anthropology, Texas A&M University, College Station.

1988a Underwater Excavations of 17th-Century Buildings at the Intersection of Lime and Queen Streets, Port Royal, Jamaica. *Underwater Archaeology Proceedings from the Society for Historical Archaeology Conference*, edited by J. P. Delgado, pp. 9-12. Reno, Nevada.

1988b Port Royal, Jamaica: A City Coming to Light. *INA Newsletter* 15(1):6-7. Institute of Nautical Archaeology, Texas A&M University, College Station.

1990a A Decade of Underwater Research at Port Royal. *INA Newsletter* 17(2):4-7. Institute of Nautical Archaeology, Texas A&M University, College Station.

1990b Port Royal 1990: The Last Excavation Season. *INA Newsletter* 17(4):14-17. Institute of Nautical Archaeology, Texas A&M University, College Station.

1991 A Decade of Excavations at Port Royal, Jamaica. *Underwater Archaeology Proceedings from the Society for Historical Archaeology Conference*, edited by J. D. Broadwater, pp. 90-94. Richmond, Virginia.

1992 Simon Benning: Pewterer of Port Royal. In *Text-Aided Archaeology*, edited by Barbara J. Little, pp. 39-54. CRC Press, Boca Raton, Florida.

1996 *Basic Methods of Conserving Underwater Archaeological Material Culture.* Legacy Resource Management Program, U.S. Department of Defense, Washington, D.C.

Hamilton, D. L., and R. Woodward
1984 A Sunken 17th-Century City: Port Royal, Jamaica. *Archaeology* 37(1):38-45.

Handler, J. S.
1963 A Historical Sketch of Pottery Manufacture in Barbados. *Journal of the Barbados Museum and Historical Society* 30(3):129-153.

Hanson, F.
1683 *The Laws of Jamaica.* Printed by H. Hills for Charles Harper, London. Microfilm, Evans Library, Texas A&M University, College Station.

Haselgrove, D.
1990 Steps Towards English Stoneware Manufacture in the 17th Century, Part 2:1650-1700. *The London Archaeologist* 6(6):152-159.

1992 Imported Pottery in the 'Book of Rates': English Customs Categories in the 16th and 17th Centuries. In *Everyday and Exotic Pottery from Europe, c. 650-1900: Studies in Honour of John G. Hurst*, edited by D. R. M. Gaimster and M. Redknap, pp. 324-335. Oxbow Books, Oxford.

Haselgrove, D., and J. van Loo
1998 Pieter van den Ancker and Imports of Frechen Stoneware Bottles and Drinking Pots in Restoration London c. 1660-67. *Post-Medieval Archaeology* 32:45-74.

Heath, B.
1991 Pots of Earth: Forms and Functions of Afro-Caribbean Ceramics. *Florida Journal of Anthropology Special Publications* 7:33-49.

Heidtke, K.
1992 Jamaican Red Clay Pipes. Unpublished M.A. thesis, Department of Anthropology, Texas A&M University, College Station.

Henretta, J. A.
1965 Economic Development and Social Structure in Colonial Boston. *William and Mary Quarterly* 22(1):75-92.

Henrickson, E. M., and M. A. McDonald
1983 Ceramic Form and Function: An Ethnographic Search and Archaeological Explanation. *American Anthropologist* 85:630-643.

Hess, K. (transcriber)
1995 *Martha Washington's Booke of Cookery.* Second edition. Columbia University Press, New York.

Hobson, R. L.
1915 *Chinese Pottery and Porcelain: An Account of the Potter's Art in China from Primitive Times to the Present Day.* Two volumes. Funk and Wagnells, New York/Cassel, London.

1923 *The Wares of the Ming Dynasty.* Charles Scribner's Sons, New York.

1948 *Handbook of the Pottery and Porcelain of the Far East in the Department of Oriental Antiquities.* Third edition. The Stationery Office/The British Museum, London.

Holman, R. G.
1975 The *Dartmouth*, a British Frigate Wrecked off Mull, 1690: Culinary and Related Items. *The International Journal of Nautical Archaeology and Underwater Exploration* 4(2):253-265.

Holme, R.
1905 [1688] *The Academy of Armory or, A Storehouse of Armory and Blazon*, Vol. 2, edited by I. H. Jeayes. The Roxburghe Club, London.

Holmes, M. R.
1951 The So-Called 'Bellarmine' Mask on Imported Rhenish Stoneware. *The Antiquaries Journal* 31:173-179.

Honey, W. B.
1927 *Guide to the Later Chinese Porcelain Periods of K'ang Hsi, Yung Cheng, and Chienlung.* Department of Ceramics, Victoria and Albert Museum, London.

1944 *The Ceramic Art of China and other Countries of the Far East.* Faber and Faber, London.

Hook, D. R.
1997 Appendix I: Provenancing Rhenish Stoneware using Neutron Activation Analysis. In *German Stoneware, 1200-1900: Archaeology and Cultural History,* by D. R. M. Gaimster, pp. 344-353. British Museum Press, London.

Horändner, E.
1986 Storing and Preserving Meat in Europe: A Historical Survey. In *Food in Change: Eating Habits from the Middle Ages to the Present Day,* edited by A. Fenton and E. Kisbán, pp. 53-59. John Donald, Edinburgh, Scotland.

Horn, J.
1994 *Adapting to a New World: English Society in the Seventeenth-Century Chesapeake.* University of North Carolina Press, Chapel Hill.

Horne, J.
1985 *A Catalogue of English Brown Stoneware from the 17th and 18th Centuries.* Jonathan Horne, London.

Howat, G. M. D.
1974 *Stuart and Cromwellian Foreign Policy.* St. Martin's Press, New York.

Hoyt, S. D.
1984 The Archaeological Survey of Pedro Bank, Jamaica 1981-1983. *The International Journal of Nautical Archaeology and Underwater Exploration* 13(2):99-114.

Husband, T.
1980 *The Wild Man: Medieval Myth and Symbolism.* The Metropolitan Museum of Art, New York.

Ingelman-Sundberg, C.
1978 Relics from the Dutch East Indiaman *Zeewijk. Western Australian Museum Publication* 10. Perth, Australia.

Island Record Office
1672-1720 *Grantors Old Series: Wills, Port Royal, Jamaica.* 15 volumes. Spanish Town, Jamaica. Microfilm, Department of Anthropology, Texas A&M University, College Station.

1683 Record of Transaction between George Nicholas and Nathaniel Cooke, 15 October 1683. *Grantors Old Series: Deeds Records, Port Royal, Jamaica* 15:166. Spanish Town, Jamaica. Microfilm, Department of Anthropology, Texas A&M University, College Station.

Jamaica Public Archives (JPA)
1674-1716 *Household Probate Inventory Records, Port Royal, Jamaica.* 10 volumes. Spanish Town, Jamaica. Microfilm, Department of Anthropology, Texas A&M University, College Station.

James, S. R.
1985 The Analysis of the Conde de Tolosá and the Nuestra Señora de Guadalupe Olive Jar Assemblage. Unpublished M.A. thesis, Department of Anthropology, Texas A&M University, College Station.

1988 A Reassessment of the Chronological and Typological Framework of the Spanish Olive Jar. *Historical Archaeology* 22(1):43-66.

Janssen, H. L.
1983 Later Medieval Pottery Production in the Netherlands. In *Ceramics and Trade*, edited by P. Davey and R. Hodges, pp. 121-186. Department of Prehistory and Archaeology, University of Sheffield, Sheffield, England.

Jenyns, S.
1955 Wares of the Transitional Period Between the Ming and Ch'ing, 1620-1683. *Archives of the Chinese Art Society of America* 9:20-42.

Kelso, W. M.
1984 *Kingsmill Plantations, 1619-1800: Archaeology of Country Life in Colonial Virginia.* Academic Press, Orlando, Florida.

Kelso, W. M., and E. A. Campbell
1974 Excavation of a Seventeenth-Century Pottery Kiln at Glebe Harbor, Westmoreland County, Virginia. *Historical Archaeology* 30(1):53-62.

Kisbán, E
1986 Food Habits in Change: The Example of Europe. In *Food in Change: Eating Habits from the Middle Ages to the Present Day,* edited by A. Fenton and E. Kisbán, pp. 2-10. John Donald, Edinburgh, Scotland.

Lessman, A. W.
1997 The Rhenish Stoneware from the Monte Cristi Shipwreck, Dominican Republic. Unpublished M.A. thesis, Department of Anthropology, Texas A&M University, College Station.

Link, M. C.
1960 Exploring the Drowned City of Port Royal. *National Geographic Magazine* 117(2):151-182.

Lion-Goldschmidt, D.
1978 *Ming Porcelain.* Rizzoli, New York/Paris.

Lipski, L., and M. Archer
1984 *Dated English Delftware: Tin-Glazed Earthenware, 1600-1800.* Sotheby, London/Scranton, Pennsylvania.

Lister, F. C, and R. H. Lister
1976a Italian Presence in Tin Glazed Ceramics of Spanish America. *Historical Archaeology* 10:28-41.

1976b A Descriptive Dictionary for 500 Years of Spanish-Tradition Ceramics (13th through 18th Centuries). *Society for Historical Archaeology Special Publication* 1. John D. Combes, editor.

1978 The First Mexican Maiolicas: Imported and Locally Produced. *Historical Archaeology* 12:1-24.

1982 *Sixteenth-Century Majolica Pottery in the Valley of Mexico.* University of Arizona Press, Tucson.

1987 *Andalusian Ceramics in Spain and New Spain: A Cultural Register from the Third Century B.C. to 1700.* University of Arizona Press, Tucson.

Long, E.
1970 [1774] *The History of Jamaica or General Survey of the Antient and Modern State of that Island With Reflections on its Situations, Settlements, Inhabitants, Climate, Products, Commerce, Laws and Government.* Three volumes. Frank Cass, London.

Longacre, W. A. (editor)
1991 *Ceramic Ethnoarchaeology.* University of Arizona Press, Tucson.

Longe, S.
1999 [1610] Mrs. Sarah Longe Her Receipt Book, c. 1610. Transcribed by R. Doggett and E. Dever Powell. In *Fooles and Fricassees: Food in Shakespeare's England,* edited by M. Caton, pp. 102-119. Folger Shakespeare Library, Washington, D.C.

McCarthy, M. R., and C. M. Brooks
1988 *Medieval Pottery in Britain A.D. 900-1600.* Leicester University Press, Leicester, England.

McClenaghan, P. E.
1988 Drinking Glasses from Port Royal, Jamaica, c. 1630-1840: A Study of Styles and Usage. Unpublished M.A. thesis, Department of Anthropology, Texas A&M University, College Station.

McCracken, G.
1988 *Culture and Consumption: New Approaches to the Symbolic Character of Consumer Goods and Activities.* Indiana University Press, Bloomington.

McDonald, R. A.
1993 *The Economy and Material Culture of Slaves: Goods and Chattels on the Sugar Plantations of Jamaica and Louisiana.* Louisiana State University Press, Baton Rouge.

Main, G. L.
1982 *Tobacco Colony: Life in Early Maryland, 1650-1720.* Princeton University Press, New Jersey.

Marken, M. W.
1994 *Pottery from Spanish Shipwrecks, 1500-1800.* University Press of Florida, Gainesville.

Markham, G.
1986 [1615] *The English House-Wife,* edited by M. R. Best. McGill-Queen's University Press, Kingston /Montreal, Canada.

Marx, R. F.
1967 *Excavation of the Sunken City of Port Royal, December 1965-December 1966: A Preliminary Report.* Institute of Jamaica, Kingston.

1968a *Excavation of the Sunken City of Port Royal, January 1967-March 1968: A Preliminary Report.* Jamaica National Trust Commission, Kingston.

1968b *Clay Smoking Pipes Recovered from the Sunken City of Port Royal, October 1, 1967-March 31, 1968.* Jamaica National Trust Commission, Kingston.

1968c *Brass and Copper Items Recovered from the Sunken City of Port Royal, May 1, 1966-March 31, 1968.* Jamaica National Trust Commission, Kingston.

1969 *Glass Bottles Recovered from the Sunken City of Port Royal, May 1, 1966-March 31, 1968.* Caribbean Research Institute, College of the Virgin Islands, St. Thomas.

1971 *Silver and Pewter Items Recovered from the Sunken City of Port Royal, May 1, 1966-March 31, 1968.* Caribbean Research Institute, College of the Virgin Islands, St. Thomas.

Marryat, J.
1850 *Collections Towards a History of Pottery and Porcelain in the 15th, 16th, 17th and 18th Centuries.* John Murray, London.

Mathewson, R. D.
1972 Jamaican Ceramics: An Introduction to 18th-Century Folk Pottery in the West African Tradition. *Jamaica Journal* 6(2):54-56.

1973 Archaeological Analysis of Material Culture as a Reflection of Sub-Cultural Differentiation in 18th Century Jamaica. *Jamaica Journal* 7(1-2):25-29.

Matson, F. R. (editor)
1965a *Ceramics and Man.* Aldine, Chicago.

1965b Ceramic Ecology: An Approach to the Study of Early Cultures of the Near East. In *Ceramics and Man,* edited by F. R. Matson, pp. 202-217. Aldine, Chicago.

Mayes, P.
1972 *Port Royal, Jamaica: Excavations 1969-70.* Jamaica National Trust Commission, Kingston.

Mead, W. E.
1931 *The English Medieval Feast.* Houghton, Boston.

Merrifield, R.
1987 *The Archaeology of Ritual and Magic.* B. T. Batsford, London.

Meyers, A. D.
1998 Ethnic Distinctions Among Colonial Jamaican Merchants, 1685-1716. *Social Science History* 22(1): 47-81.

1999 West African Tradition in the Decoration of Colonial Jamaican Folk Pottery. *International Journal of Historical Archaeology* 3(4):201-223.

Miller, G. L.
1980 Classification and Economic Scaling of 19th Century Ceramics. *Historical Archaeology* 14:1-40.

1991 A Revised Set of CC Index Values for Classification and Scaling of English Ceramics from 1787-1880. *Historical Archaeology* 25(1):1-25.

Mintz, S. W.
1985 *Sweetness and Power: The Place of Sugar in Modern History.* Viking Penguin, New York/London.

Monteiro, J. P.
1994 *Oriental Influences on Seventeenth Century Portuguese Ceramics.* Museu Nacional do Azulejo, Lisbon, Portugal.

Morley-Fletcher, H., and R. McIlroy
1984 *Christie's Pictorial History of European Pottery.* Phaidon, Oxford.

Mountford, A. R.
1971 *The Illustrated Guide to Staffordshire Salt-Glazed Stoneware.* Barrie and Jenkins, London.

Mountford, A. R., and F. Celoria
1968 Some Examples of Sources in the History of 17th-Century Ceramics. *Journal of Ceramic History* 1. George Street Press, Stafford, England.

Mudge, J. M.
1986 *Chinese Export Porcelain in North America.* Clarkson N. Potter, New York.

Muffett, T.
1655 *Health's Improvement.* Printed by T. Newcomb for S. Thomson, London. Microfilm, Evans Library, Texas A&M University, College Station.

Multhauf, R. P.
1978 *Neptune's Gift: A History of Common Salt.* Johns Hopkins University Press, Baltimore, Maryland.

Noël Hume, I.
1958 German Stoneware Bellarmines: An Introduction. *Antiques* 74(5):439-441.

1962 An Indian Ware of the Colonial Period. *Archaeological Society of Virginia Bulletin* 17(1):2-14.

1963 A Late Seventeenth-Century Pottery Kiln near Jamestown. *Antiques* 83(5):550-552.

1967 Rhenish Gray Stoneware in Colonial America. *Antiques* 92(3):349-353.

1970 *A Guide to Artifacts of Colonial America.* Reprint of 1969 edition. Alfred A. Knopf, New York.

1977 Early English Delftware from London and Virginia. *Colonial Williamsburg Occasional Papers in Archaeology* 2. Williamsburg, Virginia.

Orser, C. E., and B. Fagan
1995 *Historical Archaeology.* Harper Collins, New York.

Orton, C. R., and J. E. Pearce
1984 The Pottery. In Excavations at Aldgate, 1974, by A. Thompson, F. Grew, and J. Schofield, pp. 34-68. *Post Medieval Archaeology* 18:1-148.

Oswald, A., R. J. C. Hildyard, and R. G. Hughes
1982 *English Brown Stoneware.* Faber and Faber, London.

Outlaw, A.
1989 *Governor's Land: Archaeology of Early Seventeenth-Century Settlements.* University Press of Virginia, Richmond.

Owen. N. C.
1988 HMS *Hazardous* Wrecked 1706: Pre-Disturbance Survey Report 1987. *The International Journal of Nautical Archaeology and Underwater Exploration* 17(4):285-293.

Palmer, R.
1973 *The Water Closet: A New History.* David and Charles, Newton Abbot, England.

Pawson, M., and D. Buisseret
1975 *Port Royal, Jamaica.* Clarendon Press, Oxford.

Pearce, J. E.
1992 *Post Medieval Pottery in London, 1500-1700.* Vol. 1, *Border Wares.* Her Majesty's Stationery Office, London.

1998 A Rare Delftware Hebrew Plate and Associated Assemblage from an Excavation in Mitre Street, City of London. *Post-Medieval Archaeology* 32:95-112.

Pendery, S.
1987 Symbols of Community: Status Differences and the Archaeological Record in Charlestown, Massachusetts, 1630-1760. Unpublished Ph.D. dissertation, Department of Anthropology, Harvard University, Cambridge, Massachusetts.

1999 Portuguese Tin-Glazed Earthenware in Seventeenth-Century New England: A Preliminary Study. *Historical Archaeology* 33(4):58-77.

Philadelphia Museum of Art
1984 *Masters of Seventeenth-Century Dutch Genre Painting.* (Exhibition 18 March-13 May.) Catalogue by Christopher Brown, Jan Kelch, Otto Naumann, William Robinson, Peter C. Sutton, and Cynthia von Bogendorf-Rupprath. Edited by J. I. Watkins. Distributed by University of Pennsylvania Press, Philadelphia.

Phillippo, J. M.
1970 [1843] *Jamaica: Its Past and Present State.* Negro Universities Press, Westport, Connecticut.

Pittman, W. E.
1993 A Survey of 17th-Century Chesapeake Ceramic Typologies. In The Archaeology of 17th-Century Virginia, pp. 357-369, edited by T. R. Reinhart and D. J. Pogue. *Special Publications* 30. Archaeological Society of Virginia, Richmond.

Plot, R.
1686 *The Natural History of Stafford-shire.* Printed at the Theater, Oxford. Microfilm, Evans Library, Texas A&M University, College Station.

Poole, J.
1995 *English Pottery.* Cambridge University Press, Cambridge.

Price, B.
1978 *The Arthur Negus Guide to English Pottery and Porcelain.* Hamlyn, London.

Priddy, A. J.
1975 The 17th- and 18th-Century Settlement Pattern of Port Royal. *Jamaica Journal* 9(2, 3):8-10.

Pryor, S., and K. Blockley
1978 A 17th-Century Kiln Site at Woolwich. *Post-Medieval Archaeology* 12:30-85.

Rice, P.
1987 *Pottery Analysis: A Sourcebook.* University of Chicago Press, Chicago.

Rice, P. (editor)
1984 *Pots and Potters: Current Approaches in Ceramic Archaeology.* Institute of Archaeology, University of California, Los Angeles.

Quinn, K.
1971 London Pottery 1565-1636. *Transactions of the English Ceramic Circle* 8(1):57-78.

Rackham, B.
1926 *Early Netherlands Maiolica.* G. Bles, London.

1987 *Catalogue of the Glaisher Collection of Pottery and Porcelain in the Fitzwilliam Museum, Cambridge.* Reprint of 1935 edition. Two volumes. Antique Collector's Club, Suffolk, England.

Rawson, J.
1984 *Chinese Ornament: The Lotus and the Dragon.* The British Museum, London.

Ray, A.
1968 *English Delftware Pottery in the Robert Hall Warren Collection, Ashmolean Museum, Oxford.* Faber, London.

Reineking von Bock, G.
1966 Die Entwicklung der Bartmaske an Rheinischem Steinzug. *Keramos* 34:30-43.

1986 *Steinzug.* Third edition. Katalogue des Kunstgewerbemuseums Köln IV, Cologne, Germany.

Root, W.
1958 *The Food of France.* Knopf, New York.

Root, W., and R. de Rochemont
1976 *Eating in America: A History.* William Morrow, New York.

Rule, M.
1982 *The Mary Rose: The Excavation and Raising of Henry VIII's Flagship.* Naval Institute Press, Annapolis, Maryland.

Sassoon, H.
1981 Ceramics from the Wreck of a Portuguese Ship at Mombasa. *Azania* 16:97-130.

Schaefer, R. G.
1998 A Typology of Seventeenth-Century Dutch Ceramics and its Implications for American Historical Archaeology. *British Archaeological Reports International Series* 702. Oxford.

Schiffer, M. B.
1987 *Formation Processes of the Archaeological Record.* University of New Mexico Press, Albuquerque.

Shackel, P. A.
1993 *Personal Discipline and Material Culture: An Archaeology of Annapolis, Maryland, 1695-1870.* University of Tennessee Press, Knoxville.

1994 Town Plans and Everyday Material Culture: An Archaeology of Social Relations in Colonial Maryland's Capital Cities. In *Historical Archaeology of the Chesapeake*, edited by P. A. Shackel and B. J. Little, pp. 85-96. Smithsonian Institution Press, Washington D.C.

Shammas, C.
1990 *The Pre-industrial Consumer in England and America.* Clarendon Press, Oxford.

Skibo, J. M.
1992 *Pottery Function: A Use-Alteration Perspective.* Plenum Press, New York.

Sloane, H.
1707 *A Voyage to the Islands Madera, Barbados, Nieves, S. Christophers and Jamaica with the Natural History of the Last of those Islands.*, Vol. 1. The British Museum, London. Microfilm, Evans Library, Texas A&M University, College Station.

Smart Martin, A
1994 Fashionable Sugar Dishes, Latest Fashion Ware: The Creamware Revolution in the Eighteenth-Century Chesapeake. In *Historical Archaeology of the Chesapeake*, edited by P. A. Shackel and B. J. Little, pp. 169-187. Smithsonian Institution Press, Washington D.C.

Smith, M. F.
1983 *The Study of Ceramic Function from Artifact Size and Shape.* Ph. D. dissertation, Department of Anthropology, University of Oregon, Eugene. University Microfilms International, Ann Arbor, Michigan.

Smith, W. C.
1995 *Analysis of the Weight Assemblage of Port Royal, Jamaica.* Ph.D. dissertation, Department of Anthropology, Texas A&M University, College Station. University Microfilms International, Ann Arbor, Michigan.

South, S.
1977 *Method and Theory in Historical Archaeology.* Academic Press, New York.

Spencer-Wood, S. M.
1987 Miller's Indices and Consumer-Choice Profiles. In *Consumer Choice in Historical Archaeology*, edited by S. M. Spencer-Wood, pp. 321-358. Plenum Press, New York.

Stanbury, M.
1974 *Batavia Catalogue.* Department of Maritime Archaeology, Western Australia Museum, Perth.

Sténuit, R.
1974 Early Relics of the VOC Trade from Shetland: The Wreck of the Fluit *Lastdrager* Lost off Yell 1653. *The International Journal of Nautical Archaeology and Underwater Exploration* 3(2):213-256.

1977 The Wreck of the VOC Fluit *Lastdrager* Lost off Yell (Shetland) 1653. In The Loss of the Verenigde Oostindische Compagnie Jacht *Vergulde Draeck*, Western Australia 1656: An Historical Background and Excavation Report, Vol. 1, edited by J. N. Green, pp. 403-468. *British Archaeological Reports Supplementary Series* 36. Oxford.

Stephan, H. -G.
1983 The Development and Production of Medieval Stoneware in Germany. In *Ceramics and Trade: The Production and Distribution of Later Medieval Pottery in North-West Europe*, edited by P. Davey and R. Hodges, pp. 95-120. University of Sheffield Press, Sheffield, England.

Steponaitis, V. P.
1984 Technological Studies of Prehistoric Pottery from Alabama: Physical Properties and Vessel Function. In *The Many Dimensions of Pottery: Ceramics in Archaeology and Anthropology*, edited by S. E. van der Leeuw and A. C. Pritchard, pp. 79-122. Institute for Pre- and Proto-History, University of Amsterdam, The Netherlands.

Stone, G. W.
1970 Ceramics in Suffolk County, Massachusetts Inventories 1680-1775: A Preliminary Study with Divers Comments Thereon, and Sundry Suggestions. *The Conference on Historic Site Archaeology Papers 1968*, 3(2):73-90.

Stubbe, H.
1662 *The Indian Nectar, Or A Discourse Concerning Chocolata.* Printed by F. C. for Andrew Crook, London.

Swift, J.
1958 [1726] *Gulliver's Travels and Other Writings.* Random House, New York.

Tait, H.
1960 Southwark (Alias Lambeth) Delftware and the Potter Christian Wilhelm, Part 1. *Connoisseur* 146 (585):34-42.

1961 Southwark (Alias Lambeth) Delftware and the Potter Christian Wilhelm, Part 2. *Connoisseur* 147 (591):22-29

1962 *Porcelain.* Hamlyn, London/New York.

Taylor, J.
1688 Second Part of the Historie of his Life and Travels in America. Manuscript on file, Institute of Jamaica, Kingston.

Thirsk, J.
1999 Food in Shakespeare's England. In *Fooles and Fricassees: Food in Shakespeare's England,* edited by M. Caton, pp. 13-25. Folger Shakespeare Library, Washington, D.C.

Thompson, A., F. Grew, and J. Schofield
1984 Excavations at Aldgate, 1974. *Post Medieval Archaeology* 18:1-148.

Thornton, D.
1992 The Probate Inventories of Port Royal, Jamaica. Unpublished M.A. thesis, Department of Anthropology, Texas A&M University, College Station.

Thwaite, A.
1973 The Chronology of the Bellarmine Jug. *Connoisseur* 182 (734):255-262.

Troy, J.
1977 *Salt-Glazed Ceramics.* Watson-Guptill, New York/Pitman, London.

Tryon, T.
1691 *A New Art of Brewing Beer, Ale, and Other Sorts of Liquors.* Second edition. Printed for Thomas Salisbury, London. Microfilm, Evans Library, Texas A&M University, College Station.

Turnbaugh, S. P.
1977 Ideo-Cultural Variation and Change in the Massachusetts Bay Colony. *Conference on Historic Site Archaeology Papers* 2:169-235.

1983 17th and 18th Century Lead-Glazed Redwares in the Massachusetts Bay Colony. *Historical Archaeology* 17(1):3-17.

Turnbaugh, S. P. (editor)
1985 *Domestic Pottery of the Northeastern United States.* Academic Press, New York.

Turner, J. S. T.
1971 An Augustan Election: The 1710 Election in the County of Surrey. *Surrey Archaeological Collection* 68:131-151.

Ukers, W. H.
1935 *All About Coffee.* The Tea and Coffee Trade Journal, New York.

Vaisey, D. G., and F. Celoria
1974 Inventory of George Ecton of Abingdon, Berks, 1696. *Journal of Ceramic History* 7:22-42. George Street Press, Stafford, England.

Van der Leeuw, S. E., and A. C. Pritchard (editors)
1984 *The Many Dimensions of Pottery: Ceramics in Archaeology and Anthropology.* Institute for Pre- and Proto-History, University of Amsterdam, The Netherlands.

Van der Pijl-Ketel C. L. (editor)
1982 *The Ceramic Load of the* Witte Leeuw *1613.* Rijksmuseum, Amsterdam.

Vigarello, G.
1988 *Concepts of Cleanliness: Changing Attitudes in France since the Middle Ages.* Cambridge University Press, Cambridge.

Wadley, C. A.
1985 Historical Analysis of Pewter Spoons Recovered from the Sunken City of Port Royal, Jamaica. Unpublished M.A. thesis, Department of Anthropology, Texas A&M University, College Station.

Walsh, L. S
1983 Urban Amenities and Rural Sufficiency: Living Standards and Consumer Behavior in the Colonial Chesapeake, 1643-1777. *The Journal of Economic History* 43(1):109-117.

Ward, E.
1933 [1700] *A Trip to Jamaica: With a True Character of the People and the Island.* Reproduced by The Facsimile Text Society, Columbia University Press, New York.

Watkins, C. M.
1973 Ceramics Used in America: Comparisons. In *Ceramics in America,* edited by I. M. G. Quimby, pp. 191-196. University Press of Virginia, Charlottesville.

Watkins, J.
1792 *An Essay towards a History of Bideford in the County of Devon.* Published by the author, Exeter, England.

Watkins, L. W.
1950 *Early New England Potters and their Wares.* Harvard University Press, Cambridge, Massachusetts.

Weatherill, L.
1971 *The Pottery Trade and North Staffordshire, 1660-1760.* Augustus M. Kelly, New York.

1986 *The Growth of the Pottery Industry in England, 1660-1815.* Garland, New York.

1988 *Consumer Behavior and Material Culture in Britain, 1660-1760.* Routledge, London/ New York.

Weatherill, L., and R. Edwards
1971 Pottery-Making in London and Whitehaven in the Late Seventeenth Century. *Post-Medieval Archaeology* 5:160-181.

Wheaton, B. K.
1983 *Savoring the Past: The French Kitchen and Table from 1300-1789.* University of Pennsylvania Press, Philadelphia.

Wheaton, T. R., and P. H. Garrow
1985 Acculturation and the Archaeological Record in the Carolina Low Country. In *The Archaeology of Slavery and Plantation Life*, edited by T. A. Singleton, pp. 239-260. Academic Press, San Diego.

Wilcoxen, C.
1987 *Dutch Trade and Ceramics in America in the Seventeenth Century.* Albany Institute of History and Art, Albany, New York.

1992a A Preliminary Analysis of Ceramics at Port Royal. Manuscript on file, Department of Anthropology, Texas A&M University, College Station.

1992b Tin-Glazed Pottery from Port Royal: Analytical Inventory. Manuscript on file, Department of Anthropology, Texas A&M University, College Station.

Willey, G. R., and J. A. Sabloff
1983 *A History of American Archaeology.* Third edition. W. H. Freeman, New York.

Wills, G.
1967 English Pottery in 1696: An Unpublished Document. *Apollo* June 1967:436-443.

Wondrausch, M.
1986 *Mary Wondrausch on Slipware: A Potter's Approach.* A&C Black, London.

Wright, L.
1960 *Clean and Decent: The Fascinating History of the Bathroom and the Water Closet.* Routledge and Kegan Paul, London.

Yentsch, A. E.
1990 Minimum Vessel Lists as Evidence of Change in Folk and Courtly Traditions of Food Use. *Historical Archaeology* 4(3):24-53.

1991 Chesapeake Artefacts and Their Cultural Context: Pottery and the Food Domain. *Post-Medieval Archaeology* 25:25-72.

Y-Worth, W.
1692 *Cerevisiarii Comes, Or, The New and True Art of Brewing.* Printed for J. Taylor and S. Clement, London. Microfilm, Evans Library, Texas A&M University, College Station.

Zahedieh, N.
1986a The Merchants of Port Royal, Jamaica, and the Spanish Contraband Trade, 1655-1692. *William and Mary Quarterly* 43:570-593.

1986b Trade, Plunder, and Economic Development in Early English Jamaica, 1655-1689. *Economic History Review* 39:205-222.

APPENDIX A

CERAMIC INVENTORY FROM LAYER 3 AT BUILDINGS 4 AND 5

BUILDING 4

Location	Ceramic Type and Form	Quantity
Room 4A	Staffordshire slipware chocolate/coffee cup	1
	English/Dutch undecorated tin-glazed earthenware plate	1
	English/Dutch undecorated tin-glazed drinking pot	1
	English decorated tin-glazed bowl	1
	English decorated tin-glazed earthenware flower vase	1
Subtotal		**5**
Room 4B	Coarse ware storage pots (unidentified source)	2
	African-slave-made Yabba bowl	1
	English/Dutch undecorated tin-glazed earthenware apothecary pot	1
	English (?) decorated tin-glazed earthenware chocolate/coffee cup	1
	Mexican majolica plate	1
	English stoneware mug	1
Subtotal		**7**
Yard 4A	English Redware cooking pot	1
	English Redware storage pots	3
	Red Border ware storage pot	1
	Red Border ware porringer	1
	Red Border ware chamber pot	1
	African-slave-made Yabba cooking pots	4
	African-slave-made Yabba bowls	2
	Staffordshire slipware bowl	1
	English undecorated tin-glazed earthenware caudle cups	2
	English/Dutch undecorated tin-glazed earthenware drinking pot	1
	English undecorated tin-glazed earthenware apothecary pots	2
	Italianate-Spanish majolica saucer	1
	Mexican majolica plate	1
	Westerwald stoneware mugs	3
	Iberian (?) white-slipped stoneware storage pot	1
	Dutch/South East Indian (?) stoneware costrel	1
	Chinese export porcelain (blue-and-white) bowl	1
Subtotal		**27**
Yard 4B	English Redware bowl	1
	African-slave-made Yabba bowls	2
	North Devon Sgraffito bowl	1
	English/Dutch undecorated tin-glazed earthenware plate	1
	English undecorated tin-glazed earthenware apothecary pot	1
	Spanish/Italianate-Spanish majolica plate	1
	Chinese export porcelain (undecorated white) saucer	1
Subtotal		**8**
Alley	English Redware cooking pot	1
	English Redware chamber pot	1
	White Border ware porringer	1
	Staffordshire slipware chocolate/coffee cup	1
	North Holland slipware saucer	1
	English (?) decorated tin-glazed earthenware plate	1
	Portuguese majolica plate	1
	Rhenish *Bartmannkrüge*	2
Subtotal		**9**
Total		**56**

BUILDING 5

Location	Ceramic Type and Form	Quantity
Room 1	English Redware cooking pot	1
	English Redware storage pots	4
	North Devon Plain ware sherd (unknown vessel–storage pot?)	1
	Iberian unglazed white coarse ware ewer (cross-mended with Room 2)	1
	African-slave-made Yabba bowl	1
	Staffordshire slipware saucer	1
	North Italian slipware costrel	1
	Rhenish *Bartmannkrug*	1
	English stoneware bottle	1
	English undecorated tin-glazed earthenware plate	1
	English undecorated (?) tin-glazed earthenware salt holder (cross-mended with Room 3 and Yard 5)	1
	English/Dutch undecorated tin-glazed earthenware drinking pot	1
	English undecorated (?) tin-glazed earthenware apothecary pot	1
	English undecorated tin-glazed earthenware chamber pot	1
	English decorated tin-glazed earthenware bottle	1
	Dutch decorated tin-glazed earthenware serving dish (cross-mended with Room 2)	1
	English decorated tin-glazed earthenware chocolate/coffee cup	1
	English (?) decorated tin-glazed earthenware washbasin (cross-mended with Yard 5)	1
Subtotal		**21**
Room 2	English Redware colander	1
	English Redware storage pot	1
	North Devon Gravel-Tempered ware storage pot	1
	White Border ware pipkin (cross-mended with Yard 5)	1
	Staffordshire Mottled slipware pan	1
	English undecorated tin-glazed earthenware caudle cup	1
	English undecorated (?) tin-glazed earthenware apothecary pots	2
	English (?) undecorated (?) tin-glazed earthenware washbasin	1
	English decorated tin-glazed earthenware saucer	1
	Dutch (?) decorated tin-glazed earthenware chocolate/coffee cups	2
	English/Dutch decorated tin-glazed earthenware punch bowl	1
	Dutch (?) decorated tin-glazed earthenware lid (unknown vessel)	1
	Rhenish miniature stoneware bottle	1
	Chinese export porcelain (Batavia ware) bowl (cross-mended with Yard 5)	1
Subtotal		**16**
Room 3	English Redware storage pot	1
	English/Dutch undecorated tin-glazed earthenware plates	2
	English undecorated (?) tin-glazed earthenware apothecary pots	3
	English (?) decorated tin-glazed earthenware chocolate/coffee cup	1
	English (?) decorated tin-glazed earthenware cup (use unknown)	1
	English decorated tin-glazed earthenware drinking pot (cross-mended with Room 2 and Yard 4A)	1
	Spanish/Italianate-Spanish majolica plate	1
Subtotal		**10**
Room 4	African-slave-made Yabba bowl	1
	English/Dutch undecorated tin-glazed earthenware drinking pot	1
Subtotal		**2**

BUILDING 5 (continued)

Location	Ceramic Type and Form	Quantity
Yard 5	English Redware storage pots	2
	English Redware bowl	1
	English Redware chamber pot	1
	North Devon Gravel-Tempered ware storage pot	1
	Red Border ware storage pot	1
	White Border ware chamber pot	1
	'Spanish Storage Jars'	2
	Iberian olive jars	2
	Staffordshire slipware porringer	1
	Staffordshire slipware chocolate/coffee cup	1
	Staffordshire slipware cup (use unknown)	1
	English/Dutch undecorated tin-glazed earthenware plate	1
	English undecorated tin-glazed earthenware porringer	1
	English undecorated (?) tin-glazed earthenware apothecary pots	4
	English undecorated tin-glazed earthenware chamber pot	1
	Mexican majolica plate	1
	Rhenish miniature stoneware bottle	1
	Chinese export porcelain (undecorated *blanc de chine*) cup	1
Subtotal		**24**
Sidewalk	Italian (?) coarse ware costrel	1
	Spanish majolica bowl	1
	Rhenish *Bartmannkrug*	1
	Chinese export porcelain (undecorated *blanc de chine*) cup	1
	Chinese export porcelain (Batavia ware) tea bowl	1
Subtotal		**5**
Yard 6	English Redware cooking pot	1
	English decorated tin-glazed earthenware apothecary pots	2
	English stoneware mug	1
Subtotal		**4**
Yard 7	African-slave-made Yabba storage pots	2
	African-slave-made Yabba bowl	1
	English undecorated tin-glazed earthenware porringer	1
	English stoneware storage pot	1
	Dutch/South East Indian (?) stoneware costrel	1
Subtotal		**6**
Total		**88**

APPENDIX B

CERAMIC WARES IN PRODUCTION IN 1692, AS FOUND IN ALL LAYERS AT BUILDING 4/5 (THE DATA BASE SHERDS)

This appendix includes the ceramic data base compiled by the author for the 'data base sherds' recovered from Building 4/5 at Port Royal, Jamaica. It includes information on site provenance, ceramic ware, type, form, sherd size, decoration, etc. (see below for keys to the field codes).

Year = year excavated

Unit/No. = site provenance

Form = whole vessel form:

APOT	apothecary pot
BASN	basin
BOTT	bottle
BOWL	bowl (incl. tea and punch bowls)
CHAM	chamber pot
COLN	colander
COOK	cooking pot
COST	costrel/flask
CUP	cup
DISH	dish
DRNK	drinking pot
EWER	ewer
MUG	mug
PAN	pan
PIP	pipkin
PLAT	plate
PORR	porringer
SAUC	saucer
SALT	salt stand
STOR	storage pot
VASE	vase
UNKN	unknown (small sherd size)

Part = whole vessel part:

RIM	rim (may be too small to estimate diameter)
BDY	body
BSE	base (may be too small/from a central part of base to estimate diameter)
ATT	attachment (handle, spout, lid, knob, or medallion)
R/B/B	rim/body/base attached
WHLE	complete vessel

Ware = ceramic ware:

CRW	coarse ware
SLW	slipware
TIN	tin-glazed earthenware
STW	stoneware
POR	porcelain

Type = ceramic type and variety (see Ceramic Typology Coding Key)

Thick = sherd thickness:
1 = 0-5 mm
2 = 5+ -10 mm
3 = 10+ -15 mm
4 = 15+ -20 mm
5 = 20+ mm

Diam = approx. diam. of whole rim, base, or both (e.g., 2/3 = 5+ -10 cm rim/10+ -15 cm base):
1 = 0-5 cm
2 = 5+ -10 cm
3 = 10+ -15 cm
4 = 15+ -20 cm
5 = 20+ -25 cm
6 = 25+ cm

Ht = height (near-intact/complete vessels only): numbered key as for 'diameter' above

Location = context of sherd within Building 4/5

Qty = number of sherds

Assoc = cross-mended sherds

Notes = description, e.g., glaze color, attrition, etc.

CERAMIC TYPOLOGY
This numbered typology is to be used with the ceramic data base. The coding system is straightforward. A White Border ware sherd, for example, is recorded in the 'Type' field as 1.11; an unglazed/undecorated English Redware sherd is recorded as 1.31; a plain/undecorated English/Dutch tin-glazed earthenware sherd is recorded as 3.12, etc.

1. COARSE WARE (CRW)
10. LEAD GLAZED (ENGLISH)
- 11 = White Border Ware
- 12 = Red Border Ware
- 13 = North Devon Plain Ware
- 14 = North Devon Gravel Tempered Ware
- 15 = Buckley Ware

20. UNGLAZED/GLAZED (CONTINENTAL EUROPEAN)
- 21 = Olive Jar (Iberian)
- 22 = Spanish Storage Jar Type
- 23 = Rey Ware (Iberian)

30. REDWARE (ENGLISH, WHEEL-THROWN)
- 31 = No Glaze/No Decoration
- 32 = No Glaze/Decoration
- 33 = Interior Glaze/No Decoration
- 34 = Interior Glaze/Decoration
- 35 = Exterior Glaze/No Decoration
- 36 = Exterior Glaze/Decoration
- 37 = Glazed Both Surfaces/No Decoration
- 38 = Glazed Both Surfaces/Decoration

40. YABBA WARE (AFRICAN-JAMAICAN, HAND-COILED)
- 41 = No Glaze/No Decoration
- 42 = No Glaze/Decoration
- 43 = Interior Glaze/No Decoration
- 44 = Interior Glaze/Decoration
- 45 = Exterior Glaze/No Decoration
- 46 = Exterior Glaze/Decoration
- 47 = Glazed Both Surfaces/No Decoration
- 48 = Glazed Both Surfaces/Decoration

2. SLIPWARE (SLW)
10. ENGLISH
- 11 = Wrotham Ware
- 12 = Metropolitan Ware
- 13 = North Devon Slip/Sgraffito
- 14 = Combed/Feathered/Trailed
- 15 = Annular
- 16 = Dot
- 17 = Reverse Slip
- 18 = Staffordshire Mottled

20. CONTINENTAL EUROPEAN
- 21 = Wanfried Ware (German)
- 22 = North Holland (Dutch)
- 23 = North Italian Red Marbleized

3. TIN-GLAZED EARTHENWARE (TIN)
10. DELFTWARE (ENGLISH/DUTCH)
- 11 = De-enameled/Unidentified (sherd too small)
- 12 = Plain/Undecorated
- 13 = Blue
- 14 = Polychrome
- 15 = Bianca Sopra Bianca
- 16 = Bianca Sopra Azzuro
- 17 = Mimosa Pattern
- 18 = Frazackerly

20. MAJOLICA (EUROPEAN/MEXICAN)
 21 = Morisco Ware (Spanish)
 22 = Sevilla Ware (Italianate-Spanish)
 23 = Montelupo Majolica (Italian)
 24 = Ligurian Majolica (Italian)
 25 = Faenza Majolica (Italian)
 26 = Mexico City Ware (Mexican)
 27 = Puebla Ware (Mexican)

30. FAÏENCE (FRENCH/PORTUGUESE)
 31 = Nevers Blue (French)
 32 = Rouen (French)
 33 = Portuguese

4. STONEWARE (STW)

10. GLAZED, BROWN (EUROPEAN)
 11 = Rhenish Brown
 12 = Bellarmine (salt-glazed) (Rhenish)
 13 = Debased Bellarmine (salt-glazed) (Rhenish)
 14 = Fulham Tavern Ware (salt-glazed) (English)
 15 = Burslem Tavern Ware (salt-glazed) (English)
 16 = Staffordshire Brown (salt-glazed) (English)
 17 = Nottingham Lustered (salt-glazed) (English)

20. GLAZED, BLUE-GRAY (EUROPEAN)
 21 = Rhenish Gray
 22 = Westerwald (sprig-molded, blue and purple)
 23 = Westerwald (stamped/incised, floral devices)
 24 = Debased Westerwald
 25 = Embellished Hohr Gray Rhenish

30. GLAZED, WHITE (ENGLISH)
 31 = Scratch Brown
 32 = Slip Dipped
 33 = Molded
 34 = Scratch Blue
 35 = Debased Scratch Blue
 36 = Littler's Blue
 37 = Transfer Printed

40. OTHER
 41 = Refined Red (unglazed) (English)
 42 = Astbury-type (English)
 43 = Shaw Ware (English)
 44 = Jackfield-type (English)
 45 = Black Basaltes (unglazed) (English)
 46 = Elers Ware (unglazed) (English)
 47 = Jasper Ware (unglazed) (English)
 48 = Lustreware (English)

5. PORCELAIN (POR)

10. CHINESE
 11 = Undecorated White
 12 = Late Ming
 13 = Batavia Ware
 14 = Blanc de Chine
 15 = Underglaze Blue
 16 = Overglaze Enamel
 17 = Famille Rose
 18 = Armorial Ware
 19 = Blue Willow

20. NON-CHINESE
 21 = Polychrome Overglaze (English)
 22 = Hand-Painted Blue Underglaze (English)
 23 = Bone China (European, American)
 24 = Porcellaneous Ware (European, American)

Year	Unit	No.	Layer	Form	Part	Ware	Type	Thick	Diam	Ht	Qty	Location	Assoc	Notes
87	574	1	3	SAUC	RIM	TIN	3.11	1	2	0	1	ROOM 1		
87	574	2	3	PLAT	RIM	TIN	3.11	1	4	0	1	ROOM 1		
87	575	3	3	UNKN	BDY	TIN	3.11	1	0	0	1	ROOM 1		
87	575	3	3	UNKN	RIM	TIN	3.11	1	0	0	1	ROOM 1		
87	575	3	3	UNKN	BDY	TIN	3.11	1	0	0	2	ROOM 1		
87	404	3	3	APOT	RIM	TIN	3.11	1	4	0	2	YARD 4B		
87	575	3	3	CUP	RIM	TIN	3.11	1	2	0	1	ROOM 1		
87	575	3	3	CUP	BSE	TIN	3.13	1	1	0	1	ROOM 1		low footring; design on exterior
87	575	2	3	UNKN	BDY	TIN	3.11	1	0	0	1	ROOM 1		
89	642		2	CUP	BDY	TIN	3.11	1	0	0	1	ROOM 1		
89	646	3	3	UNKN	BDY	TIN	3.11	1	0	0	1	ROOM 1		
89	646	4	3	UNKN	BDY	TIN	3.11	1	0	0	2	ROOM 1		
87	404	4	3	UNKN	BSE	TIN	3.13	1	2	0	1	YARD 4B		flatware/shallow bowl? design on obverse
89	599		3	UNKN	BDY	TIN	3.11	1	0	0	1	ROOM 1		
89	599		3	UNKN	BDY	TIN	3.11	1	0	0	1	ROOM 1		
89	599		3	APOT	RIM	TIN	3.11	1	2	0	1	ROOM 1		ointment pot?
89	599		3	UNKN	BDY	TIN	3.11	1	0	0	1	ROOM 1		
89	599		3	UNKN	BDY	TIN	3.20	1	0	0	1	ROOM 1		pristine glaze; red
89	599		3	APOT	BSE	TIN	3.11	1	1	0	1	ROOM 1		ointment pot; low footring; melted glaze (fire damage)
89	599		3	PORR	BDY	TIN	3.11	1	0	0	1	ROOM 1		
89	599		3	BOWL	BSE	TIN	3.13	1	2	0	1	ROOM 1		low footring; design on obverse
89	599		3	DRNK	BDY	TIN	3.11	1	0	0	1	ROOM 1		
89	599		3	UNKN	BDY	TIN	3.11	1	0	0	1	ROOM 1		
89	599		3	APOT	RIM	TIN	3.12	1	2	0	1	ROOM 1		drug jar
89	599		3	APOT	RIM	TIN	3.12	3	3	0	1	ROOM 1		
89	599		3	DRNK	ATT	TIN	3.11	2	0	0	1	ROOM 1		spout; see similar spouts 87-375-1, 89-734-2
89	654	3	3	BOWL	BSE	TIN	3.12	1	3	0	2	ROOM 1		low footring
97	404	2	3	UNKN	RIM	TIN	3.11	3	0	0	1	YARD 4B		
89	654	4	3	UNKN	BDY	TIN	3.11	1	0	0	1	ROOM 1		
89	656	4	3	UNKN	BDY	TIN	3.11	2	0	0	1	ROOM 1		storage pot?
89	656	3	3	UNKN	BDY	TIN	3.13	1	0	0	1	ROOM 1		flatware/bowl; floral design on interior
89	599		3	UNKN	BDY	TIN	3.11	1	0	0	1	ROOM 1		
89	654	4	3	CUP	BDY	TIN	3.14	1	0	0	1	ROOM 1		blue/purple design on exterior
90	5910		3	UNKN	BDY	TIN	3.11	1	0	0	1	ROOM 1		
89	656	3	3	SALT	RIM	TIN	3.12	1	3	0	1	ROOM 1	TIN 1	salt holder
89	646	4	3	UNKN	BDY	TIN	3.11	1	0	0	1	ROOM 1		
89	656	7	3	DRNK	ATT	TIN	3.12	1	4	0	12	ROOM 1		reconstructed lid
89	682		2	APOT	RIM	TIN	3.11	1	3	0	2	ROOM 2		
89	682		2	UNKN	BDY	TIN	3.11	1	0	0	3	ROOM 2		
89	606		3	UNKN	BDY	CRW	1.35	1	0	0	1	SIDEWALK		greenish brown glaze; inclusions
89	683		3	PLAT	BSE	TIN	3.11	1	4	0	1	ROOM 2		low footring
89	683		3	UNKN	BDY	TIN	3.11	1	0	0	1	ROOM 2		
89	683		3	CUP	BDY	TIN	3.14	1	0	0	1	ROOM 2		blue/purple design on exterior
89	683	3	3	UNKN	RIM	TIN	3.11	1	2	0	1	ROOM 2		drug jar/storage pot?
89	683	9	3	UNKN	BDY	TIN	3.14	1	0	0	1	ROOM 2		
89	683		3	UNKN	BDY	TIN	3.14	1	0	0	1	ROOM 2		blue/purple geometric design on exterior
89	685	4	3	UNKN	BDY	TIN	3.11	1	0	0	1	ROOM 2		
89	685	4	3	UNKN	BDY	TIN	3.11	1	0	0	1	ROOM 2		
89	683	9	3	UNKN	RIM	TIN	3.14	1	3	0	1	ROOM 2		blue/purple design on exterior
89	685	4	3	UNKN	BDY	TIN	3.11	1	0	0	1	ROOM 2		

Year	Unit	No.	Layer	Form	Part	Ware	Type	Thick	Diam	Ht	Qty	Location	Assoc	Notes
89	670	2	1	UNKN	BDY	TIN	3.13	1	0	0	1	ROOM 2		design on exterior
89	6710	2	3	UNKN	BDY	TIN	3.11	1	0	0	1	ROOM 2		
89	688	4	3	APOT	RIM	TIN	3.11	1	5	0	1	ROOM 2		drug jar
89	670	2	1	CUP	RIM	TIN	3.11	1	2	0	1	ROOM 2		three thin lines edge rim exterior; unidentified design below rim
89	673		3	APOT	RIM	TIN	3.11	1	3	0	3	ROOM 2		
89	745		3	APOT	RIM	TIN	3.11	1	3	0	1	YARD 4A		drug jar
89	674	12	3	BASN	RIM	TIN	3.12	1	6	0	1	ROOM 2		washbasin?
89	683		3	CUP	BSE	TIN	3.13	1	1	0	1	ROOM 2		tea/coffee; low footring; thin line around base exterior
89	688	4	3	CUP	BSE	TIN	3.14	1	1	0	1	ROOM 2		tea/coffee
89	599		3	DRNK	ATT	TIN	3.12	1	0	0	1	ROOM 2	TIN 2	lid
89	678	4	3	PLAT	RIM	TIN	3.12	1	5	0	1	ROOM 2	TIN 3	
89	6710	2	3	UNKN	BDY	TIN	3.12	1	0	0	1	ROOM 2		
89	6710	2	3	DRNK	RIM	TIN	3.12	1	3	0	1	ROOM 2	TIN 4	part reconstructed
89	688	2	3	CUP	RIM	TIN	3.14	1	2	0	2	ROOM 2	TIN 5	tea/coffee; blue/purple floral design on exterior; see similar cups 89-678-11, 89-662-4
89	685	10	3	CUP	RIM	TIN	3.14	1	2	0	1	ROOM 2	TIN 5	tea/coffee; blue/purple floral design on exterior
89	734	1	3	CUP	RIM	TIN	3.14	1	2	0	1	ROOM 3	TIN 5	tea/coffee; blue/purple floral design on exterior
89	683		3	CUP	BSE	TIN	3.13	1	3	0	3	ROOM 2	TIN 6	caudle
89	688	2	3	CUP	BDY	TIN	3.13	1	0	0	1	ROOM 2	TIN 6	caudle
89	6710	2	3	CUP	RIM	TIN	3.13	1	3	0	1	ROOM 2	TIN 6	caudle
89	683		3	CUP	ATT	TIN	3.13	1	0	0	4	ROOM 2	TIN 6	part handle; caudle
89	685	4	3	CUP	BDY	TIN	3.13	1	0	0	1	ROOM 2	TIN 6	caudle
89	753	3	3	CUP	RIM	TIN	3.13	1	3	0	2	YARD 4A	TIN 6	caudle
89	734	1	3	CUP	BDY	TIN	3.13	1	0	0	1	ROOM 3	TIN 6	caudle
89	738	5	3	UNKN	BDY	TIN	3.11	1	0	0	4	ROOM 3		
89	732		2	UNKN	BDY	TIN	3.11	1	0	0	1	ROOM 3		pink
89	738	2	3	UNKN	BDY	TIN	3.11	1	0	0	1	ROOM 3		
89	7310	1	3	UNKN	BDY	TIN	3.11	1	0	0	1	ROOM 3		
89	732		2	UNKN	BDY	TIN	3.13	1	0	0	1	ROOM 3		design over light blue exterior
89	733	1	3	CUP	BDY	TIN	3.11	1	0	0	1	ROOM 3		caudle?
89	734	2	3	CUP	BDY	TIN	3.11	1	0	0	1	ROOM 3		caudle?
89	734	4	3	BOWL	RIM	TIN	3.12	1	4	0	1	ROOM 3		
89	734	2	3	DRNK	ATT	TIN	3.11	2	0	0	1	ROOM 3		spout; see similar spouts 87-375-1, 89-599
89	736	2	3	APOT	RIM	TIN	3.11	3	3	0	1	ROOM 3		
89	734	2	3	UNKN	BSE	TIN	3.11	1	2	0	1	ROOM 3		small bowl/caudle cup/ ug? low footring
89	737	6	3	SALT	ATT	TIN	3.12	1	0	0	1	ROOM 3	TIN 1	lug
89	737	6	3	SALT	R/B/B	TIN	3.12	1	2 / 3	2	9	ROOM 3	TIN 1	part reconstructed salt holder
89	738	2	3	DRNK	ATT	TIN	3.12	1	3	0	1	ROOM 3	TIN 2	lid
89	734	4	3	DRNK	ATT	TIN	3.12	1	3	0	1	ROOM 3	TIN 2	lid
89	738	5	3	PLAT	R/B/B	TIN	3.11	1	5 / 4	3	8	ROOM 3	TIN 3	
89	734	2	3	PLAT	BDY	TIN	3.11	1	0	0	3	ROOM 3	TIN 3	
89	858	2	3	UNKN	BDY	TIN	3.11	1	0	0	1	ROOM 4		
89	852		2	DRNK	RIM	TIN	3.13	1	3	0	1	ROOM 4	TIN 7	
89	857	1	3	UNKN	BDY	TIN	3.11	1	0	0	1	ROOM 4		
89	857	1	3	DRNK	BDY	TIN	3.13	1	0	0	1	ROOM 4	TIN 7	strap handle attached
89	748	3	3	DRNK	RIM	TIN	3.13	1	3	0	1	YARD 4A	TIN 7	
89	745	4	3	DRNK	BDY	TIN	3.13	1	0	0	1	YARD 4A	TIN 7	
89	853		3	DRNK	BDY	TIN	3.13	1	0	0	2	YARD 4A	TIN 7	
89	858	2	3	UNKN	BDY	TIN	3.13	1	0	0	1	ROOM 4		
90	922		2	DRNK	BDY	TIN	3.13	1	0	0	1	YARD 6	TIN 7	
89	858	2	3	UNKN	BDY	TIN	3.11	1	0	0	2	ROOM 4		
89	854		3	CUP	BSE	TIN	3.12	2	2	0	7	ROOM 4	TIN 6	caudle

Year	Unit	No.	Layer	Form	Part	Ware	Type	Thick	Diam	Ht	Qty	Location	Assoc	Notes
89	858	2	3	CUP	RIM	TIN	3.13	1	2	0	1	ROOM 4		tea/coffee? design on exterior
89	852		2	DRNK	RIM	TIN	3.12	1	3	0	1	ROOM 4	TIN 2	lid
89	656	13	3	BASN	RIM	TIN	3.13	1	6	0	4	ROOM 1	TIN 8	washbasin; chinoiserie design on interior
89	656		3	BASN	RIM	TIN	3.13	1	6	0	2	ROOM 1	TIN 8	washbasin; chinoiserie design on interior
89	656	3	3	BASN	RIM	TIN	3.13	1	6	0	2	ROOM 1	TIN 8	washbasin; chinoiserie design on interior
90	944	27	3	BASN	RIM	TIN	3.13	1	6	0	1	YARD 5	TIN 8	washbasin; chinoiserie design on interior
90	944	27	3	BASN	R/B/B	TIN	3.13	1	6 / 2	0	3	YARD 5	TIN 8	washbasin; chinoiserie design on interior
89	764	3	3	PLAT	RIM	TIN	3.11	1	5	0	1	YARD 4B		
89	763	3	3	UNKN	BSE	TIN	3.11	1	2	0	1	YARD 4B		high footring
90	960	5	1	APOT	RIM	TIN	3.13	1	2	0	1	YARD 6		drug jar; chinoiserie design on exterior
89	715	8	3	VASE	WHLE	TIN	3.13	1	3 / 2	4	75	ROOM 4A		flower vase; design on exterior
89	654	4	3	DISH	RIM	TIN	3.13	1	6	0	1	ROOM 1	TIN 9	chinoiserie (panneled) design on obverse; see TIN 19, 25 (same dish?)
89	656	3	3	DISH	RIM	TIN	3.13	1	6	0	1	ROOM 1	TIN 9	chinoiserie (panneled) design on obverse; see TIN 19, 25 (same dish?)
89	656	3	3	DISH	BSE	TIN	3.13	1	3	0	1	ROOM 1	TIN 9	low footring ; chinoiserie (panneled) design on obverse; see TIN 19, 25 (same dish?)
89	947	4	3	DISH	RIM	TIN	3.13	1	6	0	1	YARD 5	TIN 25	chinoiserie (panneled) design on obverse; see TIN 9,19 (same dish?)
87	404	4	3	UNKN	BSE	TIN	3.11	1	2	0	1	YARD 4B		design on one surface (too little to identify pattern)
87	404	3	3	APOT	RIM	TIN	3.11	1	4	0	2	YARD 4B		
89	689	2	3	DISH	RIM	TIN	3.13	1	6	0	1	ROOM 2	TIN 9	chinoiserie (panneled) design on obverse ; see TIN 19, 25 (same dish?)
89	678	11	3	CUP	R/B/B	TIN	3.14	1	2 / 1	2	1	ROOM 2		tall coffee; blue/purple/yellow floral/insect design on exterior; see 89-662-2 (matched set?)
89	662	2	2	CUP	BSE	TIN	3.14	1	2	2	1	SIDEWALK		tall coffee; blue/purple/yellow floral/insect design on exterior ; see 89-678-11 (matched set?)
89	6810	4	3	APOT	ATT	TIN	3.14	1	2	0	1	ROOM 2	TIN 10	reconstructed lid; blue/purple/yellow chinoiserie design on exterior
89	775	3	3	APOT	ATT	TIN	3.14	1	2	0	1	ROOM 4B	TIN 10	reconstructed lid; blue/purple/yellow chinoiserie design on exterior
89	739	4	3	CUP	RIM	TIN	3.14	1	2	0	1	ROOM 3	TIN 11	reconstructed quatrefoil
89	738	4	3	CUP	RIM	TIN	3.14	1	2	0	1	ROOM 3	TIN 11	reconstructed quatrefoil
89	7310	1	3	CUP	R/B/B	TIN	3.14	1	2 / 1	1	3	ROOM 3	TIN 11	reconstructed quatrefoil; low footring; vestiges of a handle; tulip design on interior
87	446	4	3	CUP	R/B/B	TIN	3.14	1	2 / 1	2	1	ROOM 4B		tall coffee; high footring; blue/purple floral design on exterior
90	941		1	BOWL	RIM	CRW	1.31	2	4	0	1	YARD 5		fire damage; yellowish red; core; English
89	786	7	3	DISH	BSE	TIN	3.13	1	6	0	6	ROOM 4A		low footring; chinoiserie design on obverse
89	602		2	CUP	BSE	TIN	3.13	1	1	0	1	SIDEWALK		coffee; low footring; design on exterior (too little to identify pattern)
89	646	4	3	BOTT	ATT	TIN	3.13	4	0	0	1	ROOM 1		part rounded handle; cloud scroll design
89	656	2	3	PLAT	RIM	TIN	3.13	1	4	0	1	ROOM 1		
89	687	1	3	BOWL	BDY	TIN	3.11	1	0	0	2	ROOM 3	TIN 12	
89	582		2	BOWL	BDY	TIN	3.11	1	0	0	1	SIDEWALK	TIN 12	
89	683		3	BOWL	RIM	TIN	3.11	1	5	0	1	ROOM 2	TIN 12	
89	688	2	3	BOWL	BDY	TIN	3.11	1	0	0	1	ROOM 2	TIN 12	
89	676	4	3	BOWL	BDY	TIN	3.11	1	0	0	3	ROOM 2	TIN 12	
89	676	4	3	BOWL	RIM	TIN	3.11	1	5	0	1	ROOM 2	TIN 12	
89	670	4	1	BOWL	BDY	TIN	3.11	1	0	0	1	ROOM 2	TIN 12	
89	670	4	1	BOWL	RIM	TIN	3.11	1	5	0	1	ROOM 2	TIN 12	
89	683	9	3	BOWL	BSE	TIN	3.13	1	3	0	4	ROOM 2		high footring; floral design on obverse (very faint - could be polychrome) on face
89	689	4	3	SAUC	RIM	TIN	3.13	1	2	0	1	ROOM 2		design on obverse (too little of it remains to identify pattern)
89	734	10	3	APOT	WHLE	TIN	3.11	1	1 / 1	1	1	ROOM 3		ointment pot
89	852		2	APOT	WHLE	TIN	3.11	1	1 / 1	1	1	ROOM 4		ointment pot
89	735	1	3	APOT	R/B/B	TIN	3.11	1	2 / 1	1	1	ROOM 3	TIN 13	ointment pot
89	739	1	3	APOT	RIM	TIN	3.11	1	2	0	1	ROOM 3	TIN 13	ointment pot
89	735	8	3	CUP	BSE	TIN	3.14	1	1	0	1	ROOM 3		coffee; high footring; blue/purple design on exterior (too little identify pattern)
89	713	4	3	PLAT	BSE	TIN	3.11	1	3	0	3	ROOM 4A		low footring
87	404	2	3	UNKN	RIM	TIN	3.11	1	3	0	1	YARD 4B		
89	727	2	3	UNKN	BDY	TIN	3.11	1	0	0	1	ROOM 4A		
89	786	2	3	UNKN	BDY	TIN	3.11	1	0	0	1	ROOM 4A		storage pot/drug jar?
89	794	4	3	UNKN	BDY	TIN	3.11	1	0	0	1	ROOM 4A		

Year	Unit	No.	Layer	Form	Part	Ware	Type	Thick	Diam	Ht	Qty	Location	Assoc	Notes
89	812		2	UNKN	BDY	TIN	3.11	1	0	0	1	ROOM 4B		
89	825		3	UNKN	BDY	TIN	3.11	1	0	0	1	ROOM 4B		
89	835	4	3	PLAT	RIM	TIN	3.13	1	4	0	1	ROOM 4B		two lines edge rim, one line edges well; possibly polychrome (floral?) design across obverse; pink
89	836	1	3	BOWL	BSE	TIN	3.12	1	2	0	1	ROOM 4B		high footring
89	839	1	3	UNKN	RIM	TIN	3.11	1	0	0	1	ROOM 4B		
89	839	3	3	UNKN	BDY	TIN	3.11	1	0	0	1	ROOM 4B		
89	833		3	PLAT	RIM	TIN	3.13	1	4	0	1	ROOM 4B		chinoiserie design on obverse
89	776	3	3	CUP	BDY	TIN	3.14	1	0	0	1	ROOM 4B		coffee; blue/purple floral design on exterior
89	803	2	3	CUP	BDY	TIN	3.14	1	0	0	1	ROOM 4B		coffee; part scrolled handle attached; blue/purple design on exterior
89	714	2	3	DRNK	ATT	TIN	3.11	1	0	0	1	ROOM 4A	TIN 2	lid
89	714		3	DRNK	ATT	TIN	3.11	1	0	0	1	ROOM 4A	TIN 2	lid
87	745	4	3	BOWL	BDY	TIN	3.13	1	0	0	1	YARD 4A		floral design on obverse
87	7410	4	3	UNKN	BDY	TIN	3.11	1	0	0	1	YARD 4A		
87	371		1	UNKN	BSE	TIN	3.11	1	0	0	2	YARD 4A		pink
87	375	3	3	UNKN	RIM	TIN	3.11	1	2	0	1	YARD 4A		
87	375	3	3	UNKN	BDY	TIN	3.11	1	0	0	1	YARD 4A		
87	376	1	3	UNKN	RIM	TIN	3.11	1	0	0	1	YARD 4A		
87	384	2	3	UNKN	BDY	TIN	3.11	1	0	0	1	YARD 4A		
87	372		2	UNKN	BDY	TIN	3.11	1	0	0	1	YARD 4A		
87	373	3	3	UNKN	RIM	TIN	3.11	1	3	0	1	YARD 4A		
89	743	4	3	APOT	RIM	TIN	3.11	1	2	0	1	YARD 4A		ointment pot
89	743	4	3	CUP	BDY	TIN	3.11	1	0	0	14	YARD 4A	TIN 14	caudle
89	743	4	3	BOWL	RIM	TIN	3.11	1	2	0	1	YARD 4A		
89	743	1	3	CUP	BDY	TIN	3.11	1	0	0	7	YARD 4A	TIN 14	caudle
89	743	1	3	PLAT	RIM	TIN	3.11	1	4	0	1	YARD 4A		
89	742		2	CUP	BDY	TIN	3.11	1	0	0	1	YARD 4A	TIN 14	caudle
89	752		2	CUP	RIM	TIN	3.11	1	3	0	1	YARD 4A	TIN 14	caudle
89	745	3	3	CUP	BDY	TIN	3.11	1	0	0	1	YARD 4A	TIN 14	caudle
89	745	3	3	CUP	RIM	TIN	3.11	1	4	0	3	YARD 4A	TIN 14	caudle
89	743	3	3	CHAM	RIM	TIN	3.11	1	5	0	1	YARD 4A	TIN 14	reconstructed
89	743	4	3	CUP	R/B/B	TIN	3.13	1	3 / 2	0	3	YARD 4A	TIN 14	caudle; part handle attached; low footring
89	752		2	SAUC	RIM	TIN	3.13	1	3	0	1	YARD 4A		
89	754	5	3	CUP	ATT	TIN	3.13	2	0	0	1	YARD 4A	TIN 14	handle; caudle
89	869	2	3	UNKN	BSE	TIN	3.11	1	0	0	1	YARD 4A		
89	869	3	3	UNKN	BDY	TIN	3.11	1	0	0	1	YARD 4A		
89	863	3	3	SAUC	RIM	TIN	3.11	1	2	0	1	YARD 4A		
89	864	1	3	APOT	RIM	TIN	3.11	1	2	0	1	YARD 4A	TIN 15	ointment pot
89	868	3	3	APOT	RIM	TIN	3.11	1	2	0	1	YARD 4A	TIN 15	ointment pot
89	863	3	3	APOT	BSE	TIN	3.11	1	1	0	1	YARD 4A		ointment pot; pedestal foot (post 1700)
89	864	1	3	UNKN	BDY	TIN	3.11	1	0	0	1	YARD 4A		
89	863	1	3	CUP	RIM	TIN	3.13	1	2	0	1	YARD 4A		coffee; floral design on exterior
90	891		1	PLAT	BSE	TIN	3.12	1	4	0	1	YARD 4A	TIN 16	blue stains
90	894	4	3	PLAT	BSE	TIN	3.12	1	4	0	1	YARD 4A	TIN 16	
90	460		1	PLAT	BDY	TIN	3.11	1	0	0	1	YARD 4A		
90	463	1	3	UNKN	BDY	TIN	3.11	1	0	0	2	YARD 4A		
90	463	3	3	UNKN	BDY	TIN	3.11	1	0	0	2	YARD 4A		
90	464	1	3	UNKN	BDY	TIN	3.11	1	0	0	1	YARD 4A		
90	466	3	3	UNKN	BDY	TIN	3.11	1	0	0	1	YARD 4A		
90	465	1	3	SAUC	BSE	TIN	3.11	1	2	0	1	YARD 4A		low footring; de-enameled on underside
90	866		3	PLAT	BDY	TIN	3.13	1	0	0	1	YARD 4A		lines on obverse
90	864	4	3	UNKN	BDY	TIN	3.11	1	0	0	2	YARD 4A		

Year	Unit	No.	Layer	Form	Part	Ware	Type	Thick	Diam	Ht	Qty	Location	Assoc	Notes
90	861		1	APOT	RIM	TIN	3.11	1	2	0	2	YARD 4A		ointment pot
90	866	4	3	UNKN	BDY	TIN	3.11	1	0	0	1	YARD 4A		pink
90	865	3	3	UNKN	BDY	TIN	3.13	1	0	0	1	YARD 4A		pristine design (too little to identify pattern)
89	745	3	3	CHAM	R/B/B	TIN	3.12	1	5 / 3	3	8	YARD 4A	TIN 17	reconstructed
89	599		3	CHAM	BDY	TIN	3.12	1	0	0	1	ROOM 1	TIN 17	reconstructed
89	674		3	CHAM	BDY	TIN	3.12	1	0	0	2	ROOM 2	TIN 17	reconstructed
89	678	2	3	CHAM	BDY	TIN	3.12	1	0	0	1	ROOM 2	TIN 17	reconstructed
89	6710	2	3	CHAM	BDY	TIN	3.12	1	0	0	1	ROOM 2	TIN 17	reconstructed
89	689	4	3	CHAM	BDY	TIN	3.12	1	0	0	1	ROOM 2	TIN 17	reconstructed
89	686	5	3	CHAM	BSE	TIN	3.12	1	3	0	2	ROOM 2	TIN 17	reconstructed
87	403	2	3	UNKN	BDY	TIN	3.11	1	0	0	1	YARD 4B		
87	405	2	3	APOT	RIM	TIN	3.11	1	2	0	1	YARD 4B		ointment pot
87	404	3	3	UNKN	BSE	TIN	3.11	1	3	0	1	YARD 4B		
87	404	2	3	UNKN	BDY	TIN	3.11	1	0	0	1	YARD 4B		
87	404	3	3	UNKN	BDY	TIN	3.11	1	0	0	1	YARD 4B		
87	403	2	3	UNKN	BDY	TIN	3.11	1	0	0	1	YARD 4B		
87	406	4	3	UNKN	BDY	TIN	3.11	1	0	0	1	YARD 4B		
87	400		1	UNKN	BDY	TIN	3.14	1	0	0	1	YARD 4B		blue/purple design on one surface (too little to identify pattern)
87	403	2	3	UNKN	RIM	TIN	3.11	1	2	0	1	YARD 4B		
87	404	3	3	UNKN	BDY	TIN	3.11	1	0	0	1	YARD 4B		
87	403	2	3	APOT	RIM	TIN	3.11	1	2	0	1	YARD 4B		drug jar
87	404	2	3	UNKN	BSE	TIN	3.11	1	0	0	1	YARD 4B		
87	400		1	UNKN	BDY	TIN	3.13	1	0	0	1	YARD 4B		two thin parallel lines on one surface
87	405	3	3	UNKN	BDY	TIN	3.14	1	0	0	1	YARD 4B		blue/purple design on obverse (too little to identify pattern)
87	406	3	3	APOT	BSE	TIN	3.11	1	1	0	1	YARD 4B		ointment pot ; pedestal foot (post 1700)
87	446	4	3	UNKN	RIM	TIN	3.11	1	3	0	1	ROOM 4B		
87	443	3	3	APOT	RIM	TIN	3.11	1	4	0	1	ROOM 4B		drug jar
87	446	2	3	CUP	ATT	TIN	3.13	1	0	0	1	ROOM 4B		scrolled handle; tea/coffee
87	256	2	3	UNKN	BDY	TIN	3.13	1	0	0	1	YARD 4B		floral? design on one surface
87	256	2	3	UNKN	BDY	TIN	3.13	1	0	0	1	YARD 4B		design (possibly with purple) on one surface (too little to identify pattern)
87	256	2	3	UNKN	RIM	TIN	3.11	1	2	0	1	YARD 4B		
87	256	2	3	UNKN	BDY	TIN	3.11	1	0	0	1	YARD 4B		
87	256	2	3	UNKN	RIM	TIN	3.11	1	4	0	1	YARD 4B		
87	432		2	UNKN	BDY	TIN	3.11	1	0	0	1	YARD 4B		thin and flat
87	432		2	APOT	BSE	TIN	3.11	1	1	0	1	YARD 4B		
87	435	2	3	UNKN	BDY	TIN	3.11	1	0	0	1	YARD 4B		
87	435	2	3	UNKN	BSE	TIN	3.12	1	6	0	1	YARD 4B		
87	435	2	3	UNKN	RIM	TIN	3.12	1	0	0	2	YARD 4B		fluted
87	433	2	3	CUP	RIM	TIN	3.13	1	2	0	1	YARD 4B		coffee; thin blue lines edge rim exterior
87	412		2	CUP	RIM	TIN	3.11	1	2	0	1	ALLEY		coffee; greenish blue interior
87	442		2	CUP	BDY	TIN	3.13	1	0	0	1	ROOM 4B		coffee; floral design on exterior
90	469	2	3	UNKN	BSE	TIN	3.11	1	2	0	1	ROOM 4B		low footring
87	415	1	3	PLAT	RIM	TIN	3.14	1	4	0	1	ALLEY		two thin blue lines edge rim; incised circles/scrolls on purple ground
87	405	2	3	PLAT	RIM	TIN	3.12	1	4	0	1	YARD 4B		
89	633	1	3	UNKN	BDY	TIN	3.11	2	0	0	1	SIDEWALK		
89	602		2	UNKN	BDY	TIN	3.11	1	0	0	1	SIDEWALK		
89	686	5	3	UNKN	BDY	TIN	3.11	1	0	0	1	ROOM 2		
89	634	1	3	PLAT	BSE	TIN	3.14	1	4	0	1	SIDEWALK	TIN 18	high footring; blue/purple chinoiserie on obverse
89	634	1	3	PLAT	BDY	TIN	3.14	1	0	0	1	SIDEWALK	TIN 18	blue/purple chinoiserie design on obverse
89	662		2	PLAT	BDY	TIN	3.13	1	0	0	1	SIDEWALK		blue wave design on obverse
89	666	4	3	UNKN	BDY	TIN	3.11	1	0	0	1	SIDEWALK		

Year	Unit	No.	Layer	Form	Part	Ware	Type	Thick	Diam	Ht	Qty	Location	Assoc	Notes
89	704	2	3	UNKN	BDY	TIN	3.11	2	0	0	2	SIDEWALK		
90	660		1	BOWL	RIM	TIN	3.13	1	3	0	1	SIDEWALK		scroll design on obverse
90	944	4	3	STOR	BDY	CRW	1.31	1	0	0	1	YARD 5		white slip on exterior; yellowish red; English
87	353	1	3	UNKN	BDY	TIN	3.11	1	0	0	3	ALLEY		
87	353	1	3	UNKN	ATT	TIN	3.11	2	0	0	1	ALLEY		strap handle
87	353	1	3	UNKN	BDY	TIN	3.11	1	0	0	1	ALLEY		
87	353		3	UNKN	RIM	TIN	3.11	1	0	0	1	ALLEY		
87	353	1	3	UNKN	BDY	TIN	3.11	1	0	0	1	ALLEY		
87	353	2	3	UNKN	BDY	TIN	3.11	1	0	0	3	ALLEY		
87	353	3	3	APOT	RIM	TIN	3.11	1	2	0	1	ALLEY		ointment pot
87	353	2	3	UNKN	ATT	TIN	3.11	2	0	0	1	ALLEY		strap handle
87	353	3	3	UNKN	BDY	TIN	3.11	1	0	0	1	ALLEY		
87	354	1	3	UNKN	BDY	TIN	3.11	1	0	0	1	ALLEY		
87	353	1	3	UNKN	BDY	TIN	3.13	1	0	0	1	ALLEY		design on one surface (too little to identify pattern)
87	351	1	1	UNKN	BDY	TIN	3.11	1	0	0	1	ALLEY		
87	263	1	3	CUP	BSE	TIN	3.11	1	1	0	1	ALLEY		coffee; low footring
87	264	2	3	UNKN	BDY	TIN	3.11	1	0	0	1	ALLEY		
87	416	3	3	UNKN	BDY	TIN	3.11	1	0	0	1	ALLEY		
87	425	2	3	UNKN	ATT	TIN	3.11	3	0	0	1	ALLEY		rounded handle
87	413	1	3	UNKN	BDY	TIN	3.11	1	0	0	1	ALLEY		
87	453	1	3	UNKN	BSE	TIN	3.11	1	0	0	1	ALLEY		high footring
87	413	1	3	UNKN	RIM	TIN	3.14	1	0	0	1	ALLEY		blue/purple design on exterior (too little to identify pattern)
87	415	3	3	UNKN	BDY	TIN	3.13	1	0	0	1	ALLEY		flower design on on surface
87	415	3	3	UNKN	BDY	TIN	3.13	1	0	0	1	ALLEY		shades of blue on exterior; pristine white interior
87	262		2	CUP	RIM	TIN	3.13	1	2	0	1	ALLEY		coffee; two lines edge rim exterior, flower head below
87	353	1	3	PLAT	BDY	TIN	3.13	1	0	0	1	ALLEY		two lines edge rim on obverse, two lines surround well
90	452	2	2	APOT	RIM	TIN	3.11	1	2	0	1	ALLEY		ointment pot
90	918	1	3	UNKN	BDY	TIN	3.11	1	0	0	1	YARD 5		
90	912		2	UNKN	BDY	TIN	3.11	1	0	0	1	YARD 5		
90	918	1	3	UNKN	RIM	TIN	3.11	1	0	0	1	YARD 5		
90	917	10	3	UNKN	BDY	TIN	3.11	1	0	0	2	YARD 5		
90	918	1	3	UNKN	BDY	TIN	3.11	1	0	0	1	YARD 5		
90	912		2	UNKN	BDY	TIN	3.11	1	0	0	1	YARD 5		
90	913		3	UNKN	BDY	TIN	3.11	1	0	0	1	YARD 5		
90	913		3	APOT	RIM	TIN	3.11	1	2	0	1	YARD 5		drug jar
90	912	15	2	PLAT	BDY	TIN	3.11	1	0	0	1	YARD 5		
90	9110	3	3	UNKN	BDY	TIN	3.11	1	0	0	1	YARD 5		
90	918	12	3	UNKN	BDY	TIN	3.11	1	0	0	1	YARD 5		flat; thin
90	9110	4	3	UNKN	BDY	TIN	3.11	1	0	0	1	YARD 5		
90	918	1	3	UNKN	BDY	TIN	3.11	1	0	0	2	YARD 5		
90	917	3	3	UNKN	BDY	TIN	3.11	1	0	0	1	YARD 5		
90	9110	1	3	UNKN	BSE	TIN	3.11	1	0	0	1	YARD 5		
90	919	1	3	APOT	BDY	TIN	3.11	2	0	0	1	YARD 5		
90	917	3	3	UNKN	BDY	TIN	3.26	1	0	0	1	YARD 5		blue geometric design on one surface; dull white glaze
90	917	3	3	UNKN	BDY	TIN	3.11	1	0	0	1	YARD 5		
90	967	4	3	UNKN	BDY	TIN	3.11	1	0	0	1	YARD 6		
90	942		2	DISH	RIM	TIN	3.13	1	6	0	1	YARD 5	TIN 19	chinoiserie design (panneled) on obverse; see TIN 9, 25 (same dish?)
89	6710	4	3	DISH	RIM	TIN	3.13	1	6	0	1	ROOM 1	TIN 19	chinoiserie design (panneled) on obverse; see TIN 9, 25 (same dish?)
90	904	7	3	UNKN	RIM	TIN	3.12	1	4	0	2	YARD 5	TIN 20	porringer/drug jar?
90	904	7	3	UNKN	BDY	TIN	3.12	1	0	0	3	YARD 5	TIN 20	porringer/drug jar?
89	737	1	3	UNKN	BDY	TIN	3.12	1	0	0	1	ROOM 3	TIN 20	porringer/drug jar?

Year	Unit	No.	Layer	Form	Part	Ware	Type	Thick	Diam	Ht	Qty	Location	Assoc	Notes
90	9411	4	3	PLAT	RIM	TIN	3.12	1	4	0	1	YARD 5		
90	947	4	3	UNKN	BDY	TIN	3.11	1	0	0	1	YARD 5		
90	9410	3	3	PLAT	BDY	TIN	3.12	1	0	0	1	YARD 5		center/well
90	9411	4	3	UNKN	BDY	TIN	3.11	1	0	0	1	YARD 5		
90	959	4	3	UNKN	BDY	TIN	3.11	1	0	0	1	YARD 5		pink fabric
90	9412	1	3	UNKN	BDY	TIN	3.11	1	0	0	1	YARD 5		
90	944	3	3	UNKN	BDY	TIN	3.11	1	0	0	1	YARD 5		
90	949	3	3	UNKN	BDY	TIN	3.12	1	0	0	1	YARD 5		
90	9410	2	3	PLAT	BDY	TIN	3.11	1	0	0	1	YARD 5		center/well
90	944	20	3	UNKN	BDY	TIN	3.11	1	0	0	1	YARD 5		
90	9410	2	3	APOT	RIM	TIN	3.11	1	2	0	1	YARD 5	TIN 21	ointment pot
90	949	4	3	APOT	BDY	TIN	3.11	1	0	0	1	YARD 5	TIN 21	ointment pot
90	957	2	3	BOWL	BDY	TIN	3.12	1	0	0	1	YARD 5		
90	946	2	3	UNKN	BDY	TIN	3.11	1	0	0	1	YARD 5		
90	944	4	3	UNKN	BDY	TIN	3.11	1	0	0	1	YARD 5		
90	944	1	3	PLAT	BDY	TIN	3.11	1	0	0	1	YARD 5		
90	947	2	3	UNKN	BDY	TIN	3.11	1	0	0	1	YARD 5		
90	944	2	3	PLAT	RIM	TIN	3.11	1	4	0	1	YARD 5		
90	945	2	3	UNKN	ATT	TIN	3.11	1	0	0	1	YARD 5		handle stub
90	946		3	UNKN	BDY	TIN	3.11	1	0	0	1	YARD 5		
90	959	4	3	UNKN	BDY	TIN	3.11	1	0	0	1	YARD 5		
90	919	4	3	UNKN	BDY	TIN	3.14	1	0	0	1	YARD 5		blue/purple design on one surface (too little to identify pattern)
90	904	2	3	CUP	BDY	TIN	3.14	1	0	0	1	YARD 5		coffee; blue/purple design (tendril / branch) on exterior
90	9412	3	3	UNKN	RIM	TIN	3.11	1	0	0	1	YARD 5		scalloped; greenish blue ground
89	745	1	3	UNKN	BDY	TIN	3.13	1	0	0	1	YARD 4A		amorphous shape on exterior (too little to identify pattern)
90	918	1	3	PLAT	R/B/B	TIN	3.11	1	5 / 4	3	2	YARD 5	TIN 3	low footring
90	942		2	PLAT	BDY	TIN	3.11	1	0	0	1	YARD 5	TIN 3	
90	913		3	PLAT	RIM	TIN	3.11	1	5	0	1	YARD 5	TIN 3	
90	9410	3	3	PLAT	BDY	TIN	3.11	1	0	0	1	YARD 5	TIN 3	well
89	852		2	CUP	BSE	TIN	3.13	1	3	0	1	ROOM 4	TIN 7	
89	738	5	3	CUP	BDY	TIN	3.13	1	0	0	1	ROOM 3	TIN 7	
90	9183		3	DRNK	ATT	TIN	3.11	2	0	0	1	YARD 5	TIN 2	finial knob for lid
90	919	1	3	DRNK	ATT	TIN	3.11	1	0	0	1	YARD 5	TIN 2	lid
90	9110	1	3	DRNK	ATT	TIN	3.11	1	0	0	1	YARD 5	TIN 2	lid
89	9610	2	3	DRNK	ATT	TIN	3.11	1	0	0	1	YARD 6	TIN 2	lid
90	944	1	3	DRNK	ATT	TIN	3.11	1	0	0	1	YARD 5	TIN 2	lid
90	9411	4	3	SALT	RIM	TIN	3.11	1	2	0	1	YARD 5	TIN 1	salt holder
90	944	1	3	PORR	R/B/B	TIN	3.12	1	4 / 4	3	14	YARD 5		part 3-holed horizontal handle attached
90	918	12	3	DRNK	RIM	TIN	3.12	1	3	0	1	YARD 5	TIN 4	part reconstructed
90	918	13	3	DRNK	BDY	TIN	3.12	1	0	0	1	YARD 5		scrolled handle attached
90	912		2	DRNK	BDY	TIN	3.12	1	0	0	1	YARD 5	TIN 4	
90	917	10	3	DRNK	RIM	TIN	3.12	1	3	0	1	YARD 5	TIN 4	
90	917	10	3	DRNK	BDY	TIN	3.12	1	0	0	1	YARD 5	TIN 4	
90	969		3	APOT	WHLE	TIN	3.13	1	3 / 2	3	1	YARD 6		drug jar; entwined/plain lines in horizontal figure-eight pattern around body
90	916	1	3	APOT	R/B/B	TIN	3.11	1	1 / 1	1	2	YARD 5		ointment pot; pedestal foot (post 1700)
90	9410	2	3	APOT	R/B/B	TIN	3.11	1	2 / 2	1	1	YARD 5		ointment pot; pedestal foot (post 1700)
90	903	6	3	APOT	R/B/B	TIN	3.11	1	2 / 2	2	2	YARD 5	TIN 22	drug jar; flared base
90	917	7	3	APOT	RIM	TIN	3.11	1	2	0	1	YARD 5	TIN 22	drug jar
90	947	12	3	CHAM	BDY	TIN	3.12	1	0	0	1	YARD 5	TIN 23	
90	944	22	3	CHAM	RIM	TIN	3.12	1	4		1	YARD 5	TIN 23	
90	944	23	3	CHAM	BDY	TIN	3.12	1	0	0	3	YARD 5	TIN 23	

Year	Unit	No.	Layer	Form	Part	Ware	Type	Thick	Diam	Ht	Qty	Location	Assoc	Notes
90	944	2	3	CHAM	BDY	TIN	3.12	1	0	0	1	YARD 5	TIN 23	
90	944	7	3	CHAM	BDY	TIN	3.12	1	0	0	1	YARD 5	TIN 23	
90	944	1	3	CHAM	R/B/B	TIN	3.12	1	4 / 4	3	10	YARD 5	TIN 23	
90	944	19	3	CHAM	BDY	TIN	3.12	1	0	0	1	YARD 5		attached to grooved / ridged handle; similar to handle on TIN 19
90	944	28	3	CHAM	BDY	TIN	3.12	1	0	0	1	YARD 5	TIN 23	
90	944	28	3	CHAM	RIM	TIN	3.12	1	4	0	1	YARD 5	TIN 23	
90	944	21	3	CHAM	BDY	TIN	3.12	1	0	0	1	YARD 5	TIN 23	
90	944	21	3	CHAM	BSE	TIN	3.12	1	4	0	2	YARD 5	TIN 23	low footring
90	9612	1	3	UNKN	BDY	TIN	3.11	1	0	0	1	YARD 6		
90	9610		3	UNKN	BDY	TIN	3.11	1	0	0	2	YARD 6		
90	9612	1	3	PLAT	RIM	TIN	3.12	1	5	0	1	YARD 6		
90	962		2	UNKN	BDY	TIN	3.12	1	0	0	1	YARD 6		square; flat
90	963	1	3	UNKN	BDY	TIN	3.11	1	0	0	1	YARD 6		thin
90	961	2	1	UNKN	BDY	TIN	3.11	1	0	0	1	YARD 6		
90	9610	2	3	UNKN	BDY	TIN	3.11	1	0	0	1	YARD 6		
90	9612		3	UNKN	BSE	TIN	3.12	1	2	0	2	YARD 6		high footring; fire damage
90	965	5	3	UNKN	BSE	TIN	3.11	1	2	0	1	YARD 6		high footring
90	9610		3	SALT	RIM	TIN	3.12	1	3	0	1	YARD 6	TIN 1	salt holder
89	884	2	3	BOWL	BDY	TIN	3.12	1	0	0	1	YARD 7		
90	882		2	UNKN	BDY	TIN	3.11	1	0	0	1	YARD 7		
90	886	4	3	PLAT	RIM	TIN	3.11	1	4	0	1	YARD 7		
90	882		2	UNKN	BDY	TIN	3.14	1	0	0	1	YARD 7		attached to 87-534-4 (not in dbase) ; blue/black geometric leaf design on obverse
90	889	11	3	PORR	R/B/B	TIN	3.12	1	3 / 3	1	8	YARD 7	TIN 24	part reconstructed; handle absent; fire damage; see 90-944-1
90	889	7	3	PORR	R/B/B	TIN	3.12	1	3 / 3	1	2	YARD 7	TIN 24	part reconstructed; handle absent; fire damage; see 90-944-1
87	375	1	3	DRNK	ATT	TIN	3.11	1	1	0	1	YARD 4A		spout; part body attached; see similar 89-599, 89-734-2
87	413	4	3	DRNK	R/B/B	TIN	3.12	1	3 / 3	3	4	ALLEY		two grooved loop handles, bulbous bottom, wide collar encircles upper body
89	654	4	3	DISH	BDY	TIN	3.13	1	0	0	1	ROOM 1	TIN 25	chinoiserie (panneled) design on obverse; see TIN 9, 19 (same dish?)
89	656	3	3	DISH	BSE	TIN	3.13	1	3	0	2	ROOM 1	TIN 25	chinoiserie (panneled) design on obverse; see TIN 9,19 (same dish?)
89	656	3	3	DISH	BDY	TIN	3.13	1	0	0	1	ROOM 1	TIN 25	chinoiserie (panneled) design on obverse; see TIN 9, 19 (same dish?)
89	656	3	3	DISH	RIM	TIN	3.13	1	6	0	1	ROOM 1	TIN 25	chinoiserie (panneled) design on obverse; see TIN 9, 19 (same dish?)
89	739	18	3	CUP	RIM	TIN	3.14	1	2	0	1	ROOM 3	TIN 11	reconstructed quatrefoil
89	858	2	3	CUP	BSE	TIN	3.13	1	3	0	1	ROOM 4	TIN 7	low footring
90	100		1	PLAT	RIM	TIN	3.26	1	5	0	1	B5/LOT 100		three wide bright blue lines / one thin blue line radiate upward from center to rim edge; red
90	917	7	3	PLAT	BDY	TIN	3.26	1	0	0	1	YARD 5	TIN 26	pin holes on reverse ; blue floral / dot design on obverse (too little to identify pattern); red
89	914		3	PLAT	RIM	TIN	3.26	1	5	0	1	YARD 5	TIN 26	three blue dots on obverse (too little to identify pattern); orange red
89	794	4	3	PLAT	BDY	TIN	3.26	1	0	0	1	ROOM 4A	TIN 26	blue floral/dot design on obverse (too little to identify pattern); dull white glaze; red
89	706	4	3	PLAT	BSE	TIN	3.26	1	2	0	1	SIDEWALK	TIN 26	high footring; reddish; Mexico City Fine Grade (San Luis cobalt blue-on-White)?
87	406	3	3	PLAT	BSE	TIN	3.26	1	2	0	2	YARD 4B	TIN 26	high footring; blue (floral?) design on obverse; pristine white glaze; orange red
87	434	2	3	PLAT	BDY	TIN	3.26	1	0	0	1	YARD 4B	TIN 26	three blue dots on obverse ; pristine white glaze; red
89	645	2	3	UNKN	BSE	SLW	2.14	1	2	0	1	ROOM 1		combed exterior
87	406	3	3	PLAT	BSE	TIN	3.26	1	2	0	1	YARD 4B	TIN 27	high footring; blue dots on obverse; orange red
90	463	1	3	PLAT	BSE	TIN	3.27	1	2	0	1	YARD 4A		high footring; dark blue/green geometric circles/ lines on obverse; Abo/ Polychrome?
89	735	1	3	PLAT	RIM	TIN	3.21	1	3	0	1	ROOM 3		light blue/white (faded) lines edge rim; yellowish; Yayal Blue-on-White?
89	732		2	PLAT	BDY	TIN	3.22	1	0	0	1	ROOM 3		light yellow/brown thick glaze light cream
89	708	2	3	BOWL	BDY	TIN	3.21	3	0	0	1	SIDEWALK		light blue crude wave-like design on exterior / vestiges of design on interior
89	743	4	3	PLAT	BDY	TIN	3.22	1	0	0	1	YARD 4A		blue/purple design on obverse; light cream
90	864	4	3	SAUC	RIM	TIN	3.22	2	3	0	1	YARD 4A		vertical pin holes on reverse; blue floral design on obverse, edged on either side by two blue lines
87	254	2	3	PLAT	RIM	TIN	3.22	1	3	0	1	YARD 4B		narrow orange line edges rim; vestiges of oval forms (colors faded) in well; unglazed
87	425	4	3	PLAT	BDY	TIN	3.33	2	0	0	1	ALLEY		blue panneled design in well; buff
89	623	4	3	SAUC	BSE	SLW	2.14	2	2	0	1	ROOM 1		shallow; combed design on obverse with partial letter H; press-molded
89	654	3	3	PAN	RIM	SLW	2.18	1	3	0	1	ROOM 1	SLW 1	part reconstructed; oval shape; dark reddish brown slip

Year	Unit	No.	Layer	Form	Part	Ware	Type	Thick	Diam	Ht	Qty	Location	Assoc	Notes
89	646	4	3	PAN	RIM	SLW	2.18	1	3	0	1	ROOM 1	SLW 1	part reconstructed; oval shape; dark reddish brown slip
89	678	7	3	PAN	R/B/B	SLW	2.18	1	3 / 4	2	5	ROOM 2	SLW 1	part reconstructed; oval shape; dark reddish brown slip
89	678	7	3	UNKN	BDY	CRW	1.10	1	0	0	1	ROOM 2		
89	6710	2	3	UNKN	BDY	SLW	2.14	1	0	0	1	ROOM 2		trailed slip design on exterior
89	734	4	3	BOWL	BDY	SLW	2.14	1	0	0	1	ROOM 3	SLW 2	combed design on interior
89	852		2	UNKN	BDY	CRW	1.10	1	0	0	1	ROOM 4		olive glaze on both surfaces
90	893	1	3	UNKN	BSE	CRW	1.10	1	2	0	1	YARD 4A		olive glaze on both surfaces
90	753	1	3	BOWL	R/B/B	SLW	2.14	1	2 / 2	2	1	YARD 4A	SLW 2	shallow; combed design on interior
89	891		1	UNKN	BDY	SLW	2.14	1	0	0	1	YARD 4A		combed/feathered design on obverse
87	375	3	3	UNKN	BDY	SLW	2.14	1	0	0	1	YARD 4A		combed design on exterior
87	371		1	PLAT	RIM	SLW	2.14	1	4	0	1	YARD 4A		combed design on obverse
89	783		3	BOWL	RIM	SLW	2.14	1	2	0	2	ROOM 4A	SLW 2	combed design on interior
89	715	9	3	CUP	RIM	SLW	2.14	1	2	0	4	ROOM 4A	SLW 3	part reconstructed; trailed band edges rim exterior
89	715	9	3	CUP	BDY	SLW	2.14	1	0	0	4	ROOM 4A	SLW 3	part reconstructed; combed/feathered design on exterior
89	743	1	3	CUP	RIM	SLW	2.14	1	2	0	1	YARD 4A	SLW 3	part reconstructed; trailed band edges rim exterior
89	715	7	3	CUP	BDY	SLW	2.14	1	0	0	1	ROOM 4A	SLW 3	combed/feathered design on exterior
89	717	5	3	CUP	BDY	SLW	2.14	1	0	0	1	ROOM 1	SLW 3	combed/feathered design on exterior
89	715	7	3	CUP	BSE	SLW	2.14	1	2	0	2	ROOM 1	SLW 3	olive glaze on interior
87	405	1	3	UNKN	BSE	SLW	2.14	1	2	0	1	YARD 4B	SLW 3	olive glaze on interior
87	404	2	3	UNKN	BDY	SLW	2.14	1	0	0	2	YARD 4B		combed design on exterior
87	255	2	3	UNKN	BDY	SLW	2.14	1	0	0	1	YARD 4B		combed design on exterior
87	403	1	3	SAUC	RIM	SLW	2.15	1	3	0	1	YARD 4B		three narrow dark brown lines (perpendicular to rim) on light green glaze on interior
87	405	3	3	BOWL	BDY	SLW	2.13	2	0	0	1	YARD 4B		incised geometric design on interior ; light reddish brown
90	9610	3	3	STOR	BDY	CRW	1.10	3	0	0	1	YARD 6		light red
90	942		2	STOR	RIM	CRW	1.10	2	5	0	1	YARD 5		white slip on exterior; light red
89	606	1	3	PAN	R/B/B	SLW	2.18	1	3 / 4	2	1	SIDEWALK	SLW 1	part reconstructed; oval shape; dark reddish brown slip
87	421		1	CUP	BSE	SLW	2.14	1	2	0	1	ALLEY		footring; handle stub attached; olive glaze on interior; combed design on exterior
87	415	3	3	UNKN	BDY	CRW	1.10	1	0	0	1	ALLEY		olive glaze on both surfaces
87	415	4	3	UNKN	RIM	SLW	2.14	1	2	0	1	ALLEY		combed design on exterior
87	353	2	3	UNKN	BDY	CRW	1.10	1	0	0	1	ALLEY		dark green stained
87	353	1	3	UNKN	BDY	SLW	2.14	1	0	0	1	ALLEY		trailed design on obverse
87	243	5	3	UNKN	BDY	SLW	2.14	1	0	0	1	ALLEY		trailed lines on one surface
87	264	3	3	UNKN	BDY	SLW	2.14	1	0	0	1	ALLEY		trailed lines/combed design on one surface
90	950		1	UNKN	BDY	SLW	2.14	1	0	0	1	YARD 5		trailed lines on exterior
90	949	10	3	UNKN	RIM	SLW	2.14	1	2	0	1	YARD 5		brown trailed line edges rim exterior
90	918	2	3	UNKN	RIM	SLW	2.14	1	2	0	2	YARD 5		brown trailed line edges rim exterior
90	946	1	3	CUP	ATT	SLW	2.14	2	0	0	1	YARD 5		handle; feathered design
90	919	1	3	CUP	BSE	SLW	2.14	1	2	0	1	YARD 5	SLW 4	part handle attached; feathered/combed design on exterior
90	919	1	3	CUP	BDY	SLW	2.14	1	0	0	2	YARD 5	SLW 4	feathered/combed design on exterior
90	958	9	3	PORR	R/B/B	SLW	2.16	1	2 / 2	2	2	YARD 5	SLW 5	dark brown dots edge rim exterior, trailed line around lower exterior body
90	9410	2	3	PORR	BSE	SLW	2.16	1	2	0	1	YARD 5	SLW 5	olive glaze on interior
90	9410	8	3	CUP	WHLE	SLW	2.14	1	1 / 1	1	1	YARD 5		combed/trailed design on exterior
90	965		3	CUP	RIM	SLW	2.14	1	2	0	4	YARD 6	SLW 6	combed/ trailed design on exterior
90	965		3	CUP	BDY	SLW	2.14	1	0	0	9	YARD 6	SLW 6	part handle attached; combed/trailed design on exterior
90	9612	1	3	CUP	BDY	SLW	2.14	1	0	0	2	YARD 6	SLW 6	combed/trailed design on exterior
90	962		2	CUP	BDY	SLW	2.14	1	0	0	2	YARD 6	SLW 6	combed/trailed design on exterior
89	622		2	UNKN	BSE	SLW	2.23	1	0	0	1	ROOM 1		white slipped lines covered with green glaze; reddish orange
87	353	1	3	SAUC	BDY	SLW	2.22	1	0	0	1	ALLEY		white slipped design (too little to identify pattern) on one surface; reddish brown
87	372		2	UNKN	BDY	SLW	2.20	1	0	0	1	YARD 4A		white slip in spiral design on obverse; unglazed; fire damage
87	594	2	3	COST	BDY	SLW	2.23	2	0	0	1	ROOM 1	SLW 7	
87	574	2	3	COST	BDY	SLW	2.23	2	0	0	12	ROOM 1	SLW 7	

Year	Unit	No.	Layer	Form	Part	Ware	Type	Thick	Diam	Ht	Qty	Location	Assoc	Notes
87	574	2	3	COST	BDY	SLW	2.23	2	0	0	1	ROOM 1	SLW 7	lion/human mask handle attached
87	571		1	COST	BDY	SLW	2.23	2	0	0	1	ROOM 1	SLW 7	
87	574	1	3	BOTT	BDY	STW	4.10	2	0	0	1	ROOM 1		dull reddish brown glaze on exterior; gray; Crouch Ware? English
87	422		2	BOTT	BDY	STW	4.10	2	0	0	1	ALLEY	STW 1	glazed both surfaces; light yellow; English
87	592		2	BOTT	BDY	STW	4.10	2	0	0	1	ROOM 1	STW 1	glazed both surfaces; light yellow; English
87	592		2	UNKN	BDY	STW	4.10	2	0	0	1	ROOM 1		mottled brown on exterior, pink wash on interior; grayish white; English
87	572		2	UNKN	BDY	STW	4.10	2	0	0	1	ROOM 1		light brown glazed exterior; grayish white fabric; English
87	592		2	UNKN	BDY	STW	4.30	2	0	0	1	ROOM 1		cream / white wash on exterior; glaze on both surfaces; grayish white
87	572		2	MUG	BDY	STW	4.20	1	0	0	1	ROOM 1		blue blobs on exterior; finely ridged exterior; gray; German? see similar 90-104-2
89	625	2	3	BOTT	BDY	STW	4.13	1	0	0	1	ROOM 1		mouth/neck with mask
89	652		2	STOR	BDY	STW	4.10	2	0	0	1	ROOM 1		light brown; English
89	651		1	STOR	BSE	STW	4.10	2	6	0	1	ROOM 1		white slip wash on exterior; reddish brown glaze; English
89	681		1	BOTT	RIM	STW	4.10	2	1	0	1	ROOM 2		part mouth/neck; yellowish brown glazed both surfaces; buff; English
89	622		2	STOR	BDY	STW	4.10	1	0	0	1	ROOM 1		heavily ridged/throwing rings; brown glazed smooth exterior; English
89	642	2	2	MUG	BSE	STW	4.20	1	2	0	1	ROOM 1		Westerwald mug? glazed both surfaces
89	642		2	BOTT	RIM	STW	4.10	2	1	0	1	ROOM 1		short neck; yellowish brown glaze on exterior, cream glaze on interior; English
90	104	2	3	MUG	BDY	STW	4.20	1	0	0	1	B5/LOT 100		blue blobs on exterior; finely ridged light gray exterior; white slip on interior; cream/white; English?
89	681		1	MUG	RIM	STW	4.10	1	2	0	1	ROOM 2		narrow dark brown painted line/incised line edges rim exterior; white slip/light brown glaze; English
89	681		1	BOTT	RIM	STW	4.10	1	2	0	1	ROOM 2		mouth/neck; one incised line edges rim exterior; glaze on interior; fire use; English
89	732		2	BOTT	BSE	STW	4.30	1	2	0	1	ROOM 3		white glaze on exterior; English
89	852		2	COST	BDY	STW	4.40	1	0	0	1	ROOM 4		short ridged horizontal handle attached; light yellowish brown; Dutch /East Indian? See also STW 5
89	682		2	MUG	BDY	STW	4.24	1	0	0	1	ROOM 2		blue stamped circular medallions; crude execution; orange patches (fire damage?)
89	672		2	UNKN	BDY	STW	4.30	1	0	0	1	ROOM 2		glazed both surfaces; cream/white
87	683	5	3	BOTT	WHLE	STW	4.21	1	0	3	1	ROOM 2		grooved handle; gray/light brown freckles
89	834	3	3	MUG	ATT	STW	4.21	2	0	0	1	ROOM 4B	STW 2	rounded handle - Westerwald mug?
90	941		1	UNKN	BDY	STW	4.10	2	0	0	2	YARD 5		glazed both surfaces; inclusions; orange brown; English
90	909	4	3	UNKN	BDY	STW	4.11	1	0	0	1	YARD 5		two incised lines/brown glaze freckled exterior; light yellowish brown slipped/glazed interior
90	902		2	MUG	BSE	STW	4.30	1	2	0	2	YARD 5		low footring
90	917	4	3	UNKN	BDY	STW	4.11	1	0	0	1	YARD 5		Bartmann? mug/jug? brown glaze freckled exterior; gray
90	944	4	3	UNKN	BDY	STW	4.11	1	0	0	1	YARD 5		Bartmann? mug/jug? orange brown glaze/dense freckled exterior; light brown
90	948	5	3	BOTT	WHLE	STW	4.11	1	0	3	1	YARD 5		grooved handle; brown glazed freckled exterior
89	632		2	MUG	BDY	STW	4.22	1	0	0	1	SIDEWALK		blue/purple; stamped manganese heart/ two horizontal sprigged lines on exterior
89	632		2	MUG	BDY	STW	4.23	1	0	0	1	SIDEWALK		blue/gray; stamped circular medallion/combed lines on exterior
89	702		2	MUG	BDY	STW	4.23	1	0	0	1	SIDEWALK		blue/gray; stamped circular medallion/combed lines on exterior
89	702		2	MUG	BDY	STW	4.21	1	0	0	1	SIDEWALK	STW 2	part handle attached
89	602		2	BOTT	BDY	STW	4.10	2	0	0	1	SIDEWALK		English?
89	604	3	3	BOTT	RIM	STW	4.12	2	1	0	1	SIDEWALK		handle scar; brown glazed on both surfaces
89	637		3	BOTT	ATT	STW	4.12	3	0	0	1	SIDEWALK		medallion (armorial)
89	632		2	BOTT	RIM	STW	4.10	2	1	0	1	SIDEWALK		mouth/neck; part glaze on interior; light yellowish brown; English
87	583	3	3	MUG	BDY	STW	4.32	1	0	0	1	SIDEWALK		brown glazed freckled exterior; pale grayish white glaze on interior; see 90-923-1
87	372		2	MUG	BDY	STW	4.10	1	0	0	1	YARD 4A		three narrow horizontal incised lines on exterior; light brown glazed both surfaces; English
87	745	3	3	UNKN	BDY	STW	4.11	1	0	0	1	YARD 4A		Bartmann? mug/jug?; dark brown glaze freckled exterior
87	442		2	MUG	RIM	STW	4.32	1	2	0	2	ROOM 4B	STW 3	brown band edges rim exterior; glaze on both surfaces
87	423	1	3	MUG	BDY	STW	4.32	1	0	0	1	ALLEY	STW 3	glaze on both surfaces
87	383	3	3	UNKN	BDY	STW	4.21	1	0	0	1	YARD 4A		gray freckled exterior
87	384	2	3	MUG	RIM	STW	4.23	1	2	0	3	YARD 4A		mouth/neck/upper body; gray stamped rosettes/ horizontal blue bands on exterior
89	884	3	3	STOR	BDY	STW	4.10	2	0	0	1	YARD 7		light brown glaze on exterior, pink interior; inclusions; grayish white; English
89	756	4	3	MUG	BSE	STW	4.20	1	3	0	1	YARD 4A		Westerwald mug?
89	892		2	COST	BDY	STW	4.40	1	0	0	1	YARD 4A	STW 4	thin; light pink exterior; unglazed; gray;East Indian? Dutch?
89	7410	3	3	COST	BDY	STW	4.40	1	0	0	2	YARD 4A	STW 4	thin; light pink exterior; unglazed; gray;East Indian? Dutch?
89	747	2	3	COST	BDY	STW	4.40	1	0	0	1	YARD 4A	STW 4	thin; light pink exterior; unglazed; gray;East Indian? Dutch?

Year	Unit	No.	Layer	Form	Part	Ware	Type	Thick	Diam	Ht	Qty	Location	Assoc	Notes
89	745	4	3	COST	BDY	STW	4.40	1	0	0	5	YARD 4A	STW 4	thin; light pink exterior; unglazed; gray;East Indian? Dutch?
89	868	3	3	COST	BDY	STW	4.40	1	0	0	1	YARD 4A	STW 4	thin; light pink exterior; unglazed; gray;East Indian? Dutch?
89	884	4	3	MUG	BSE	STW	4.10	1	2	0	1	YARD 7		light brown glaze on both surfaces; orange patches (fire damage?); English
90	463	1	3	MUG	BDY	STW	4.23	1	0	0	1	YARD 4A		blue/gray stamped rosette/combed lines on exterior
89	897	2	3	MUG	BDY	STW	4.23	1	0	0	1	YARD 4A		blue/gray stamped rosette/combed lines on exterior
90	865	3	3	UNKN	BDY	STW	4.21	1	0	0	1	YARD 4A		medallion? incised chevrons on one surface
90	469	3	3	BOTT	ATT	STW	4.12	2	0	0	1	YARD 4A		medallion (armorial)
90	886	2	3	COST	BDY	STW	4.40	1	0	0	2	YARD 7	STW 5	part handle attached; light brown/ pink mottled exterior; unglazed; East Indian? Dutch?
90	886	8	3	COST	BDY	STW	4.40	1	0	0	1	YARD 7	STW 5	part handle att'd (three pressed ovals); brown/pink mottled exterior; unglazed; East Indian? Dutch?
90	892		2	COST	BDY	STW	4.40	1	0	0	1	YARD 4A	STW 5	light brown/pinkish mottled exterior; unglazed;East Indian? Dutch?
90	866	4	3	COST	BDY	STW	4.40	1	0	0	1	YARD 4A	STW 5	light brown/pinkish mottled exterior; unglazed; East Indian? Dutch?
89	753	3	3	COST	BDY	STW	4.40	1	0	0	1	YARD 4A	STW 5	light brown/pinkish mottled exterior; unglazed; East Indian ? Dutch?
89	897	2	3	COST	BDY	STW	4.40	1	0	0	1	YARD 4A	STW 5	light brown/pinkish mottled exterior; unglazed; East Indian ? Dutch?
90	373	4	3	STOR	BSE	STW	4.10	2	3	0	1	YARD 4A		white slip on exterior; orange patches (fire damage?); light gray;English?;Staffordshire Brown?
90	465	3	3	UNKN	BDY	STW	4.11	1	0	0	1	YARD 4A		dark brown glaze/dense freckled on exteior
87	433	4	3	MUG	BDY	STW	4.23	1	0	0	1	YARD 4B		two blue/gray stamped circular medallions on exterior
87	400		1	MUG	BDY	STW	4.10	2	0	0	1	YARD 4B	STW 6	mouth/neck; light brown glaze on both surfaces; English
87	415	1	3	BOTT	ATT	STW	4.12	2	0	0	1	ALLEY		medallion (floral); gray/brown glaze freckled exterior
87	405	4	3	UNKN	BDY	STW	4.21	1	0	0	1	YARD 4B		gray glaze freckled exterior
87	406	4	3	BOTT	BDY	STW	4.12	2	0	0	1	YARD 4B	STW 7	reddish brown glaze freckled exterior; gray
87	434	3	3	UNKN	BDY	STW	4.21	2	0	0	1	YARD 4B		light brown gray glaze freckled exterior; pale gray
87	415	1	3	UNKN	BDY	STW	4.21	2	0	0	1	ALLEY		light brown gray glaze freckled exterior; pale gray
87	403	1	3	UNKN	BDY	STW	4.21	2	0	0	1	YARD 4B		gray/brown glaze freckled exterior
87	246	4	3	UNKN	BDY	STW	4.10	2	0	0	1	ALLEY	STW 6	light brown glazed on both surfaces; orange patches (fire damage?); English
87	264	3	3	UNKN	BDY	STW	4.11	1	0	0	1	ALLEY		Bartmann? mug/jug? brown glaze freckled exterior; gray
87	351		1	BOTT	BDY	STW	4.10	2	0	0	1	ALLEY		light brown glaze on exterior; white/cream; English
87	416	4	3	MUG	BDY	STW	4.20	1	0	0	2	ALLEY	STW 8	thin; two embossed horizontal lines; light gray glaze on exterior; pinkish wash interior; German?
87	415	3	3	BOTT	RIM	STW	4.12	2	1	0	19	ALLEY	STW 7	neck/mouth; reddish brown glaze freckled exterior; gray
87	243	5	3	UNKN	BDY	STW	4.21	2	0	0	1	ALLEY		light brown/gray glaze freckled exterior; pale gray
87	454	3	3	BOTT	ATT	STW	4.10	2	0	0	1	ALLEY		strap handle; cream; English
90	922		2	UNKN	BDY	STW	4.40	1	0	0	1	YARD 6		pink freckled exterior; glaze on interior; cream/white
90	928	1	3	MUG	BDY	STW	4.20	1	0	0	1	YARD 6	STW 8	thin; two embossed horizontal lines; light gray glaze on exterior; pinkish wash interior; German?
90	963	1	3	UNKN	BDY	STW	4.11	1	0	0	1	YARD 6		Bartmann? orange brown glaze/dense freckled exterior; light brown
90	909	5	3	BOWL	BSE	POR	5.13	1	3	0	1	YARD 5	POR 1	brown wash on exterior; underglaze blue floral design on interior center
89	673		3	BOWL	BDY	POR	5.13	1	0	0	1	ROOM 2	POR 1	high footring; brown wash on exterior, underglaze blue floral design on interior center
90	755	90	3	BOWL	BSE	POR	5.15	1	2	0	2	YARD 4A	POR 2	high footring; blue/white floral design/borders on both surfaces
90	892		2	BOWL	BSE	POR	5.15	1	2	0	1	YARD 4A	POR 2	high footring; blue/white floral design/borders on both surfaces
90	755	90	3	BOWL	RIM	POR	5.15	1	3	0	1	YARD 4A	POR 2	high footring; blue/white floral design/borders on both surfaces
90	892		2	BOWL	BDY	POR	5.15	1	0	0	3	YARD 4A	POR 2	high footring; blue/white floral design/borders on both surfaces
90	890		1	BOWL	RIM	POR	5.15	1	3	0	1	YARD 4A	POR 2	high footring; blue/white floral design/borders on both surfaces
90	942		2	CUP	BDY	POR	5.14	1	0	0	2	YARD 5		three appliqued floral motifs around exterior; attached to 87-544-4; 534-1; 546-1
89	682		2	CUP	BDY	POR	5.13	1	0	0	1	ROOM 2		brown wash on exterior
89	687	4	3	BOWL	RIM	POR	5.13	1	3	0	1	ROOM 2		teabowl; brown wash on exterior
87	435	2	3	BOWL	BSE	POR	5.13	1	2	0	1	YARD 4B		teabowl; brown wash on exterior
89	634	1	3	COOK	BDY	CRW	1.33	2	0	0	1	SIDEWALK		straight sided; yellowish green glaze; fire use
87	584	3	3	COST	BSE	CRW	1.10	2	3	0	1	SIDEWALK		yellowish brown glaze; possibly North Italian/Spanish; see similar form SLW 8
87	592		2	STOR	ATT	CRW	1.42	2	3	0	1	ROOM 1		lid with finial knob; circular hand-molded ridges radiate outward to the rim from the knob
87	351		1	PLAT	BSE	POR	5.11	1	2	0	1	ALLEY		footing; floral design molded on center; bluish tinge
89	621		1	SAUC	R/B/B	POR	5.11	1	3 / 2	0	1	ROOM 1		footring
90	785	10	3	STOR	BDY	CRW	1.31	3	0	0	1	ROOM 4B		white slip on interior; light red; inclusions
90	100		1	STOR	BDY	CRW	1.31	2	0	0	1	B5/LOT 100		white slip wash on exterior; olive jar?

Year	Unit	No.	Layer	Form	Part	Ware	Type	Thick	Diam	Ht	Qty	Location	Assoc	Notes
89	727	5	3	STOR	BDY	CRW	1.31	3	0	0	11	ROOM 4A		fire use; English?
90	892		2	UNKN	BDY	CRW	1.31	1	0	0	1	YARD 4A		
90	894	1	3	UNKN	BDY	CRW	1.31	1	0	0	1	YARD 4A		
87	372		2	UNKN	BSE	CRW	1.10	1	2	0	3	YARD 4A	CRW 1	shallow with flared walls; light green glaze on interior, some glaze on exterior; fire use
87	381		1	UNKN	BSE	CRW	1.10	1	2	0	1	YARD 4A	CRW 1	shallow with flared walls; light green glaze on interior, some glaze on exterior; fire use
87	412		2	APOT	BSE	CRW	1.37	1	2	0	1	ALLEY		ointment pot; black glaze; English (Hertfordshire?) Spanish?
87	353	20	3	COOK	RIM	CRW	1.37	3	6	0	2	ALLEY	CRW 2	handle attached; internal seating/stepped profile (for lid); reddish brown glaze; English
87	353	20	3	COOK	BDY	CRW	1.37	3	0	0	1	ALLEY	CRW 2	reddish brown glaze
87	381		1	COOK	BDY	CRW	1.37	3	0	0	1	YARD 4A	CRW 2	reddish brown glaze
90	917	3	3	STOR	BDY	CRW	1.20	1	0	0	1	YARD 5		white slip wash on exterior; pinkish gray
90	955	2	3	STOR	BDY	CRW	1.31	2	0	0	2	YARD 5		fire use; English?
90	903	2	3	UNKN	BDY	CRW	1.10	1	0	0	1	YARD 5		olive glaze on one surface
90	913		3	STOR	BSE	CRW	1.22	2	5	0	1	YARD 5		bacin form; light yellowish brown; Spanish?
90	969	2	3	STOR	BDY	CRW	1.37	1	0	0	1	YARD 6		ointment pot? glossy black glaze; English (Hertfordshire?) Spanish?
90	9610	1	3	STOR	BSE	CRW	1.31	2	5	0	1	YARD 6	CRW 3	bacin form; inclusions; orange red; Spanish?
90	9610	2	3	STOR	BSE	CRW	1.31	2	5	0	1	YARD 6	CRW 3	bacin form; inclusions; orange red; Spanish?
90	963	7	3	STOR	BSE	CRW	1.31	2	5	0	1	YARD 6	CRW 3	bacin form; incised line on wall; inclusions; orange red; Spanish?
90	944	2	3	PIP	RIM	CRW	1.11	2	3	0	1	YARD 5	CRW 4	olive glaze on interior; internal seating/stepped profile (for lid); buff; see CRW 5
90	947	2	3	PIP	RIM	CRW	1.11	2	3	0	1	YARD 5	CRW 4	olive glaze on interior; buff; see CRW 5
90	947	2	3	PIP	BDY	CRW	1.11	1	0	0	1	YARD 5	CRW 4	olive glaze on interior; buff; see CRW 5
89	670	2	1	PIP	BDY	CRW	1.11	1	0	0	1	ROOM 2	CRW 4	olive glaze on interior; buff; see CRW 5
90	949	4	3	PIP	ATT	CRW	1.11	2	0	0	1	YARD 5	CRW 4	handle; olive glaze; buff; see CRW 5
89	822		2	PIP	BDY	CRW	1.11	1	0	0	1	ROOM 4B		olive glaze on interior; buff; part of CRW 4?
87	423	3	3	PORR	R/B/B	CRW	1.11	1	3 / 3	2	1	ALLEY		handle attached; olive glaze on interior; fire use; buff; see 87-747-5
90	904	14	3	STOR	R/B/B	CRW	1.12	2	2 / 2	2	3	YARD 5		small storage container; olive glaze on both surfaces
89	897	2	3	UNKN	BSE	CRW	1.33	1	2	0	4	YARD 4A		high footring; fire use; brown glaze
90	944	2	3	PIP	BSE	CRW	1.11	1	2	0	1	YARD 5	CRW 5	feet (2) attached; olive glaze on interior; buff; see CRW 4
89	683		3	PIP	BSE	CRW	1.11	1	2	0	1	ROOM 2	CRW 5	foot (1) attached; yellow glaze on interior; buff; see CRW 4
90	469	6	3	CHAM	WHLE	CRW	1.12	2	4 / 3	3	1	YARD 4A		olive glaze on both surfaces
90	9610	5	3	CHAM	WHLE	CRW	1.12	2	5 / 4	3	1	YARD 6		olive glaze on both surfaces
87	747	5	3	PORR	WHLE	CRW	1.12	2	2 / 2	2	1	YARD 4A		horizontal handle attached; brown glaze on both surfaces; see 87-423-3
89	857	2	3	PORR	BDY	CRW	1.11	1	0	0	1	ROOM 4		olive glaze on interior; buff
89	8510	2	3	PORR	BDY	CRW	1.11	1	0	0	1	ROOM 4	CRW 6	olive glaze on interior; buff; part of CRW 4?
89	727	2	3	PIP	BDY	CRW	1.11	1	0	0	1	ROOM 4	CRW 6	olive glaze on interior; buff; part of CRW 4?
90	908	6	3	CHAM	WHLE	CRW	1.11	2	4 / 3	3	1	YARD 5		olive glaze on interior; white wash on exterior; buff
89	623	3	3	STOR	RIM	CRW	1.14	5	5	0	1	ROOM 1	CRW 7	finger impressed rim; internal seating / stepped profile; olive glaze on interior; brown; core
90	944	35	3	STOR	RIM	CRW	1.14	5	5	0	1	YARD 5	CRW 7	finger impressed rim; internal seating / stepped profile; olive glaze on interior; brown; core
87	574	2	3	UNKN	BDY	CRW	1.13	2	0	0	1	ROOM 1		olive glaze on interior; grayish brown glaze on exteior; light gray
89	599		3	STOR	BDY	CRW	1.14	3	0	0	1	ROOM 1	CRW 8	olive glaze on interior; pink; core
89	654	1	3	STOR	BDY	CRW	1.14	3	0	0	3	ROOM 1	CRW 8	olive glaze on interior; pink; core
87	255	1	3	STOR	BDY	CRW	1.14	3	0	0	1	ROOM 1	CRW 8	olive glaze on interior; pink; core
89	862		3	STOR	BDY	CRW	1.14	3	0	0	1	ROOM 1	CRW 8	olive glaze on interior; pink; core
89	682		3	STOR	RIM	CRW	1.14	4	6	0	1	ROOM 2	CRW 8	finger impressed rim; internal seating/stepped profile; olive glaze on interior; pink; core
89	683	3	3	STOR	BDY	CRW	1.14	3	0	0	1	ROOM 2	CRW 8	olive glaze on interior; pink; core
89	683	1	3	STOR	BDY	CRW	1.14	3	0	0	5	ROOM 2	CRW 8	olive glaze on interior; pink; core
89	675		3	STOR	BDY	CRW	1.14	3	0	0	1	ROOM 2	CRW 8	olive glaze on interior; pink; core
89	683	2	3	STOR	RIM	CRW	1.14	5	6	0	1	ROOM 2	CRW 8	finger impressed rim; internal seating/stepped profile; olive glaze on interior; pink; core
89	683	2	3	STOR	BDY	CRW	1.14	3	0	0	8	ROOM 2	CRW 8	olive glaze on interior; pink; core
89	677	1	3	STOR	BDY	CRW	1.14	3	0	0	1	ROOM 2	CRW 8	olive glaze on interior; pink; core
89	683		3	STOR	BDY	CRW	1.14	3	0	0	3	ROOM 2	CRW 8	olive glaze on interior; pink; core
89	682		2	STOR	BDY	CRW	1.14	3	0	0	1	ROOM 2	CRW 8	olive glaze on interior; pink; core

Year	Unit	No.	Layer	Form	Part	Ware	Type	Thick	Diam	Ht	Qty	Location	Assoc	Notes
89	739	1	3	STOR	BDY	CRW	1.14	3	0	0	1	ROOM 2	CRW 8	olive glaze on interior; pink; core
90	912		2	STOR	BDY	CRW	1.14	3	0	0	1	YARD 5	CRW 8	olive glaze on interior; pink; core
90	944	35	3	STOR	RIM	CRW	1.14	4	6	0	1	YARD 5	CRW 8	loop handle attached; finger impressed rim; internal seating/stepped profile; pink; core
90	906	2	3	STOR	BDY	CRW	1.14	3	0	0	1	YARD 5	CRW 8	olive glaze on interior; pink; core
90	944	3	3	STOR	BDY	CRW	1.14	3	0	0	1	YARD 5	CRW 8	olive glaze on interior; pink; core
90	944	1	3	STOR	BDY	CRW	1.14	3	0	0	1	YARD 5	CRW 8	olive glaze on interio ; pink; core
90	944	5	3	STOR	BSE	CRW	1.14	5	6	0	5	YARD 5	CRW 8	olive glaze on interior; pink; core
90	942		2	STOR	BSE	CRW	1.14	5	6	0	1	YARD 5	CRW 8	olive glaze on interior; pink; core
89	6810	5	3	STOR	BDY	CRW	1.31	4	0	0	6	ROOM 2	CRW 9	reddish brown; inclusions
89	654	4	3	STOR	BDY	CRW	1.31	4	0	0	1	ROOM 1	CRW 9	reddish brown; inclusions
89	655	4	3	STOR	RIM	CRW	1.22	5	4	0	1	ROOM 1		bacin form; white slip on exterior; yellowish red/reddish brown; inclusions; Spanish?
89	737	2	3	STOR	RIM	CRW	1.31	5	5	0	1	ROOM 3	CRW 9	dark red; inclusions
89	737	4	3	STOR	BDY	CRW	1.31	4	0	0	2	ROOM 3	CRW 9	dark red; inclusions
89	737	4	3	STOR	RIM	CRW	1.31	5	5	0	1	ROOM 3	CRW 9	dark red; inclusions
89	738	4	3	STOR	BDY	CRW	1.31	4	0	0	1	ROOM 3	CRW 9	white slip on exterior; reddish brown; inclusions
89	738	4	3	STOR	RIM	CRW	1.31	5	5	0	1	ROOM 3	CRW 9	dark red; inclusions
90	947	1	3	STOR	RIM	CRW	1.31	5	5	0	1	YARD 5	CRW 9	dark red; inclusions
90	947	6	3	STOR	BDY	CRW	1.20	4	0	0	5	YARD 5	CRW 12	yellowish red fabric; inclusions
90	104	4	3	CHAM	R/B/B	CRW	1.33	2	3	4	3	B5/LOT 100	CRW 10	dark gray/olive gray glaze; vestiges of white slip on exterior; reddish brown; inclusions
90	104	5	3	CHAM	BDY	CRW	1.33	2	0	0	1	B5/LOT 100	CRW 10	dark gray/olive gray glaze; vestiges of white slip on exterior; reddish brown; inclusions
90	104	5	3	CHAM	RIM	CRW	1.33	2	3	0	1	B5/LOT 100	CRW 10	dark gray/olive gray glaze; vestiges of white slip on exterior; reddish brown; inclusions
90	103		3	CHAM	R/B/B	CRW	1.33	2	3	4	2	B5/LOT 100	CRW 10	dark gray/olive gray glaze; vestiges of white slip on exterior; reddish brown; inclusions
87	384	15	3	STOR	RIM	CRW	1.33	4	4	0	1	YARD 4A	CRW 11	olive glaze on exterior; reddish brown glaze on interior; red; inclusions
87	353	1	3	STOR	BDY	CRW	1.37	3	0	0	1	ALLEY	CRW 11	olive glaze on exterior; reddish brown glaze on interior; red; inclusions
87	384	3	3	STOR	BDY	CRW	1.37	3	0	0	2	YARD 4A	CRW 11	olive glaze on exterior reddish brown glaze on interior; red; inclusions
90	469	2	3	UNKN	BDY	CRW	1.31	2	0	0	1	YARD 4A		grayish brown wash on both surfaces; reddish brown; inclusions
87	433	2	3	STOR	BDY	CRW	1.37	2	0	0	1	YARD 4B		dark reddish brown glaze; red; inclusions
89	742		2	STOR	BDY	CRW	1.37	4	0	0	1	YARD 4A		light reddish brown glaze; red
89	732		2	COOK	RIM	CRW	1.31	5	3	0	1	ROOM 3		fire use; dark red
89	752		2	STOR	RIM	CRW	1.20	5	3	0	1	YARD 4A	CRW 12	white slip on exterior; yellowish brown; inclusions
89	748	4	3	STOR	RIM	CRW	1.20	5	3	0	1	YARD 4A	CRW 12	yellowish brown; inclusions
89	644	1	3	COOK	RIM	CRW	1.33	5	3	0	1	ROOM 1		light reddish brown glaze; red
89	642		2	STOR	RIM	CRW	1.41	5	4	0	1	ROOM 1		reddish brown; inclusions; core
89	732		2	STOR	RIM	CRW	1.35	3	4	0	1	ROOM 3		straight sided; reddish brown glaze; light reddish brown; inclusions
89	795	3	3	STOR	RIM	CRW	1.20	5	2	0	1	ROOM 4A		yellowish red
89	654	2	3	STOR	BSE	CRW	1.31	5	0	0	1	ROOM 1		reddish brown; inclusions; core
89	823	2	3	STOR	BDY	CRW	1.31	5	0	0	1	ROOM 4B		reddish brown; inclusions
89	653	2	3	STOR	BDY	CRW	1.31	3	0	0	1	ROOM 1		reddish brown
89	852		2	STOR	BDY	CRW	1.31	3	0	0	1	ROOM 4		reddish brown
89	748	2	3	STOR	BDY	CRW	1.31	3	0	0	2	YARD 4A		reddish brown
89	656	4	3	STOR	BDY	CRW	1.31	3	0	0	1	ROOM 1		reddish brown
89	673	5	3	STOR	BDY	CRW	1.31	4	0	0	1	ROOM 2		reddish yellow
89	738	3	3	STOR	BDY	CRW	1.31	3	0	0	1	ROOM 3		reddish yellow
89	734	3	3	STOR	BDY	CRW	1.31	4	0	0	1	ROOM 3		reddish yellow
89	351		1	UNKN	BDY	CRW	1.31	2	0	0	1	ALLEY		small storage container? vestiges of white slip on one surface; reddish yellow
89	734	5	3	UNKN	BDY	CRW	1.31	1	0	0	1	ROOM 3		vestiges of white slip on one surface; reddish yellow
89	794	1	3	STOR	BDY	CRW	1.20	3	0	0	1	ROOM 4A		pinkish white wash on both surfaces; light brown; core; Spanish?
90	954	2	3	STOR	BDY	CRW	1.31	4	0	0	1	YARD 5		vestiges of white slip on both surfaces ; light reddish brown; Spanish?
90	922		2	STOR	BDY	CRW	1.31	2	0	0	1	YARD 6		white wash on interior; yellow stained (fire use?); reddish brown;Spanish?
90	901		1	BOWL	BDY	CRW	1.31	2	0	0	1	YARD 5		reddish brown
90	942		2	BOWL	BDY	CRW	1.31	2	0	0	1	YARD 5		smoothed; reddish brown

Year	Unit	No.	Layer	Form	Part	Ware	Type	Thick	Diam	Ht	Qty	Location	Assoc	Notes
90	912		2	COOK	BDY	CRW	1.31	2	0	0	1	YARD 5		fire use; reddish brown
90	920		1	STOR	ATT	CRW	1.31	4	0	0	1	YARD 6		handle; reddish brown
90	101		1	STOR	BDY	CRW	1.31	2	0	0	1	B5/LOT 100		reddish brown; inclusions
90	892		2	STOR	BDY	CRW	1.31	3	0	0	1	YARD 4A		reddish brown
90	945	1	3	STOR	RIM	CRW	1.20	1	2	0	3	YARD 5		ointment pot? light brown
90	941		1	STOR	BDY	CRW	1.31	2	0	0	1	YARD 5		light red
90	373	1	3	UNKN	BSE	CRW	1.41	1	1	0	1	YARD 4B		bowl? hand molded; light red
90	902		2	STOR	BDY	CRW	1.31	2	0	0	1	YARD 5		light red
90	910		1	STOR	BDY	CRW	1.31	2	0	0	1	YARD 5		white slip wash on exterior; reddish yellow; Spanish?
90	102		2	STOR	BDY	CRW	1.41	2	0	0	1	B5/LOT 100		horizontal handle attached (not useful); 10 mm thick; reddish yellow; gray core
90	887		3	STOR	BDY	CRW	1.31	2	0	0	1	YARD 7		white wash on exterior; fire use; light red
90	9012	1	3	STOR	BSE	CRW	1.10	2	0	0	1	YARD 5		flat; glazed black both surfaces; reddish brown
87	371		1	BOWL	BDY	CRW	1.41	2	0	0	1	YARD 4A		brown fabric
87	252		2	BOWL	BDY	CRW	1.41	2	0	0	1	YARD 4B		brown fabric
90	102		2	BOWL	BDY	CRW	1.41	1	0	0	1	B5/LOT 100		eroded exterior; brown; core
87	353	1	3	BOWL	BDY	CRW	1.41	3	0	0	1	ALLEY		eroded exterior; reddish brown; core
89	762		2	COOK	BDY	CRW	1.41	2	0	0	1	YARD 4B		fire use; black
87	351		1	COOK	RIM	CRW	1.41	2	6	0	2	ALLEY		fire use; black
87	351		1	STOR	BDY	CRW	1.31	2	0	0	1	ALLEY		white wash on interior; reddish brown; inclusions; core; Spanish?
87	612		2	BOWL	BDY	CRW	1.33	2	0	0	1	ROOM 1		olive glaze; fire use; reddish brown; inclusions; core
90	892		2	BOWL	BDY	CRW	1.33	2	0	0	1	YARD 4A		olive glaze; red; inclusions
90	961		1	BOWL	BDY	CRW	1.33	1	0	0	1	YARD 6		olive glaze; red; inclusions
89	619		3	BOWL	BDY	CRW	1.33	2	0	0	1	ROOM 1		olive glaze; reddish brown; inclusions
89	718		3	BOWL	BDY	CRW	1.33	1	0	0	1	ROOM 4A		olive glaze; reddish brown; inclusions
87	405	3	3	BOWL	BDY	CRW	1.41	3	0	0	1	YARD 4B		reddish brown
90	102		2	BOWL	BDY	CRW	1.33	1	0	0	1	B5/LOT 100		olive glaze; reddish brown; inclusions; core
90	466	3	3	BOWL	BDY	CRW	1.41	1	0	0	1	YARD 4A		reddish brown
90	462		2	BOWL	BDY	CRW	1.47	1	0	0	1	YARD 4A		olive glaze; reddish brown; inclusions
90	922		2	BOWL	BDY	CRW	1.33	2	0	0	1	YARD 6		olive glaze; reddish brown; inclusions; core
87	246	3	3	BOWL	BDY	CRW	1.33	3	0	0	1	ALLEY		olive glaze; yellow brown; inclusions
90	962		2	BOWL	BDY	CRW	1.33	3	0	0	1	YARD 6		olive glaze; reddish brown; inclusions; core
89	702		2	BOWL	BDY	CRW	1.33	2	0	0	1	SIDEWALK		olive glaze; reddish brown; inclusions
87	372		2	BOWL	BDY	CRW	1.33	1	0	0	1	YARD 4A		yellow glaze; reddish brown; inclusions
89	600		1	BOWL	BDY	CRW	1.33	2	0	0	1	SIDEWALK		olive glaze; red; inclusions
89	869	2	3	BOWL	BDY	CRW	1.47	1	0	0	1	YARD 4A		olive glaze; red; inclusions
89	712		2	BOWL	BDY	CRW	1.41	2	0	0	1	ROOM 4A		eroded; red; core
87	372		2	BOWL	BDY	CRW	1.41	2	0	0	1	YARD 4A		eroded; reddish brown; core
90	902		2	BOWL	BDY	CRW	1.33	2	0	0	1	YARD 5		olive glaze; red; inclusions
87	572		2	BOWL	BDY	CRW	1.33	3	0	0	1	ROOM 1		olive glaze; reddish brown ; inclusions ; core
90	462		2	BOWL	BDY	CRW	1.33	2	0	0	1	YARD 4A		olive glaze; red; inclusions
89	602		2	BOWL	BDY	CRW	1.33	2	0	0	1	SIDEWALK		olive glaze; eroded exterior; reddish brown; inclusions; core
90	892		2	BOWL	BDY	CRW	1.33	1	0	0	1	YARD 4A		olive glaze; red; inclusions
87	415	1	3	BOWL	BDY	CRW	1.33	2	0	0	1	ALLEY		olive glaze; reddish brown; inclusions
89	852		2	BOWL	BDY	CRW	1.33	2	0	0	1	ROOM 4		olive glaze; red; inclusions
89	600			BOWL	BDY	CRW	1.33	1	0	0	1	SIDEWALK		olive glaze; reddish brown; inclusions
90	922		2	BOWL	BDY	CRW	1.47	1	0	0	1	YARD 6		olive glaze; red; inclusions
89	600		1	BOWL	BDY	CRW	1.41	2	0	0	1	SIDEWALK		reddish brown
87	372		2	BOWL	RIM	CRW	1.47	3	4	0	2	YARD 4A		olive glaze; reddish brown; inclusions; core
87	415	3	3	BOWL	BDY	CRW	1.31	1	0	0	1	ALLEY		reddish brown
90	942		2	BOWL	BDY	CRW	1.47	2	0	0	1	YARD 5		olive glaze; reddish brown; inclusions
90	102		2	BOWL	BDY	CRW	1.33	3	0	0	1	B5/LOT 100		olive glaze; red; inclusions; core

Year	Unit	No.	Layer	Form	Part	Ware	Type	Thick	Diam	Ht	Qty	Location	Assoc	Notes
87	260		1	BOWL	BDY	CRW	1.47	2	0	0	1	ALLEY		olive glaze; reddish brown; inclusions; core
89	652		2	STOR	BDY	CRW	1.38	1	0	0	1	ROOM 1		ribbing on exterior; red; core
87	422		2	BOWL	RIM	CRW	1.43	3	4	0	2	ALLEY		olive glaze; reddish brown; inclusions
89	762		2	BOWL	BDY	CRW	1.33	2	0	0	1	YARD 4B		olive/brown glaze; light red
90	940		1	BOWL	RIM	CRW	1.43	2	4	0	2	YARD 5		finger impressed; olive glaze; reddish brown; inclusions
87	572		2	BOWL	RIM	CRW	1.43	3	4	0	2	ROOM 1		olive glaze; reddish brown; inclusions; core
90	467	4	3	BOWL	BDY	CRW	1.33	2	0	0	1	YARD 4A		olive/brown glaze; light red
89	621		1	BOWL	BDY	CRW	1.43	2	0	0	1	ROOM 1		olive glaze; red; inclusions; core
87	371		1	COOK	BDY	CRW	1.37	2	0	0	1	YARD 4A		fire use; reddish brown glaze; red
89	650		1	BOWL	RIM	CRW	1.47	3	5	0	2	ROOM 1		olive glaze; red; inclusions
90	922		2	BOWL	RIM	CRW	1.43	2	4	0	2	YARD 6		olive glaze; reddish brown; inclusions; core
90	922		2	BOWL	BDY	CRW	1.43	2	0	0	1	YARD 6		olive glaze; reddish brown; inclusions; core
89	602		2	BOWL	RIM	CRW	1.43	2	6	0	2	SIDEWALK		yellow glaze; reddish brown; inclusions
90	9012	1	3	UNKN	BDY	CRW	1.20	2	0	0	1	YARD 5		white wash on both surfaces; gray; Spanish?
89	617		3	BOWL	BDY	CRW	1.34	2	0	0	1	ROOM 1		gray/olive glaze; incised lines; light reddish brown; core; Spanish?
87	592		2	BOWL	RIM	CRW	1.43	2	5	0	2	ROOM 1		olive glaze; reddish brown; inclusions; core
89	600		1	BOWL	BDY	CRW	1.43	2	0	0	1	SIDEWALK		olive glaze; red; inclusions; core
89	732		2	BOWL	BDY	CRW	1.43	3	0	0	1	ROOM 3		olive glaze; light brown; inclusions
89	652		2	BOWL	BDY	CRW	1.43	2	0	0	1	ROOM 1		olive glaze; reddish brown; inclusions
89	657		3	BOWL	BDY	CRW	1.43	3	0	0	1	ROOM 1		olive glaze; red; inclusions; core
90	469	2	3	STOR	RIM	CRW	1.31	1	2	0	2	YARD 4A		light red
90	843	2	3	BOWL	RIM	CRW	1.43	3	5	0	2	SIDEWALK		olive glaze; red; inclusions; core
87	372		2	BOWL	RIM	CRW	1.43	2	4	0	2	YARD 4A		olive glaze; reddish brown; inclusions
87	435	2	3	BOWL	BDY	CRW	1.31	2	0	0	1	YARD 4B		light red
89	732		2	BOWL	RIM	CRW	1.43	3	6	0	2	ROOM 3		olive glaze; red; inclusions
90	922		2	BOWL	RIM	CRW	1.43	2	6	0	2	YARD 6		olive glaze; red; inclusions; core
87	433	3	3	BOWL	BDY	CRW	1.31	2	0	0	1	YARD 4B		light red
89	732		2	BOWL	RIM	CRW	1.44	2	5	0	2	ROOM 3		finger impressed; olive glaze; red; inclusions; core
90	912		2	BOWL	BDY	CRW	1.43	2	0	0	1	YARD 5		olive glaze; reddish brown; inclusions
87	351		1	BOWL	RIM	CRW	1.43	3	5	0	2	ALLEY		olive glaze; reddish brown; inclusions; core
87	431		1	BOWL	RIM	CRW	1.31	2	2	0	1	YARD 4B		light red
89	651		1	BOWL	BDY	CRW	1.41	2	0	0	1	ROOM 1		reddish brown
89	863	1	3	BOWL	RIM	CRW	1.41	1	5	0	2	YARD 4A		yellow brown
90	944	1	3	STOR	BDY	CRW	1.31	2	0	0	1	YARD 5		white slip/wash on exterior; light red; Spanish?
89	833	2	3	BOWL	RIM	CRW	1.41	2	4	0	3	ROOM 4B		reddish brown
90	918	1	3	STOR	BDY	CRW	1.31	4	0	0	1	YARD 5		fire use; light red; core
89	884	2	3	BOWL	RIM	CRW	1.41	2	5	0	2	YARD 7		reddish brown; core
87	371		1	BOWL	BDY	CRW	1.41	2	0	0	1	YARD 4A		reddish brown; core
89	815	4	3	BOWL	BDY	CRW	1.41	2	0	0	1	SIDEWALK		reddish brown; core
89	812		2	BOWL	RIM	CRW	1.41	2	5	0	2	SIDEWALK		reddish brown; core
89	852		2	BOWL	BDY	CRW	1.31	2	0	0	1	ROOM 4		light red
87	371		1	BOWL	BDY	CRW	1.41	2	0	0	1	YARD 4A		reddish brown; core
90	102		2	BOWL	RIM	CRW	1.43	2	5	0	2	B5/LOT 100		olive glaze; red; inclusions; core
89	8610	3	3	STOR	BDY	CRW	1.31	2	0	0	1	YARD 4A		light red; inclusions
90	946	2	3	STOR	BDY	CRW	1.31	5	0	0	1	YARD 5		light red; inclusions
87	372		2	BOWL	RIM	CRW	1.41	2	5	0	2	YARD 4A		olive glaze; reddish brown; core
89	711		1	BOWL	RIM	CRW	1.41	2	4	0	2	ROOM 4A		olive glaze; reddish brown; core
89	434	3	3	STOR	BDY	CRW	1.31	4	0	0	1	YARD 4B		light red; inclusions
87	582		2	BOWL	BDY	CRW	1.41	2	0	0	1	SIDEWALK		reddish brown; core
87	732		2	BOWL	BDY	CRW	1.41	2	0	0	1	ROOM 3		eroded; reddish brown
89	602		2	COOK	RIM	CRW	1.41	3	5	0	2	SIDEWALK		reddish brown; core

Year	Unit	No.	Layer	Form	Part	Ware	Type	Thick	Diam	Ht	Qty	Location	Assoc	Notes
87	584	3	3	COLN	RIM	CRW	1.37	2	5	0	1	SIDEWALK		colander; four holes; olive glaze; reddish yellow
87	571		1	BOWL	BDY	CRW	1.41	1	0	0	1	ROOM 1		reddish brown; core
90	922		2	BOWL	BDY	CRW	1.41	2	0	0	1	YARD 6		red; core
89	859	1	3	BOWL	BDY	CRW	1.41	2	0	0	1	ROOM 4		red
87	405	3	3	BOWL	BDY	CRW	1.41	2	0	0	1	YARD 4B		reddish brown
87	432		2	BOWL	BDY	CRW	1.41	2	0	0	1	YARD 4B		reddish brown
87	371		1	BOWL	BDY	CRW	1.41	2	0	0	1	YARD 4A		reddish brown; core
87	371		1	COOK	BDY	CRW	1.41	1	0	0	1	YARD 4A		fire use; reddish brown; core
89	732		2	BOWL	BDY	CRW	1.41	2	0	0	1	ROOM 3		red; core
87	420		1	BOWL	BDY	CRW	1.41	2	0	0	1	ALLEY		reddish brown
87	401		1	BOWL	BDY	CRW	1.41	3	0	0	1	YARD 4B		eroded; reddish brown
89	7310		3	COOK	BDY	CRW	1.41	2	0	0	1	ROOM 3		reddish brown
89	862		2	STOR	BDY	CRW	1.41	4	0	0	1	YARD 4A		red; core
89	652		2	COOK	BDY	CRW	1.41	1	0	0	1	ROOM 1		fire use; red
89	891		1	BOWL	BDY	CRW	1.41	3	0	0	1	YARD 4A		reddish brown; core
90	912		2	COOK	BDY	CRW	1.41	3	0	0	1	YARD 5		fire use; red; core
89	622		2	BOWL	BDY	CRW	1.41	2	0	0	1	ROOM 1		reddish brown; core
87	371		1	COOK	BDY	CRW	1.41	2	0	0	1	YARD 4A		fire use; reddish brown core
90	944	4	3	COOK	BDY	CRW	1.41	1	0	0	1	YARD 5		fire use; reddish brown; core
87	371		1	BOWL	BDY	CRW	1.41	1	0	0	1	YARD 4A		eroded; reddish brown
87	374	2	3	BOWL	BDY	CRW	1.41	2	0	0	1	YARD 4A		red
87	374	2	3	STOR	BDY	CRW	1.41	3	0	0	1	YARD 4A		red
87	371		1	BOWL	BDY	CRW	1.41	2	0	0	1	YARD 4A		reddish brown
89	745	1	3	BOWL	BDY	CRW	1.41	3	0	0	1	YARD 4A		red
89	891		1	BOWL	BSE	CRW	1.41	2	0	0	1	YARD 4A		red; core
87	412		2	BOWL	BDY	CRW	1.41	3	0	0	1	ALLEY		reddish brown; core
89	891		1	BOWL	BDY	CRW	1.41	2	0	0	1	YARD 4A		red; core
89	622		2	BOWL	BDY	CRW	1.41	2	0	0	1	ROOM 1		red; core
87	575	2	3	COOK	BDY	CRW	1.41	1	0	0	1	ROOM 1		reddish brown
90	462		2	BOWL	BDY	CRW	1.41	2	0	0	1	YARD 4A		reddish brown
87	374	2	3	BOWL	BDY	CRW	1.41	3	0	0	1	YARD 4A		reddish brown; core
89	602		2	BOWL	BDY	CRW	1.41	2	0	0	1	SIDEWALK		red
89	732		2	BOWL	BDY	CRW	1.41	2	0	0	1	ROOM 3		red
89	733	2	3	BOWL	BDY	CRW	1.41	3	0	0	1	ROOM 3		red; core
87	374	2	3	BOWL	BDY	CRW	1.41	3	0	0	1	YARD 4A		red; core
89	842		2	BOWL	RIM	CRW	1.42	2	4	0	1	SIDEWALK		grooved; reddish brown
89	651		1	BOWL	BDY	CRW	1.43	2	0	0	1	ROOM 1		reddish brown; core
87	371		1	BOWL	BDY	CRW	1.41	1	0	0	1	YARD 4A		reddish brown
89	732		2	BOWL	BDY	CRW	1.41	3	0	0	1	ROOM 3		red
89	682		2	BOWL	BDY	CRW	1.41	2	0	0	1	ROOM 2		red; core
87	374	2	3	BOWL	BDY	CRW	1.41	3	0	0	1	YARD 4A		red
90	844	4	3	BOWL	RIM	CRW	1.41	2	5	0	2	SIDEWALK		reddish brown; core
87	574	3	3	UNKN	ATT	CRW	1.37	4	0	0	1	ROOM 1		handle stub; very dark gray glaze; reddish brown
89	374	2	3	BOWL	BDY	CRW	1.41	3	0	0	1	YARD 4A		red
90	963	1	3	BOWL	BDY	CRW	1.41	2	0	0	1	YARD 6		red
89	652		2	BOWL	BSE	CRW	1.41	1	0	0	1	ROOM 1		reddish brown; core
87	372		2	COOK	RIM	CRW	1.41	2	6	0	2	YARD 4A		fire use; reddish brown; core
90	653		3	BOWL	BDY	CRW	1.33	2	0	0	1	ROOM 1		dusky red glaze; white wash/slip; light reddish brown
87	352		2	COOK	BDY	CRW	1.41	2	0	0	1	ALLEY		fire use; brown; core
90	962		2	BOWL	BDY	CRW	1.41	2	0	0	1	YARD 6		reddish brown; core
87	244	4	3	COOK	BDY	CRW	1.41	2	0	0	1	ALLEY		fire use; reddish brown; core

Year	Unit	No.	Layer	Form	Part	Ware	Type	Thick	Diam	Ht	Qty	Location	Assoc	Notes
90	102		2	COOK	BDY	CRW	1.41	1	0	0	1	B5/LOT 100		fire use; brown; core
87	442		2	COOK	BDY	CRW	1.41	2	0	0	1	YARD 4B		fire use; brown
87	352		2	COOK	BDY	CRW	1.41	2	0	0	1	YARD 4A		fire use; brown; core
90	961		1	COOK	BDY	CRW	1.41	2	0	0	1	YARD 6		reddish brown
87	251		1	COOK	BDY	CRW	1.41	2	0	0	1	YARD 4B		fire use; reddish brown
87	251		1	COOK	BDY	CRW	1.41	2	0	0	1	YARD 4B		fire use; brown; core
87	420		1	BOWL	RIM	CRW	1.41	2	5	0	2	ALLEY		eroded; reddish brown; core
87	655	1	3	STOR	BDY	CRW	1.33	4	0	0	1	ROOM 1		reddish brown glaze; light red
87	572		2	BOWL	BDY	CRW	1.41	2	0	0	1	ROOM 1		reddish brown ; core
90	465	4	3	COOK	BDY	CRW	1.41	2	0	0	1	YARD 4A		eroded ; fire use ; yellow brown
89	622		2	BOWL	RIM	CRW	1.41	2	4	0	2	ROOM 1		reddish brown; core
89	654	2	3	UNKN	BDY	CRW	1.35	1	0	0	1	ROOM 1		reddish brown glaze; reddish brown
90	962		2	BOWL	BDY	CRW	1.41	1	0	0	1	YARD 6		yellow brown; core
87	243	1	3	COOK	BDY	CRW	1.41	1	0	0	1	ALLEY		fire use; brown; core
89	735	1	3	BOWL	BDY	CRW	1.41	3	0	0	1	ROOM 3		brown
87	384	4	3	COOK	BDY	CRW	1.41	1	0	0	1	YARD 4A		fire use; reddish brown; core
89	732		2	COOK	BDY	CRW	1.41	1	0	0	1	ROOM 3		fire use; brown; core
90	902		2	BOWL	BDY	CRW	1.41	2	0	0	1	YARD 5		eroded; yellow brown; core
87	371		1	COOK	BDY	CRW	1.41	1	0	0	1	YARD 4A		fire use; reddish brown; core
87	354	4	3	COOK	BDY	CRW	1.41	1	0	0	1	ALLEY		fire use; reddish brown; core
90	922		2	COOK	BDY	CRW	1.41	2	0	0	1	YARD 6		fire use; brown
90	922		2	COOK	BDY	CRW	1.41	1	0	0	1	YARD 6		fire use; yellow brown
87	592		2	COOK	BDY	CRW	1.41	2	0	0	1	ROOM 1		fire use; reddish brown; core
87	372		2	COOK	BDY	CRW	1.41	2	0	0	1	YARD 4A		fire use; reddish brown
87	433	1	3	BOWL	BDY	CRW	1.41	3	0	0	1	YARD 4B		reddish brown; core
87	433	1	3	BOWL	BDY	CRW	1.41	3	0	0	1	YARD 4B		reddish brown; core
87	372		2	COOK	BDY	CRW	1.41	2	0	0	1	YARD 4A		fire use; brown
90	952		2	COOK	BDY	CRW	1.41	3	0	0	1	YARD 5		fire use; reddish brown; core
89	867	3	3	COOK	BDY	CRW	1.41	2	0	0	1	YARD 4A		eroded; fire use; reddish brown
89	806	3	3	COOK	BDY	CRW	1.41	2	0	0	1	ROOM 4B		eroded; fire sue; reddish brown
87	350	1	1	COOK	BDY	CRW	1.41	3	0	0	1	ALLEY		eroded; fire use; brown
89	869	3	3	COOK	BDY	CRW	1.41	1	0	0	1	YARD 4A		eroded; fire use; brown; core
89	8610	3	3	COOK	BDY	CRW	1.41	1	0	0	1	YARD 4A		eroded; fire use; reddish brown; core
87	350		1	COOK	BDY	CRW	1.41	3	0	0	1	ALLEY		eroded; fire use; brown; core
87	372		2	COOK	BDY	CRW	1.41	1	0	0	1	YARD 4A		fire use; brown
87	353	1	3	COOK	BDY	CRW	1.41	1	0	0	1	ALLEY		fire use ; reddish brown; core
89	851		1	COOK	BDY	CRW	1.41	1	0	0	1	ROOM 4		brown; core
87	372		2	COOK	BDY	CRW	1.41	2	0	0	1	YARD 4A		fire use; reddish brown; core
87	372		2	COOK	BDY	CRW	1.41	2	0	0	1	YARD 4A		fire use; brown; core
89	869	3	3	COOK	BDY	CRW	1.41	3	0	0	1	YARD 4A		eroded; fire use; reddish brown; inclusions; core
89	816	3	3	COOK	BDY	CRW	1.41	2	0	0	1	SIDEWALK		fire use; reddish brown
90	917	7	3	BOWL	BDY	CRW	1.41	3	0	0	1	YARD 5		eroded; reddish brown; core
90	922		2	BOWL	BDY	CRW	1.41	2	0	0	1	YARD 6		light brown; core
87	381		1	COOK	BDY	CRW	1.41	1	0	0	1	YARD 4A		fire use; reddish brown; core
90	861		1	BOWL	BDY	CRW	1.41	3	0	0	1	YARD 4A		eroded; brown
87	384	4	3	COOK	BDY	CRW	1.41	1	0	0	1	YARD 4A		fire use; reddish brown; core
87	353	1	3	BOWL	BDY	CRW	1.41	1	0	0	1	ALLEY		reddish brown; core
89	602		2	COOK	BDY	CRW	1.41	1	0	0	1	SIDEWALK		eroded; fire use; brown; core
90	912		2	COOK	BDY	CRW	1.41	2	0	0	1	YARD 5		eroded; fire use; brown; core
87	384	2	3	BOWL	BDY	CRW	1.41	1	0	0	1	YARD 4A		eroded; reddish brown; core
89	866		3	COOK	BDY	CRW	1.41	2	0	0	1	YARD 4A		eroded; fire use; brown; core

Year	Unit	No.	Layer	Form	Part	Ware	Type	Thick	Diam	Ht	Qty	Location	Assoc	Notes
89	869	3	3	BOWL	BDY	CRW	1.41	3	0	0	1	YARD 4A		eroded; light brown ; core
90	922		2	COOK	BDY	CRW	1.41	3	0	0	1	YARD 6		red slip on interior; fire use; brown; inclusions; core
87	582		2	BOWL	BDY	CRW	1.41	2	0	0	1	SIDEWALK		handle attached; red slip on interior; brown; inclusions; core
90	922		2	BOWL	BDY	CRW	1.41	2	0	0	1	YARD 6		red slip on interior; brown; inclusions; core
90	892		2	BOWL	BDY	CRW	1.41	2	0	0	1	YARD 4A		light brown; core
89	859	3	3	BOWL	BDY	CRW	1.41	2	0	0	1	ROOM 4		light brown; core
87	251		1	COOK	BDY	CRW	1.41	2	0	0	1	YARD 4B		eroded; reddish brown; core
89	600		1	COOK	BDY	CRW	1.41	1	0	0	1	SIDEWALK		fire use; light brown; core
89	800	3	1	COOK	BDY	CRW	1.41	2	0	0	1	ROOM 4B		eroded; light brown
87	412		2	COOK	BDY	CRW	1.41	2	0	0	1	ALLEY		fire use; brown; core
87	374	2	3	BOWL	BDY	CRW	1.41	3	0	0	1	YARD 4A		red; core
89	801	3	1	BOWL	BDY	CRW	1.41	3	0	0	1	ROOM 4B		reddish brown; core
90	465	4	3	COOK	BDY	CRW	1.41	3	0	0	1	YARD 4A		eroded; fire use; brown; core
89	869	3	3	BOWL	BDY	CRW	1.41	2	0	0	1	YARD 4A		eroded; reddish brown; core
89	661		1	COOK	BDY	CRW	1.41	3	0	0	1	SIDEWALK		fire use; brown; core
89	602		2	BOWL	BDY	CRW	1.41	1	0	0	1	SIDWALK		reddish brown
90	922		2	BOWL	BDY	CRW	1.41	2	0	0	1	YARD 6		light brown; core
87	371		1	COOK	BDY	CRW	1.41	1	0	0	1	YARD 4A		brown; core
84	805		3	BOWL	BDY	CRW	1.41	1	0	0	1	ROOM 4B		eroded; reddish brown; core
89	762		2	BOWL	BDY	CRW	1.41	2	0	0	1	YARD 4B		brown; core
87	371		1	COOK	BDY	CRW	1.41	2	0	0	1	YARD 4A		fire use; reddish brown; core
90	922		2	COOK	BDY	CRW	1.41	2	0	0	1	YARD 6		fire use; brown; core
87	384	2	3	COOK	BDY	CRW	1.41	1	0	0	1	YARD 4A		eroded; fire use; reddish brown; core
87	384	2	3	COOK	BDY	CRW	1.41	1	0	0	1	YARD 4A		fire use; brown; core
90	880		1	BOWL	BDY	CRW	1.41	3	0	0	1	YARD 7		handle attached; reddish brown; core
90	958	2	3	COOK	BDY	CRW	1.41	3	0	0	1	YARD 5		fire use; reddish brown
87	351		1	COOK	BDY	CRW	1.41	2	0	0	1	ALLEY		brown
90	962		2	COOK	BDY	CRW	1.41	3	0	0	1	YARD 6		fire use; brown ; core
89	621		1	COOK	BDY	CRW	1.41	1	0	0	1	ROOM 1		fire use; brown ; core
89	852		2	COOK	BDY	CRW	1.41	1	0	0	1	ROOM 4		fire use; light brown ; core
87	371		1	COOK	BDY	CRW	1.41	2	0	0	1	YARD 4A		fire use; black
90	952		2	COOK	BDY	CRW	1.41	3	0	0	1	YARD 5		fire use; brown; core
87	572		2	COOK	BDY	CRW	1.41	2	0	0	1	ROOM 1		fire use; brown; core
90	942		2	COOK	BDY	CRW	1.41	2	0	0	1	YARD 5		fire use; brown; core
90	962		2	COOK	BDY	CRW	1.41	2	0	0	1	YARD 6		fire use; black
89	651		1	COOK	BDY	CRW	1.41	3	0	0	1	ROOM 1		fire use; yellow brown; core
90	922		2	COOK	BDY	CRW	1.41	2	0	0	1	YARD 6		fire use; black; core
87	571		1	COOK	BDY	CRW	1.41	2	0	0	1	ROOM 1		fire use; brown; core
90	927		3	COOK	BDY	CRW	1.41	1	0	0	1	YARD 6		fire use; black; core
90	102		2	BOWL	BDY	CRW	1.41	1	0	0	1	B5/LOT 100		eroded; black; core
90	466	3	3	BOWL	BDY	CRW	1.41	4	0	0	1	YARD 4A		light brown; core
87	571		[illegible]	BOWL	R/B/B	CRW	1.42	2	4	3	1	ROOM 1		grooved rim; reddish brown; core
90	922		2	BOWL	BDY	CRW	1.43	2	0	0	1	YARD 6		olive glaze ; reddish brown; inclusions
89	897	2	3	BOWL	BDY	CRW	1.44	2	0	0	1	YARD 4A		incised lines; olive glaz ; reddish brown; inclusions; core
87	592		2	BOWL	BDY	CRW	1.43	2	0	0	1	ROOM 1		olive glaze; reddish brown; inclusions; core
90	923		3	BOWL	BDY	CRW	1.43	2	0	0	1	YARD 6		olive glaze; red; inclusions
89	655	1	3	STOR	BDY	CRW	1.41	4	0	0	1	ROOM 1		eroded; light brown
87	371		1	BOWL	BDY	CRW	1.47	2	0	0	1	YARD 4A		olive glaze; reddish brown; inclusions; core
90	940		1	COOK	BDY	CRW	1.41	1	0	0	1	YARD 5		fire use; reddish brown
90	961		1	BOWL	BDY	CRW	1.43	1	0	0	1	YARD 6		yellow glaze; red; inclusions; core
89	652		2	COOK	BDY	CRW	1.41	2	0	0	1	ROOM 1		fire use; brown; core

Year	Unit	No.	Layer	Form	Part	Ware	Type	Thick	Diam	Ht	Qty	Location	Assoc	Notes
90	911		1	COOK	BDY	CRW	1.41	2	0	0	1	YARD 5		fire use; yellow
87	351		1	COOK	BDY	CRW	1.41	2	0	0	1	YARD 4A		fire use; yellow; core
90	922		2	COOK	BDY	CRW	1.41	2	0	0	1	YARD 6		fire use; yellow; core
87	592		2	COOK	BDY	CRW	1.41	3	0	0	1	ROOM 1		fire use; yellow; core
89	822		2	COOK	BDY	CRW	1.41	2	0	0	1	ROOM 4B		fire use; brown
87	261		1	BOWL	BSE	CRW	1.41	2	0	0	1	ALLEY		reddish brown; core
87	432		2	BOWL	BDY	CRW	1.41	3	0	0	1	YARD 4B		eroded; brown; inclusions; core
90	902		2	BOWL	RIM	CRW	1.41	2	5	0	2	YARD 5		brown; inclusions; core
89	622		2	UNKN	BDY	CRW	1.31	2	0	0	1	ROOM 1		reddish brown
89	681		1	BOWL	BDY	CRW	1.43	4	0	0	1	ROOM 2		olive glaze; yellow; inclusions
89	756	3	3	UNKN	BDY	CRW	1.41	3	0	0	1	ROOM 4B		red
87	261		1	BOWL	RIM	CRW	1.41	3	5	0	2	ALLEY		eroded; reddish brown; core
90	102		2	UNKN	BDY	CRW	1.20	2	0	0	1	B5/LOT 100		gray; inclusions; core
89	672		2	BOWL	RIM	CRW	1.43	3	5	0	2	ROOM 2		eroded; olive glaze; red; inclusions
87	404	4	3	UNKN	BDY	CRW	1.31	1	0	0	1	YARD 4B		white wash/slip on both surfaces; red
89	671		1	BOWL	RIM	CRW	1.41	3	5	0	2	ROOM 2		eroded; reddish brown; core
87	382		2	BOWL	BDY	CRW	1.33	2	0	0	1	YARD 4A		olive glaze; reddish brown; inclusions
89	891		1	BOWL	BDY	CRW	1.41	3	0	0	1	YARD 4A		eroded; brown; core
90	570		1	BOWL	BDY	CRW	1.41	1	0	0	1	ROOM 1		reddish brown; core
87	371		1	STOR	BSE	CRW	1.42	2	0	0	1	YARD 4A		grooved reddish brown; core
87	412		2	BOWL	RIM	CRW	1.44	1	5	0	2	ALLEY		grooved; olive glaze; reddish brown; core
87	254	3	3	STOR	BDY	CRW	1.31	3	0	0	1	ALLEY		white slip/wash on exterior; inclusions
89	602		2	BOWL	BDY	CRW	1.43	2	0	0	1	SIDEWALK		olive glaze; red; inclusions; core
87	571		1	BOWL	RIM	CRW	1.43	3	6	0	2	ROOM 1		olive glaze; red; inclusions; core
87	412		2	UNKN	BDY	CRW	1.31	2	0	0	1	ALLEY		white wash on interior reddish yellow
89	602		2	BOWL	RIM	CRW	1.44	3	4	0	4	SIDEWALK	CRW 13	finger pressed; olive glaze; reddish brown; inclusions; core
89	602		2	BOWL	BDY	CRW	1.44	3	0	0	2	SIDEWALK	CRW 13	olive glaze; reddish brown; inclusions; core
89	792		2	BOWL	BDY	CRW	1.43	1	0	0	1	ROOM 4A		olive glaze; reddish brown; inclusions; core
87	384	4	3	COOK	RIM	CRW	1.41	4	6	0	1	YARD 4A	CRW 14	fire use? brown; core
87	405	2	3	COOK	RIM	CRW	1.41	4	6	0	1	YARD 4B	CRW 14	fire use? brown; core
89	681		1	COOK	RIM	CRW	1.43	5	5	0	2	ROOM 2		olive glaze; reddish brown; inclusion ; core
87	353	1	3	STOR	BDY	CRW	1.31	2	0	0	1	ALLEY		white wash on exterior; reddish yellow
89	646	1	3	BOWL	BDY	CRW	1.41	4	0	0	1	ROOM 1		brown slip on exterior; reddish brown; core
90	916	4	3	BOWL	BSE	CRW	1.41	3	0	0	1	YARD 5		reddish brown; core
89	752		2	BOWL	BSE	CRW	1.41	3	0	0	1	YARD 4A		red; core
89	882		2	STOR	BSE	CRW	1.41	3	0	0	1	YARD 7		reddish brown; core
89	883	2	3	STOR	BSE	CRW	1.41	3	0	0	1	YARD 7		reddish brown; core
90	962		2	BOWL	BSE	CRW	1.43	3	0	0	1	YARD 6		olive glaze; reddish brown; inclusions; core
89	600		1	UNKN	ATT	CRW	1.47	2	0	0	1	SIDEWALK		handle; olive glaze; red; inclusions
89	752	2	2	STOR	RIM	CRW	1.41	5	7	0	2	YARD 4A		reddish brown; core
87	401		1	BOWL	BDY	CRW	1.31	2	0	0	1	YARD 4B		white wash on interior; yellowish red; inclusions; core
87	422		2	BOWL	RIM	CRW	1.43	3	6	0	3	ALLEY		reddish brown; inclusions; core
89	633	2	3	BOWL	BDY	CRW	1.43	4	0	0	1	SIDEWALK		olive glaze; red; inclusions; core
90	942		2	BOWL	BDY	CRW	1.43	3	0	0	1	YARD 5		olive glaze; red; inclusions; core
89	734	3	3	STOR	BDY	CRW	1.41	5	0	0	1	ROOM 3		light brown
87	582		2	STOR	BDY	CRW	1.43	4	0	0	1	SIDEWALK		olive glaze; reddish brown; inclusions; core
89	869	3	3	COOK	BDY	CRW	1.43	1	0	0	1	YARD 4A		olive glaze; red; inclusions; core
90	942		2	BOWL	RIM	CRW	1.43	3	5	0	3	YARD 5		olive glaze; red; inclusions; core
89	646	4	3	BOWL	BDY	CRW	1.43	6	0	0	1	ROOM 1		olive glaze; red; inclusions; core
90	952		2	UNKN	ATT	CRW	1.41	6	0	0	1	YARD 5		handle ; red; core
87	422		2	BOWL	RIM	CRW	1.43	3	4	0	3	ALLEY		olive glaze; red; inclusions

Year	Unit	No.	Layer	Form	Part	Ware	Type	Thick	Diam	Ht	Qty	Location	Assoc	Notes
87	845	3	3	STOR	BDY	CRW	1.10	6	0	0	1	SIDEWALK		olive glaze; reddish brown; Spanish?
90	922		2	COOK	RIM	CRW	1.41	3	4	0	2	YARD 6		brown; core
90	924	4	3	UNKN	BDY	CRW	1.41	2	0	0	1	YARD 6		reddish brow ; core
90	901		1	COOK	RIM	CRW	1.41	3	5	0	2	YARD 5		brown
90	910		1	STOR	BSE	CRW	1.20	3	0	0	1	YARD 5		gray; inclusions
87	372		2	COOK	BDY	CRW	1.41	3	0	0	1	YARD 4A		brown; inclusions; core
87	372		2	COOK	RIM	CRW	1.45	3	5	0	2	YARD 4A		olive glaze; yellow brown; inclusions; core
87	252		2	STOR	BDY	CRW	1.31	3	0	0	1	YARD 4B		reddish brown
90	101		1	BOWL	BDY	CRW	1.41	3	0	0	1	B5/LOT 100		brown; core
87	400		1	COOK	RIM	CRW	1.41	2	3	0	2	YARD 4B		eroded; brown
87	254	4	3	COOK	RIM	CRW	1.41	1	4	0	2	YARD 4B		brown; core
87	403	2	3	STOR	BDY	CRW	1.35	3	0	0	1	YARD 4B		light red
90	941		1	COOK	RIM	CRW	1.41	2	4	0	3	YARD 5		fire use; interior buildup; brown; core
89	712		2	COOK	BDY	CRW	1.41	2	0	0	1	ROOM 4A		eroded; reddish brown; core
87	372		2	COOK	RIM	CRW	1.47	4	4	0	1	YARD 4A		reddish brown; inclusions
87	353	2	3	UNKN	BDY	CRW	1.31	2	0	0	1	ALLEY		reddish brown
89	749	4	3	COOK	RIM	CRW	1.41	4	6	0	1	YARD 4A		eroded; yellow brown ; core
89	915	4	3	COOK	BDY	CRW	1.33	2	0	0	1	YARD 5		olive glaze; dark reddish brown
87	412		2	COOK	RIM	CRW	1.41	3	6	0	1	ALLEY		eroded; yellow brown; core
89	762		2	STOR	BDY	CRW	1.31	3	0	0	1	YARD 4B		light reddish brown
87	381		1	BOWL	BDY	CRW	1.42	4	0	0	1	YARD 4A		fire use; reddish brown
89	702		2	BOWL	BDY	CRW	1.41	3	0	0	1	SIDEWALK		reddish brown
89	671		1	BOWL	RIM	CRW	1.41	3	0	0	1	ROOM 2		red slip on exterior; eroded; black; inclusions
87	405	1	3	BOWL	RIM	CRW	1.42	3	6	0	3	YARD 4B		finger impressed handle attached; reddish brown; core
89	702		2	VASE	BSE	CRW	1.41	3	0	0	1	SIDEWALK		flower pot; hole in base center; reddish brown; core
87	351		1	BOWL	RIM	CRW	1.41	4	5	0	3	ALLEY		reddish brown; core
87	412		2	BOWL	ATT	CRW	1.42	5	0	0	1	ALLEY		handle; stamped design; eroded; reddish brown; inclusions; core
89	883	2	3	BOWL	RIM	CRW	1.42	4	5	0	3	YARD 7		stamped fan design; fire use; reddish brown; inclusions; core
89	891		1	STOR	ATT	CRW	1.42	4	0	0	1	YARD 4A		lid; stamped design; reddish brown; inclusions; core
90	892		2	BOWL	RIM	CRW	1.41	3	5	0	3	YARD 4A		handle attached; reddish brown; core
90	466		3	UNKN	BSE	CRW	1.42	4	0	0	1	YARD 4A		grooved with hole; reddish brown; inclusions; core
90	904	3	3	STOR	RIM	CRW	1.33	2	2	0	1	YARD 5	CRW 15	reddish brown glaze; light red
90	904	3	3	STOR	BDY	CRW	1.33	2	0	0	1	YARD 5	CRW 15	reddish brown glaze; light red
87	446	4	3	STOR	BDY	CRW	1.33	2	0	0	1	ROOM 4B	CRW 15	reddish brown glaze; light red
89	599		3	STOR	BDY	CRW	1.31	3	0	0	1	ROOM 1		light reddish brown
90	928		3	STOR	BDY	CRW	1.31	2	0	0	1	YARD 6		light reddish brown
89	642		2	BOWL	RIM	CRW	1.33	2	3	0	1	ROOM 1		olive glaze; light red
90	892		2	BOWL	RIM	CRW	1.37	2	3	0	1	YARD 4A		olive glaze; reddish brown
90	464	4	3	BOWL	BDY	CRW	1.37	2	0	0	1	YARD 4A		olive glaze; reddish brown
90	904	3	3	BOWL	BDY	CRW	1.37	2	0	0	1	YARD 5		olive glaze; reddish brown
87	420		1	BOWL	RIM	CRW	1.34	2	3	0	1	ALLEY		light dusky red glaze; cream band edges rim interior; light red fabric
87	254	3	3	UNKN	BDY	CRW	1.31	2	0	0	1	YARD 4B		reddish brown
90	403	2	3	UNKN	BDY	CRW	1.33	2	0	0	1	YARD 4B		light reddish brown glaze; reddish brown
87	571		1	UNKN	BDY	CRW	1.42	2	0	0	1	ROOM 1		hand molded stamped design on exterior; reddish yellow
87	425	3	3	COOK	BDY	CRW	1.43	2	0	0	1	ALLEY		hand molded handle attached; yellow glaze; reddish yellow
89	804		3	UNKN	BDY	CRW	1.33	2	0	0	1	ROOM 4B		light reddish brown glaze; reddish brown; inclusions
89	447		3	BOWL	BDY	CRW	1.33	3	0	0	1	ROOM 4B		yellow glaze; reddish brown
87	403	1	3	UNKN	BDY	CRW	1.37	2	0	0	1	YARD 4B		reddish brown glaze; reddish yellow
90	917	3	3	STOR	RIM	CRW	1.37	1	1	0	1	YARD 5		small storage container (drug jar?); olive glaze; reddish brown
89	622		2	STOR	RIM	CRW	1.37	2	4	0	1	ROOM 1	CRW 16	dusky red glaze; yellowish red
89	624		3	STOR	RIM	CRW	1.37	2	4	0	1	ROOM 1	CRW 16	dusky red glaze; yellowish red

Year	Unit	No.	Layer	Form	Part	Ware	Type	Thick	Diam	Ht	Qty	Location	Assoc	Notes
87	242		2	STOR	BDY	CRW	1.37	2	0	0	1	ALLEY		dusky red glaze; yellowish red
89	863	2	3	STOR	BDY	CRW	1.37	2	0	0	1	YARD 4A		dusky red glaze; yellowish red
90	904	3	3	BOWL	BDY	CRW	1.33	2	0	0	3	YARD 5	CRW 17	dusky red/olive glaze; light red
90	925	1	3	BOWL	BDY	CRW	1.33	2	0	0	1	YARD 6	CRW 17	dusky red/olive glaze; light red
90	905	1	3	BOWL	BDY	CRW	1.33	2	0	0	1	YARD 5	CRW 17	dusky red/olive glaze; light red
90	907	2	3	BOWL	BDY	CRW	1.33	2	0	0	1	YARD 5	CRW 17	dusky red/olive glaze; light red
89	654	3	3	STOR	RIM	CRW	1.33	2	3	0	1	ROOM 1		internal seating for lid; gray glaze; light red
89	646		3	STOR	BDY	CRW	1.33	2	0	0	1	ROOM 1		gray glaze; light red
87	453	3	3	STOR	BDY	CRW	1.33	2	0	0	1	ALLEY	CRW 18	gray glaze; light red
87	406	1	3	STOR	BDY	CRW	1.33	2	0	0	1	YARD 4B	CRW 18	gray glaze; light red
89	714		3	STOR	BDY	CRW	1.33	3	0	0	1	ROOM 4A		gray glaze; light red
89	654	2	3	STOR	BSE	CRW	1.33	3	2	0	1	ROOM 1		gray glaze; light red
87	592		2	STOR	BDY	CRW	1.33	2	0	0	1	ROOM 1		gray glaze; light red
89	654	2	3	STOR	BSE	CRW	1.33	2	3	0	1	ROOM 1		gray glaze; light red
89	732		2	BOWL	RIM	CRW	1.12	2	5	0	2	ROOM 3		olive/gray glaze
87	571		1	BOWL	RIM	CRW	1.12	2	2	0	1	ROOM 1		olive glaze on interior
90	942		2	STOR	RIM	CRW	1.38	2	3	0	1	YARD 5		olive glaze; incised lines on exterior; reddish brown; core
90	942		2	STOR	BSE	CRW	1.12	2	3	0	1	YARD 5	CRW 19	light green/yellow glaze on interior
90	969		3	STOR	BSE	CRW	1.12	2	3	0	1	YARD 6	CRW 19	light green/yellow glaze on interior
87	583		3	PORR	BDY	CRW	1.12	2	0	0	1	SIDEWALK		handle attached; yellow glaze
89	756	4	3	UNKN	BDY	CRW	1.33	2	0	0	1	YARD 4A		olive glaze; reddish brown; inclusions
90	952		2	BOWL	BDY	CRW	1.12	2	0	0	1	YARD 5		yellow/olive glaze on interior
87	254	3	3	BOWL	BDY	CRW	1.12	2	0	0	2	YARD 4B		yellow/olive glaze on interior
87	405	3	3	UNKN	BDY	CRW	1.12	2	0	0	1	YARD 4B		yellow glaze on both surfaces
89	618	3	3	UNKN	BDY	CRW	1.12	2	0	0	1	ROOM 1		yellow/light green glaze on interior
89	897	2	3	UNKN	BDY	CRW	1.12	2	0	0	1	YARD 4A		yellow/light green glaze on interior
90	904	3	3	UNKN	BDY	CRW	1.33	2	0	0	1	YARD 5		olive glaze; light red
87	582		2	UNKN	BDY	CRW	1.35	2	0	0	1	SIDEWALK		yellow glaze; reddish brown
89	893	3	3	UNKN	BDY	CRW	1.35	1	0	0	1	YARD 4A		yellow glaze; reddish brown
87	405	2	3	UNKN	BDY	CRW	1.35	1	0	0	1	YARD 4B		yellow glaze; reddish brown
89	679	5	3	EWER	RIM	CRW	1.20	1	1	0	1	ROOM 2	CRW 20	loop handle attached; cream/white; Spanish
89	654	2	3	EWER	BDY	CRW	1.20	1	0	0	1	ROOM 1	CRW 20	cream/white; Spanish
89	599		3	EWER	BDY	CRW	1.20	1	0	0	1	ROOM 1	CRW 20	cream/white; Spanish
87	606	5	3	EWER	BDY	CRW	1.20	1	0	0	1	SIDEWALK	CRW 20	cream/white; Spanish
87	622		2	EWER	BDY	CRW	1.20	1	0	0	1	ROOM 1	CRW 20	cream/white; Spanish
90	902		2	EWER	BDY	CRW	1.20	1	0	0	1	YARD 5	CRW 20	cream/white; Spanish
89	670		1	EWER	BDY	CRW	1.20	1	0	0	1	ROOM 2	CRW 20	cream/white; Spanish
89	679		3	EWER	BDY	CRW	1.20	1	0	0	1	ROOM 2	CRW 20	cream/white; Spanish
89	688	4	3	EWER	BDY	CRW	1.20	1	0	0	1	ROOM 2	CRW 20	cream/white; Spanish
89	679	5	3	EWER	BDY	CRW	1.20	1	0	0	1	ROOM 2	CRW 20	cream/white; Spanish
87	415	10	3	CHAM	WHLE	CRW	1.33	2	4 / 3	3	1	ALLEY		handle absent
87	745	3	3	STOR	BSE	CRW	1.31	3	3	0	1	YARD 4A		reddish yellow
89	815	4	3	STOR	BSE	CRW	1.31	5	3	0	1	SIDEWALK		reddish yellow
89	652		2	STOR	BDY	CRW	1.31	4	0	0	1	ROOM 1		reddish brown
87	582		2	STOR	BDY	CRW	1.31	3	0	0	1	SIDEWALK		reddish brown
89	897	1	3	STOR	RIM	CRW	1.31	3	2	0	1	YARD 4A		reddish yellow
87	376	4	3	BOWL	BDY	CRW	1.37	3	0	0	1	YARD 4A		black glaze; reddish brown
87	262		2	COOK	RIM	CRW	1.37	2	2	0	1	ALLEY		dusky red glaze ; red
89	683		3	COOK	RIM	CRW	1.33	4	4	0	2	ROOM 1		internal seating (for lid); dusky red glaze; red
87	373	2	3	BOWL	BDY	CRW	1.33	2	0	0	1	YARD 4A		olive glaze; red
87	372		2	BOWL	BDY	CRW	1.33	2	0	0	1	YARD 4A		olive glaze; red

Year	Unit	No.	Layer	Form	Part	Ware	Type	Thick	Diam	Ht	Qty	Location	Assoc	Notes
87	384	2	3	COOK	RIM	CRW	1.33	2	4	0	1	YARD 4A		olive glaze; reddish brown
89	602		2	STOR	BDY	CRW	1.21	2	0	0	1	SIDEWALK	olive	white slip/wash on exterior; pink
90	462		2	STOR	RIM	CRW	1.21	2	4	0	1	YARD 4A	olive	white slip/wash on both surfaces; pink; core; Spanish?
89	897	2	3	STOR	BDY	CRW	1.21	2	0	0	1	YARD 4A	olive	white slip/wash on exterior; pink; core
87	404	3	3	STOR	BDY	CRW	1.21	2	0	0	1	YARD 4B	olive	white slip/wash on exterior; pink; core
89	5810	4	3	COOK	BDY	CRW	1.47	3	0	0	1	SIDEWALK		reddish brown
90	104	2	3	STOR	BDY	CRW	1.21	2	0	0	1	B5/LOT 100		white slip/wash on exterior; pink; core
90	104	6	3	STOR	BDY	CRW	1.21	2	0	0	1	B5/LOT 100		white slip/wash on exterior; pink; core
89	725	3	3	STOR	RIM	CRW	1.21	2	2	0	1	ROOM 4A		white slip wash on exterior; pink; core
87	255	4	3	STOR	BDY	CRW	1.31	3	0	0	1	YARD 4A		red; inclusions
87	244	3	3	STOR	BDY	CRW	1.31	2	0	0	2	ALLEY		red
90	462		2	UNKN	BDY	CRW	1.31	2	0	0	1	YARD 4A		red
89	869	3	3	UNKN	BDY	CRW	1.31	2	0	0	1	YARD 4A		reddish yellow
89	678	2	3	BOWL	BDY	CRW	1.31	2	0	0	1	ROOM 2		reddish yellow
89	866	4	3	STOR	BDY	CRW	1.31	2	0	0	1	YARD 4A		reddish yellow
89	852		2	STOR	BDY	CRW	1.31	2	0	0	1	ROOM 4		reddish yellow
89	599		3	UNKN	BDY	CRW	1.31	2	0	0	2	ROOM 1		red
90	465	3	3	STOR	BDY	CRW	1.31	3	0	0	1	YARD 4A		red
89	861		1	UNKN	BDY	CRW	1.31	2	0	0	1	YARD 4A		reddish yellow
89	640		1	STOR	BDY	CRW	1.31	3	0	0	1	ROOM 1		red
89	654	2	3	STOR	BDY	CRW	1.31	3	0	0	1	ROOM 1		red
90	904	2	3	UNKN	BDY	CRW	1.31	2	0	0	1	YARD 5		reddish brown
87	405	3	3	UNKN	BDY	CRW	1.31	2	0	0	1	YARD 4B		reddish brown
89	867	3	3	UNKN	BDY	CRW	1.31	2	0	0	1	YARD 4A		reddish brown
89	642		2	STOR	BSE	CRW	1.31	4	4	0	1	ROOM 1		reddish brown
90	943	4	3	STOR	RIM	CRW	1.31	3	3	0	1	YARD 5	CRW 21	light red
90	943	2	3	STOR	RIM	CRW	1.31	3	3	0	1	YARD 5	CRW 21	light red
90	104	2	3	STOR	RIM	CRW	1.31	3	3	0	1	B5/LOT 100	CRW 21	light red
90	941		1	STOR	RIM	CRW	1.31	3	3	0	1	YARD 5	CRW 21	light red
90	943	2	3	STOR	BDY	CRW	1.31	3	0	0	2	YARD 5	CRW 21	light red
90	947	3	3	STOR	BDY	CRW	1.31	3	0	0	1	YARD 5	CRW 21	light red
90	954	2	3	STOR	BDY	CRW	1.21	3	0	0	1	YARD 5	CRW 22	white slip/wash on exterior; yellow
90	9612		3	STOR	RIM	CRW	1.21	3	3	0	1	YARD 6	CRW 22	complete rim; white slip/wash on exterior; yellow
90	9612		3	STOR	BDY	CRW	1.21	3	0	0	15	YARD 6	CRW 22	white slip/wash on exterior; yellow
90	948	2	3	STOR	BDY	CRW	1.21	3	0	0	1	YARD 5	CRW 22	white slip/wash on exterior; yellow
90	918	2	3	STOR	BDY	CRW	1.21	3	0	0	1	YARD 5	CRW 22	white slip/wash on exterior; yellow
90	374	2	3	STOR	BDY	CRW	1.21	3	0	0	1	YARD 4A	CRW 22	white slip/wash on exterior; yellow
90	650		1	STOR	BDY	CRW	1.21	3	0	0	4	ROOM 1	CRW 22	white slip/wash on exterior; yellow
90	954	2	3	STOR	BDY	CRW	1.21	3	0	0	1	YARD 5	CRW 22	white slip/wash on exterior; yellow
90	884		3	CUP	RIM	POR	5.11	1	2	0	1	YARD 7		
87	352		2	UNKN	BDY	POR	5.11	1	0	0	1	ALLEY		
89	632		2	BOWL	BDY	POR	5.13	1	0	0	2	SIDEWALK		blue floral design on interior; brown wash on exterior
87	587	2	3	BOWL	RIM	POR	5.13	1	3	0	1	SIDEWALK	POR 4	teabowl; blue decoration edges interior rim; brown wash on exterior
87	584	1	3	BOWL	RIM	POR	5.13	1	3	0	1	SIDEWALK	POR 4	teabowl; blue decoration edges interior rim; brown wash on exterior
89	633	2	3	CUP	RIM	POR	5.14	1	2	0	2	SIDEWALK		
87	246	2	3	SAUC	R/B/B	POR	5.11	1	3 / 2	0	1	ALLEY		
90	952		2	CUP	ATT	POR	5.11	1	O	O	1	YARD 5		handle
89	652		2	UNKN	BDY	POR	5.11	1	0	0	1	ROOM 1		
87	255	3	3	SAUC	BSE	POR	5.11	1	2	0	1	YARD 4B		
89	673		3	UNKN	BDY	POR	5.11	1	0	0	1	ROOM 2		
87	372		2	UNKN	BDY	POR	5.11	1	0	0	1	YARD 4A		

Year	Unit	No.	Layer	Form	Part	Ware	Type	Thick	Diam	Ht	Qty	Location	Assoc	Notes
89	602		2	UNKN	BDY	POR	5.11	1	0	0	2	SIDEWALK		
90	465	10	3	STOR	RIM	CRW	1.41	2	4	0	3	YARD 4A	CRW 23	reconstructed; reddish brown
90	375	1	3	STOR	BDY	CRW	1.41	2	0	0	1	YARD 4A	CRW 23	reconstructed; reddish brown
89	869	4	3	STOR	BDY	CRW	1.41	2	0	0	1	YARD 4A	CRW 23	reconstructed; reddish brown
89	888	5	3	STOR	BDY	CRW	1.41	2	0	0	14	YARD 7	CRW 23	reconstructed; reddish brown
89	891		1	STOR	BDY	CRW	1.41	2	0	0	1	YARD 4A	CRW 23	reconstructed; reddish brown
89	888	4	3	STOR	BDY	CRW	1.41	2	0	0	1	YARD 7	CRW 23	reconstructed; reddish brown
89	898	5	3	STOR	BDY	CRW	1.41	2	0	0	1	YARD 4A	CRW 23	reconstructed; reddish brown
89	893	1	3	STOR	BDY	CRW	1.41	2	0	0	1	YARD 4A	CRW 23	reconstructed; reddish brown
89	865	10	3	COOK	BDY	CRW	1.41	2	0	0	1	YARD 4A	CRW 24	reconstructed; fire use reddish brown
89	863	2	3	COOK	BDY	CRW	1.41	2	0	0	1	YARD 4A	CRW 24	reconstructed; fire use; reddish brown
89	863	4	3	COOK	BDY	CRW	1.41	2	0	0	2	YARD 4A	CRW 24	reconstructed; fire use; reddish brown
89	864	1	3	COOK	BDY	CRW	1.41	2	0	0	1	YARD 4A	CRW 24	reconstructed; fire use; reddish brown
89	863	21	3	COOK	RIM	CRW	1.41	2	4	0	1	YARD 4A	CRW 24	reconstructed; fire use; reddish brown
89	863	2	3	COOK	RIM	CRW	1.41	2	4	0	2	YARD 4A	CRW 24	reconstructed; fire use; reddish brown
89	864	3	3	COOK	BDY	CRW	1.41	2	0	0	2	YARD 4A	CRW 24	reconstructed; fire use; reddish brown

www.ingramcontent.com/pod-product-compliance
Lightning Source LLC
LaVergne TN
LVHW071212210926
848482LV00015B/791